Dynamic Coordination in the Brain

From Neurons to Mind

Strüngmann Forum Reports

Julia Lupp, series editor

The Ernst Strüngmann Forum is made possible through the generous support of the Ernst Strüngmann Foundation, inaugurated by Dr. Andreas and Dr. Thomas Strüngmann.

This Forum was supported by funds from the Deutsche Forschungsgemeinschaft (German Science Foundation)

Dynamic Coordination in the Brain

From Neurons to Mind

Edited by

Christoph von der Malsburg, William A. Phillips,
and Wolf Singer

Program Advisory Committee:
Sten Grillner, William A. Phillips, Steven M. Silverstein,
Wolf Singer, Olaf Sporns, and Christoph von der Malsburg

The MIT Press

Cambridge, Massachusetts
London, England

© 2010 Massachusetts Institute of Technology and
the Frankfurt Institute for Advanced Studies

Series Editor: J. Lupp
Assistant Editor: M. Turner
Photographs: U. Dettmar
Typeset by BerlinScienceWorks

All rights reserved. No part of this book may be reproduced in any form by electronic or mechanical means (including photocopying, recording, or information storage and retrieval) without permission in writing from the publisher.

MIT Press books may be purchased at special quantity discounts for business or sales promotional use. For information, please email special_sales@mitpress.mit.edu or write to Special Sales Department, The MIT Press, 55 Hayward Street, Cambridge, MA 02142.

The book was set in TimesNewRoman and Arial.
Printed and bound in the United States of America.

Library of Congress Cataloging-in-Publication Data

Ernst Strüngmann Forum (5th : 2009 : Frankfurt am Main, Germany)
  Dynamic coordination in the brain : from neurons to mind / edited by
  Christoph von der Malsburg, William A. Phillips, and Wolf Singer.
     p. ; cm. — (Strüngmann Forum reports)
  Fifth Ernst Strüngmann Forum held Aug. 16–21, 2009, Frankfurt am Main.
  Includes bibliographical references and index.
  ISBN 978-0-262-01471-7 (hardcover : alk. paper)
  1. Neural circuitry—Congresses. 2. Brain—Localization of functions—
  Congresses. 3. Cognitive neuroscience—Congresses. I. Malsburg,
  Christoph von der, 1942– II. Phillips, William A., 1936– III. Singer, W.
  (Wolf) IV. Title. V. Series: Strüngmann Forum reports.
  [DNLM: 1. Brain—physiology—Congresses. 2. Brain—physiopathology—
  Congresses. 3. Cognition—physiology—Congresses. 4. Neural Pathways—
  physiology—Congresses. 5. Neurons—physiology—Congresses. 6.
  Psychophysiology—Congresses.   WL 300 E727d 2010]
QP363.3.E76   2009
612.8'2—dc22
                                        2010014715
10 9 8 7 6 5 4 3 2 1

Contents

The Ernst Strüngmann Forum		vii
List of Contributors		ix
Preface		xiii
1	**Dynamic Coordination in Brain and Mind** *William A. Phillips, Christoph von der Malsburg, and Wolf Singer*	1
2	**Cortical Circuits: Consistency and Variability across Cortical Areas and Species** *Jon H. Kaas*	25
3	**Sequence Coding and Learning** *Gilles Laurent*	35
4	**What Can Studies of Comparative Cognition Teach Us about the Evolution of Dynamic Coordination?** *Uri Grodzinski and Nicola S. Clayton*	43
5	**Evolution of Dynamic Coordination** *Evan Balaban, Shimon Edelman, Sten Grillner, Uri Grodzinski, Erich D. Jarvis, Jon H. Kaas, Gilles Laurent, and Gordon Pipa*	59
6	**Modeling Coordination in the Neocortex at the Microcircuit and Global Network Level** *Anders Lansner and Mikael Lundqvist*	83
7	**Oscillation-supported Information Processing and Transfer at the Hippocampus–Entorhinal–Neocortical Interface** *György Buzsáki and Kamran Diba*	101
8	**What Are the Local Circuit Design Features Concerned with Coordinating Rhythms?** *Miles A. Whittington, Nancy J. Kopell, and Roger D. Traub*	115
9	**Coordination in Circuits** *Mayank Mehta, György Buzsáki, Andreas Kreiter, Anders Lansner, Jörg Lücke, Kevan Martin, Bita Moghaddam, May-Britt Moser, Danko Nikolić, and Terrence J. Sejnowski*	133
10	**Coordination: What It Is and Why We Need It** *Christoph von der Malsburg*	149
11	**Neocortical Rhythms: An Overview** *Wolf Singer*	159
12	**Stimulus-driven Coordination of Cortical Cell Assemblies and Propagation of Gestalt Belief in V1** *Yves Frégnac, Pedro V. Carelli, Marc Pananceau, and Cyril Monier*	169

| 13 | **Coordination in Brain Systems** | 193 |

Edvard I. Moser, Maurizio Corbetta, Robert Desimone, Yves Frégnac, Pascal Fries, Ann M. Graybiel, John-Dylan Haynes, Laurent Itti, Lucia Melloni, Hannah Monyer, Wolf Singer, Christoph von der Malsburg, and Matthew A. Wilson

| 14 | **"Hot Spots" and Dynamic Coordination in Gestalt Perception** | 215 |

Ilona Kovács

| 15 | **Coordination in Sensory Integration** | 229 |

Jochen Triesch, Constantin Rothkopf, and Thomas Weisswange

| 16 | **Neural Coordination and Human Cognition** | 235 |

Catherine Tallon-Baudry

| 17 | **Failures of Dynamic Coordination in Disease States and Their Implications for Normal Brain Function** | 245 |

Steven M. Silverstein

| 18 | **Coordination in Behavior and Cognition** | 267 |

Andreas K. Engel, Karl Friston, J. A. Scott Kelso, Peter König, Ilona Kovács, Angus MacDonald III, Earl K. Miller, William A. Phillips, Steven M. Silverstein, Catherine Tallon-Baudry, Jochen Triesch, and Peter Uhlhaas

Bibliography 301

Subject Index 345

The Ernst Strüngmann Forum

Founded on the tenets of scientific independence and the inquisitive nature of the human mind, the Ernst Strüngmann Forum is dedicated to the continual expansion of knowledge. Through its innovative communication process, the Ernst Strüngmann Forum provides a creative environment within which experts scrutinize high-priority issues from multiple vantage points.

This process begins with the identification of themes. By nature, a theme constitutes a problem area that transcends classic disciplinary boundaries. It is of high-priority interest and requires concentrated, multidisciplinary input to address the issues involved. Proposals are received from leading scientists active in their field and are selected by an independent Scientific Advisory Board. Once approved, a steering committee is convened to refine the scientific parameters of the proposal and select the participants. Approximately one year later, a focal meeting is held to which circa forty experts are invited.

Preliminary discussion around this theme began in the hallways of FIAS in 2007, shortly after Christoph von der Malsburg and I arrived in Frankfurt. Our talks expanded to include Bill Phillips, who was visiting FIAS at the time, and Wolf Singer. In August, 2008, the steering committee—comprised of Sten Grillner, Bill Phillips, Steve Silverstein, Wolf Singer, Olaf Sporns, and Christoph von der Malsburg—met to identify the key issues for debate and select the participants for the focal meeting, which was held in Frankfurt am Main, Germany, from August 16–21, 2009.

The activities and discourse surrounding a Forum begin well before participants arrive in Frankfurt and conclude with the publication of this volume. Throughout each stage, focused dialog is the means by which participants examine the issues anew. Often, this requires relinquishing long-established ideas and overcoming disciplinary idiosyncrasies that might otherwise inhibit joint examination. However, when this is accomplished, a unique synergism results and new insights emerge.

This volume conveys the synergy that arose from a group of diverse experts, each of whom assumed an active role, and is comprised of two types of contributions. The first provides background information on key aspects of the overall theme. These chapters have been extensively reviewed and revised to provide current understanding of the topics. The second (Chapters 5, 9, 13, and 18) summarizes the extensive discussions that transpired. These chapters should not be viewed as consensus documents nor are they proceedings; they transfer the essence of the discussions, expose the open questions that still remain, and highlight areas for future research.

An endeavor of this kind creates its own unique group dynamics and puts demands on everyone who participates. Each invitee contributed not only their time and congenial personality, but a willingness to probe beyond that which is evident, and I wish to extend my sincere gratitude to all. A special word of thanks goes to the steering committee, the authors of the background papers, the reviewers of the papers, and the moderators of the individual working groups (Sten Grillner, Terry Sejnowski, Bob Desimone, and Steve Silverstein). To draft a report during the Forum and bring it to its final form in the months thereafter is no simple matter, and for their efforts, we are especially grateful to Evan Balaban, Mayank Mehta, Edvard Moser, and Andreas Engel. Most importantly, I wish to extend my appreciation to the chairpersons, Christoph von der Malsburg, Bill Phillips, and Wolf Singer, whose support throughout this project was invaluable.

A communication process of this nature relies on institutional stability and an environment that encourages free thought. The generous support of the Ernst Strüngmann Foundation, established by Dr. Andreas and Dr. Thomas Strüngmann in honor of their father, enables the Ernst Strüngmann Forum to conduct its work in the service of science. In addition, the following valuable partnerships are gratefully acknowledged: the Scientific Advisory Board, which ensures the scientific independence of the Forum; the German Science Foundation, which provided financial support for this theme; and the Frankfurt Institute for Advanced Studies, which shares its vibrant intellectual setting with the Forum.

Long-held views are never easy to put aside. Yet, when this is achieved, when the edges of the unknown begin to appear and gaps in knowledge are able to be defined, the act of formulating strategies to fill these becomes a most invigorating exercise. It is our hope that this volume will convey a sense of this lively exercise and extend the inquiry further, as scientists continue to explore the mechanisms and manifestations of distributed dynamic coordination in the brain and mind across species and levels of organization.

Julia Lupp, Program Director
Ernst Strüngmann Forum
Frankfurt Institute for Advanced Studies (FIAS)
Ruth-Moufang-Str. 1, 60438 Frankfurt am Main, Germany
http://fias.uni-frankfurt.de/esforum/

List of Contributors

Evan Balaban Psychology Department, McGill University, Stewart Biological Sciences Bldg., 1205 Dr. Penfield Avenue, Montreal, QC H3A 1B1, Canada
György Buzsáki Center for Molecular and Behavioral Neuroscience, Rutgers, The State University of New Jersey, 197 University Avenue, Newark, NJ 07102, U.S.A.
Pedro V. Carelli Unité de Neurosciences, Information et Complexité (U.N.I.C.), UPR 3293, CNRS, Gif-sur-Yvette, France
Nicola S. Clayton Department of Experimental Psychology, University of Cambridge, Downing Street, Cambridge CB2 3EB, U.K.
Maurizio Corbetta Department of Neurology, Washington University, St. Louis, Campus Box 8225, 4525 Scott Avenue, St. Louis, MO 63110, U.S.A.
Robert Desimone McGovern Institute, Massachusetts Institute of Technology, MIT Bldg 46-3160, Cambridge, MA 02139, U.S.A.
Kamran Diba Center for Molecular and Behavioral Neuroscience, Rutgers, The State University of New Jersey, 197 University Avenue, Newark, NJ 07102, U.S.A.
Shimon Edelman Department of Psychology, Cornell University, 232 Uris Hall, Ithaca, NY 14853, U.S.A.
Andreas K. Engel Department of Neurophysiology and Pathophysiology, University Medical Center Hamburg-Eppendorf, Martinistr. 52, 20246 Hamburg, Germany
Yves Frégnac Unité de Neurosciences, Information et Complexité (U.N.I.C.), UPR 3293, CNRS, Gif-sur-Yvette, France
Pascal Fries Ernst Strüngmann Institute, c/o Max-Planck-Society, Deutschordenstr. 46, 60528 Frankfurt am Main, Germany
Karl Friston Wellcome Trust Centre for Neuroimaging, 12 Queen Square, London WC1N 3BG, U.K.
Ann M. Graybiel Department of Brain and Cognitive Sciences, Massachusetts Institute of Technology, McGovern Institute, 77 Massachusetts Avenue, Cambridge, MA 02139, U.S.A.
Sten Grillner Karolinska Institute, Department of Neuroscience, 17177 Stockholm, Sweden
Uri Grodzinski Department of Experimental Psychology, University of Cambridge, Downing Street, Cambridge CB2 3EB, U.K.
John-Dylan Haynes Bernstein Center for Computational Neuroscience Berlin, Haus 6, Phillipstr. 13, 10115 Berlin, Germany

Laurent Itti Viterbi School of Engineering, Computer Science Department, University of Southern California, Hedco Neuroscience Building, 3641 Watt Way, Los Angeles, CA 90089–2520, U.S.A.

Erich D. Jarvis Department of Neurobiology, Duke University Medical Center, Box 3209, Durham, NC 27710, U.S.A.

Jon H. Kaas Vanderbilt University, Department of Psychology, 301 Wilson Hall, Nashville, TN 37203, U.S.A.

J. A. Scott Kelso Center for Complex Systems and Brain Sciences, Florida Atlantic University, 777 Glades Road, Boca Raton, FL 33431, U.S.A.

Peter König Institute of Cognitive Science, University Osnabrück, Albrechtstr. 28, 49076 Osnabrück, Germany

Nancy J. Kopell Institute of Neuroscience, The Medical School, Newcastle University, Framlington Place, Newcastle NE2 4HH, U.K.

Ilona Kovács Department of Cognitive Science, Faculty of Economics and Social Science, Budapest University of Technology and Economics, ST–311, 1111, Budapest, Hungary

Andreas Kreiter Zentrum für Kognitionswissenschaften, Universität Bremen, FB 2, Postfach 33 04 40, 28334 Bremen, Germany

Anders Lansner Department of Computational Biology, Stockholm University and Royal Institute of Technology, School of Computer Science and Communication, 10044 Stockholm, Sweden

Gilles Laurent Max-Plank-Institute for Brain Research, Deutschordenstr. 46, 60528 Frankfurt am Main, Germany

Jörg Lücke Frankfurt Institute for Advanced Studies, Ruth-Moufang-Str. 1, 60438 Frankfurt am Main, Germany

Mikael Lundqvist Computational Biology and Neurocomputing, School of Computer Science and Communication, 10044 Stockholm, Sweden

Angus MacDonald III University of Minnesota, Department of Psychology, N426 Elliott Hall, Minneapolis, MN 55455, U.S.A.

Kevan Martin Institute of Neuroinformatics, ETH/UZH, Winterthurerstrasse 190, 8057 Zurich, Switzerland

Mayank Mehta Department of Neurology, Department of Physics and Astronomy, University of California, Los Angeles, CA 90095, U.S.A.

Lucia Melloni Max-Planck-Institut für Hirnforschung, Deutschordenstraße 46, 60528 Frankfurt am Main, Germany

Earl K. Miller The Picower Institute for Learning and Memory, Department of Brain and Cognitive Sciences, Massachusetts Institute of Technology, 77 Massachusetts Avenue, Cambridge, MA 02139, U.S.A.

Bita Moghaddam Department of Neuroscience, University of Pittsburgh, A210 Langley Hall, Pittsburgh, PA 15238, U.S.A.

Cyril Monier Unité de Neurosciences, Information et Complexité (U.N.I.C.), UPR 3293, CNRS, Gif-sur-Yvette, France

Hannah Monyer Clinical Neurobiology, University of Heidelberg, Im Neuenheimer Feld 364, 69120 Heidelberg, Germany

List of Contributors

Edvard I. Moser Centre for the Biology of Memory, Kavli Institute for Systems Neuroscience, Medical-Technical Research Centre, 7489 Trondheim, Norway

May-Britt Moser Centre for Biology of Memory, Kavli Institute for Systems Neuroscience, Medical-Technical Research Centre, 7489 Trondheim, Norway

Danko Nikolić Department of Neurophysiology, Max Planck Institute for Brain Research, Deutschordenstraße 46, 60528 Frankfurt am Main, Germany

Marc Pananceau Unité de Neurosciences, Information et Complexité (U.N.I.C.), UPR 3293, CNRS, Gif-sur-Yvette, France

William A. Phillips Department of Psychology, University of Stirling, Stirling, FK9 4LA, Scotland, U.K.

Gordon Pipa Max Planck Institute for Brain Research, Deutschordenstr. 46, 60528 Frankfurt am Main, Germany

Constantin Rothkopf Frankfurt Institute for Advanced Studies, Ruth-Moufang-Str.1, 60438 Frankfurt am Main, Germany

Terrence J. Sejnowski Salk Institute for Biological Studies, 100/0 N. Torrey Pines Rd., La Jolla, CA 92037, U.S.A.

Steven M. Silverstein Division of Schizophrenia Research, University of Medicine and Dentistry of New Jersey, 151 Centennial Avenue, Piscataway, NJ 08854, U.S.A.

Wolf Singer Max Planck Institute for Brain Research, Deutschordenstr. 46, 60528 Frankfurt am Main, Germany

Olaf Sporns Department of Psychology, Indiana University, 1230 York Avenue, Bloomington, IN 47405,U.S.A.

Catherine Tallon-Baudry CRICM CNRS UMR 7225, 47 Bd de l'Hôpital, 75013 Paris, France

Roger D. Traub Department of Physical Sciences, IBM T. J. Watson Research Center, Yorktown Heights, NY 10598, U.S.A.

Jochen Triesch Frankfurt Institute for Advanced Studies, Ruth-Moufang-Str. 1, 60438 Frankfurt am Main, Germany

Peter Uhlhaas Department of Neurophysiology, Max Planck Institute for Brain Research, Deutschordenstr. 46, 60528 Frankfurt am Main, Germany

Christoph von der Malsburg Frankfurt Institute for Advanced Studies, Ruth-Moufang-Str. 1, 60438 Frankfurt am Main, Germany

Thomas Weisswange Frankfurt Institute for Advanced Studies, Ruth-Moufang-Str. 1, 60438 Frankfurt am Main, Germany

Miles A. Whittington Institute of Neuroscience, The Medical School, Newcastle University, Framlington Place, Newcastle NE2 4HH, U.K.

Matthew A. Wilson The Picower Institute for Learning and Memory, Department of Brain and Cognitive Sciences, Massachusetts Institute of Technology, 43 Vassar Street, Building 46-5233, Cambridge, MA 02139, U.S.A.

Preface

Brain and mind have been studied by an exponentially growing number of scientists in the fields of cognitive science, neuroscience, and molecular biology. As a consequence, tremendous treasures of knowledge have been accumulated. Given this exuberance, individual scientists are driven to ever finer specializations. This can be a double-edged sword, because it is then too easy to lose sight of the grand picture. Thus, periodically, the attempt must be made to revisit the essential questions concerning the functions and mechanisms of brain and mind. The Strüngmann Forum Series is an ideal platform in which to accomplish this.

Supported by generous donors and a superb team of organizers-cum-undercover book editors, the Ernst Strüngmann Forum has revitalized the tradition once associated with the Dahlem Workshops and has the clout to assemble a quorum of the leading scientists in a given field. Equally important to this end is the format developed by the Forum: an intense week devoted entirely to discussion, avoiding time normally eaten up by talks, presentations, and statements of opinion and diverting detail. The essence of these concentrated discussions is distilled in strenuous nightly sessions into reports and, together with background papers written by a select group of exponents of the field, are published in book form, with this one being the fifth in the series.

Dynamic coordination is a crucial aspect of what brain and mind are all about: it is the ability to flexibly apply skills, knowledge, and situation awareness to the achievement of current goals. In planning this Forum, we were well aware that this capability and the mechanisms behind it are still in the dark to a very large extent. The purpose of the Forum was therefore not seen as bringing in a rich harvest in the fall but rather as plowing and sowing the field in spring. Turning points in science occur when an issue has reached maturity, when a critical mass of new facts cries out for a new synthesis, and when there is a general sense of urgency to reconsider basic assumptions. Developments in the last decade or two have powerfully stirred up interest in context sensitivity, in binding mechanisms, in the dynamical organization of temporal patterns and synchrony relations of neural signals, and in the agile adaptation of synaptic efficacy on all timescales. These developments alone, we felt, would be sufficient to shake up our collective thoughts at this Forum. It is evident that the longed-for new and generally acknowledged synthesis has not yet grown out of our efforts. In our opinion, however, this Forum and its report provide an authoritative account of current views of the pressing problems surrounding the issue of dynamic coordination in the brain. We hope that they will help drive the neural and cognitive sciences toward the revolution that is necessary if we are to finally understand the fundamental functions and mechanisms of mind and brain.

Christoph von der Malsburg, William A. Phillips, and Wolf Singer

1

Dynamic Coordination in Brain and Mind

William A. Phillips, Christoph von der Malsburg, and Wolf Singer

Abstract

This chapter discusses the concept of dynamic coordination and the major issues that it raises for the cognitive neurosciences. In general, coordinating interactions are those that produce coherent and relevant overall patterns of activity, while preserving the essential individual identities and functions of the activities coordinated. Dynamic coordination is the coordination that is created on a moment-by-moment basis so as to deal effectively with unpredictable aspects of the current situation. Different computational goals for dynamic coordination are distinguished and issues discussed that arise concerning local cortical circuits, brain systems, cognition, and evolution. Primary focus is on dynamic coordination resulting from widely distributed processes of self-organization, but the role of central executive processes is also discussed.

Introduction: Basic Concepts, Hypotheses, and Issues

How Is Flexibility Combined with Reliability?

The universe is lawful but unpredictable. Regularities make life possible, but unpredictability requires it to be flexible, so, biological systems must combine reliability with flexibility. Neural activity must reliably convey sensory information, cognitive contents, and motor commands, but it must do so flexibly and creatively if it is to generate novel but useful percepts, thoughts, and actions in novel circumstances. Neural activity, however, is widely distributed, which suggests that activity is dynamically coordinated so as to produce coherent patterns of macroscopic activity that are adapted to the current context, without corrupting the information that is transmitted by the local signals.

It is clear that, both within and between brain regions, different cells or groups of cells convey information about different things. New techniques and findings continue to add to our knowledge of this local selectivity and its gradual adaptation to different environments and tasks. Much of cognitive neuroscience is therefore primarily concerned with finding out what is dealt with where; that is, with localizing brain functions. We need now to understand better how these diverse activities are coordinated. Most locally specialized streams of activity can affect most others, either directly or indirectly. In that sense, the brain operates as an integrated whole. It is also clear that percepts, thoughts, and actions are normally, in some sense, both "coherent" and "relevant" to the current context, and psychological studies provide much information on the underlying processes. However, we do not yet have widely agreed conceptions of "coherence" and "relevance," nor of the dynamic coordinating interactions by which they are achieved. A central goal of this Forum was therefore to review the relevant evidence, to assess the underlying concepts, and to evaluate possible ways of exploring the major issues that arise.

Prima facie, the need to combine flexibility with reliability raises difficulties because of contrasting requirements. To be reliable, neural codes must code for the same thing when used at different times and in different contexts; however, to be flexible, codes must be used in different ways at different times and in different contexts. The notion of dynamic coordination directly raises this issue because it emphasizes the need to combine and sequence activities in a context-sensitive way that changes from moment to moment, while implying that the activities coordinated maintain their distinct identities.

Thus, fundamental issues arise for neurobiology, neurocomputational theory, cognitive psychology, and psychopathology. Does dynamic coordination depend on specific synaptic or local circuit mechanisms? Does it depend on distinct neural pathways? Can the capabilities and requirements of dynamic coordination be clarified by neurocomputational theory? What are the cognitive consequences of dynamic coordination, and what are the psychopathological consequences of its malfunction?

How Is Holism Combined with Localism?

Localism and holism have been contrasting themes throughout the history of neuroscience, with the evidence for localism being clearly dominant (Finger 1994). Emphasis upon holistic organization has waxed and waned within both neuroscience and psychology. The form of holism known as Gestalt psychology was strong for a few decades from the 1920s, but has had limited impact since then. It is now ready for a renaissance in which its phenomenological methods are complemented by rigorous psychophysical methods, and its speculations concerning electrophysiological fields are replaced by better founded neural network theories (Kovács 1996; Watt and Phillips 2000).

The apparent conflict between localism and holism is clearly shown by studies of the visual cortex. The classic studies of Kuffler, Hubel, Wiesel, and many others provide strong support for localism. They show that individual cells or small populations of cells convey information about very specific aspects of visual input. Higher levels within the hierarchy of processing compute more abstract descriptions, but local selectivity is maintained up to at least object recognition levels, and probably beyond. Analogous forms of local selectivity have been found throughout all perceptual and motor modalities. In contrast, many other experiments clearly show that activity in the visual cortex is strongly affected by a surrounding context that extends far beyond the "classical receptive field" (Gilbert 1992; Kovács 1996). There is now evidence for many forms of contextual modulation, dynamic organization, and task dependence throughout all levels of the visual system (Albright and Stoner 2002; Lamme and Spekreijse 2000). Such findings provide clear support for the holistic tradition and have led some to doubt the value of the classical localist tradition, emphasizing instead the rich nonlinear dynamics of highly connected networks. As it is clear that the arguments and evidence are very strong for both localism and holism, a better understanding is needed of how they are combined.

What Is Dynamic Coordination?

Though various aspects of the notion of dynamic coordination have long been implicit in neuroscience, it is still in need of a clear explicit expression. Here, we provide an informal perspective and briefly discuss more formal views. For further discussion of coordination and its functions, see von der Malsburg (this volume).

Simply put, dynamic coordination refers jointly to fundamental neurocomputational functions such as contextual gain modulation (Phillips and Silverstein 2003; Tiesinga et al. 2005), dynamic grouping (e.g., Singer 2004), dynamic linking (e.g., Wolfrum et al. 2008), and dynamic routing (e.g., Fries 2005). All imply that novel context-sensitive patterns of macroscopic activity can be created by modulating the strength and precise timing of local neural signals without corrupting the information that these local signals transmit. For example, consider attention, which is a paradigmatic case of gain modulation. There is ample evidence that attention amplifies or suppresses signals conveyed by local processors, but with little or no effect on the receptive field selectivity of the cells generating those signals (Deco and Thiele 2009; Reynolds and Heeger 2009; Maunsell 2004). There is also ample evidence that concurrent contextual inputs from far beyond the classical receptive field can produce similar modulatory effects on the strength and timing of local signals (Gilbert 1992; Lamme 2004). Evidence concerning contextual modulation, grouping, linking, and routing was discussed in depth at the Forum and is reviewed in the following chapters.

The proposed distinction between coding and coordinating interactions can be clarified by considering coordination at a molecular level. The genetic code that was first discovered in the structure of DNA concerns the mapping of codons (i.e., sequences of three bases along the DNA chain) to amino acids.

This coding has remained essentially constant throughout the history of life on Earth. We now know that expression of those codes is controlled in a highly context-sensitive way by noncoding sequences and epigenetic mechanisms to ensure effective adaptation to current circumstances. Furthermore, we know that only a small percentage of codons code for amino acids and that the majority are involved in coordinating their expression. Thus, the effects of coding sequences, noncoding sequences, and epigenetic mechanisms are seamlessly intertwined, but coding and coordinating mechanisms have nevertheless been clearly distinguished.

Our working assumption is that the distinction between coding and coordinating interactions can also be applied to brain function. At each level of hierarchical processing, single-unit or population codes are used to represent many diverse entities. At each level, expression of those codes is flexibly controlled in a highly context-sensitive way by various coordinating interactions to ensure effective adaptation to current circumstances. The effects of coding and coordinating interactions are seamlessly intertwined in all neuronal activity and behavior, but they are clearly distinguishable. Under natural conditions the selection, timing, and amplification or suppression of signals is coordinated with concurrent stimuli and current goals. The contribution of these coordinating interactions can be greatly reduced by presenting isolated stimuli to anesthetized animals. When this is done we find a highly selective and hierarchical functional architecture, as revealed by much of neuroscience since the 1950s. The selectivity thus discovered has led to fundamental concepts such as that of receptive field selectivity and regional specialization. In stark contrast to the genetic code, however, receptive field coding is not fixed; it adapts to the environment in which the organism finds itself. This adaptation is usually gradual, as when visual cortex adapts to the statistical structure of visual input, but it may also operate more rapidly. When it does, the new codes must remain reliable over the time span of their use, and their rate of creation must not outrun the ability of projective sites to interpret them correctly.

As our primary focus is on dynamic coordination rather than on local selectivity, we must first make clear that local selectivity does make a crucial contribution to neural activity in awake-behaving animals in natural environments. First, any system that learns from examples must find some way to overcome the "curse of dimensionality" (Bellman 1961; Edelman 2008a); that is, an exponential increase in the number of samples required with dimensionality. This can be done by first mapping the input into low-dimensional subspaces or manifolds. Fortunately, neural systems can do this because the natural input they receive has a hierarchical structure that can be exploited to reduce dimensionality. Second, there is plenty of evidence showing that neural systems discover and exploit this hierarchical structure. For example, at the macroscopic level of neuroimaging, cognitive neuroscience is replete with evidence for reliable local selectivity within and between cortical regions in awake-behaving subjects, as well as across subjects within and between species. Much remains to be discovered, however, concerning this local selectivity. *Prima facie* evidence that "context" modifies local selectivity may therefore be more easily interpreted by refining notions of the local selectivity involved.

Thus, in arguing for a central role for coordinating interactions, we are not arguing against a role for local selectivity. Rather, we take that as a given and then argue that the selectivity is, and must be, so great that coordination is inevitably required. Figure 1.1 demonstrates the modulatory role of context in vision to emphasize our central concern with the Gestalt level of organization created by these coordinating interactions. Contextual modulation is common in perception and can be seen as using Bayesian inference or prior probabilities to interpret input (Schwartz et al. 2007). In response to discussion of the concept of dynamic coordination (e.g., Moser et al., this volume), we must

Figure 1.1 Demonstration of the strength and validity of contextual disambiguation, based on a computer-generated demonstration of lightness constancy by Edward Adelson (MIT) and on a painting by René Magritte. The upper panel is a photograph (taken by Ben Craven, http://bencraven.org.uk/) of a real checkerboard and pipe. The central light-gray square of the checkerboard actually transmits less light to the camera (or eye) than the arrowed dark-gray square because of the pipe's shadow. The lower panel, where all but the arrowed black square and the central white square of the checkerboard are deleted, demonstrates this; cues that the visual system uses to disambiguate reflectance from incident light intensity have been removed. Most people will see the central square as being much lighter in the upper panel than in the lower. The central squares are, however, identical and are in both panels darker (brightness about 27, as measured by Photoshop) than the arrowed square (brightness about 33). Such contextual disambiguation is ubiquitous throughout perception, demonstrating that the capabilities of attention are highly constrained. We cannot voluntarily ignore the surrounding context (e.g., when looking at the central square in the upper panel). We think that we can attend to independent elements of sensory awareness, but we cannot. This is but one of several common, but false, beliefs about the nature of consciousness.

emphasize that the concept does not imply that coordinating interactions have no effect on local circuit activities. On the contrary, they can have significant effects. The point is that these effects preserve local selectivity. This obviously applies to dynamic grouping, linking, and routing, and in relation to contextual disambiguation by gain modulation. Tiesinga, Fellous, and Sejnowski (2008:106) state that "multiplicative gain modulation is important because it increases or decreases the overall strength of the neuron's response while preserving the stimulus preference of the neuron."

To some it might seem that, in principle, coding and coordinating interactions cannot be distinguished, because signals necessarily transmit information about everything that affects them. This intuition is misleading. The information that is transmitted specifically about a modulatory or coordinating input given the coding, or receptive field, tends to be negligible, even when that modulating input has a large effect on the transmission of receptive field information (Kay et al. 1998). Therefore, conditional mutual information measures can be used to distinguish coordinating from coding interactions (Smyth et al. 1996). By applying these measures to two alternative forced-choice data obtained in a texture segregation task with multiple cues, it was found—as predicted—that cue fusion is a coding rather than coordinating interaction, whereas attention involves coordinating rather than coding interactions (Phillips and Craven 2000).

Three formal conceptions are discussed by Engel et al. (this volume): coherent infomax (e.g., Kay et al. 1998), coordination dynamics (e.g., Kelso 1995), and predictive coding under the free energy principle (e.g., Friston 2009; Friston and Keibel 2009). They were found to have much in common, including emphasis upon the necessity for dimensionality reduction, correlations between dimensions, and the distinction between driving and modulating interactions. A major difference in emphasis between coordination dynamics and coherent infomax is that the former is concerned with coordination in general, whereas the latter has been predominantly used in relation to dynamic coordination in particular. Another difference is that the formal analyses of coordination dynamics have been applied mainly to sensorimotor coordination, whereas coherent infomax has been applied mainly to perception. Optimization under the free energy principle is the most general and extends beyond coherent infomax by formally including a loop back from motor output to sensory input, thus enabling the system to achieve its objectives both by adapting itself to its world, and its world to itself.

Above we used the notion of dynamic coordination to mean coordination that is generated on a moment-by-moment basis so as to deal effectively with the current situation. It contrasts with prespecified coordination, i.e., those activity patterns that are explicitly specified and available before the current tasks and stimulus conditions are known. Highly practiced or stereotypical patterns of coordinated activity do not have to be created dynamically as they are already available. For example, the coordinated motions of the limbs that produce normal locomotion are so well specified prior to their use that they do not need to be created anew each time. Such skills may be acquired or refined by learning, but once acquired they function as prespecified components from which appropriate patterns of activity can be built. Stereotypical patterns of coordination can be highly stable and can act as strong attractors in the dynamic landscape (Kelso 1995). Dynamic coordination, by contrast, is that which cannot be prespecified because the relevant stimulus conditions or tasks are not known until they arise. For example, in figure–ground segregation, the particular set of low-level features that are to be grouped to form a Gestalt can often be determined only after the stimulus is available (Watt and Phillips 2000). Prespecified and dynamic coordination may be seen as the contrasting ends of a spectrum, rather than as a dichotomy, because novel patterns of coordination are usually built from familiar components. In some cases the novelty may be great; in others small. Possible ways of characterizing this spectrum and of placing various cognition functions on it, such as Gestalt perception, attention, and cognitive control, were major topics at this Forum.

Finally, we wish to emphasize three points concerning population coding, hierarchy, and downward causation. First, we assume that coding is more reliable at the level of the local population, than at the level of single cells. Second, interpretations computed at higher levels of the hierarchy depend upon coordinating interactions at lower levels, and higher-level interpretations serve as a context that modulates activity at lower levels. Third, as holistic Gestalt organization can have large effects on local activity, it provides clear examples of downward causation from macroscopic to microscopic levels.

What Neural Mechanisms Express Dynamic Coordination?

The dilemma mentioned earlier—how to guarantee stability in the representation of local information transmission while permitting flexibility in building situation-dependent representations—can only be solved by clearly discriminating between local information transmission and relational Gestalt organization. One possibility is temporal binding, according to which the neurons to be bound into a group synchronize their signals in time, such that neuronal identity and rate of firing in a given interval signals local meaning, whereas the temporal fine structure expresses relations to other neurons. Neurons would thus always convey two orthogonal messages in parallel: they would first signal whether the feature for which they serve as a symbol is present and then they would communicate with which other neurons they cooperate in this very moment to form a holistic representation.

One common view proposes that feature identity and salience are conveyed by labeled line and rate codes. The binding-by-synchrony hypothesis proposes that, in addition, grouping information is conveyed by the temporal fine structure of the discharge sequence which allows for the definition of precise temporal relations between simultaneously active neurons, especially if they engage in oscillatory activity (Singer, this volume). In this way, graded relations can be encoded by varying degrees of synchrony and/or phase shifts and nested relations by phase locking among oscillations of different frequencies, known as n:m locking. Such dynamically organized spatiotemporal patterns self-organize on the backbone of networks with fixed anatomical connections and change rapidly in a context-dependent way without requiring changes of the backbone, because effective coupling among anatomically connected neurons is not only determined by synaptic efficiency but also by the temporal relations between the respective discharges, and in dynamic systems these can change rapidly (Womelsdorf et al. 2007). A second hypothetical mechanism would be based on rapidly and reversibly switching synapses, such that groups of neurons that are to express integration into a pattern do so by momentarily activating a network of mutual connections. These two mechanisms are not mutually exclusive and may indeed be like two sides of the same coin: signal correlations switching synaptic strengths, synaptic strengths modulating signal correlations (von der Malsburg 1981/1994). Alternately, synapses could be actively controlled by dedicated "control units" (Lücke et al. 2008). A third mechanism would require that for each local feature to be signaled (such as a visual edge at a given position and orientation), there must be several alternate neurons, each carrying their own distinct connection patterns. By selecting among these neurons, the system could express integration of the signaled fact into this or that network. Figure–ground separation could thus be expressed by representing each possible visual element by two neurons: an f-neuron and a g-neuron. Within the figure, f-neurons would be active; within the ground, g-neurons would be.

Mechanisms for contextual gain modulation, including those underlying attention, play a major role in dynamic coordination; they amplify activity relevant to current stimulus and task contexts while suppressing activity that is irrelevant. One mechanism discussed at this Forum concerns the "windows of opportunity" for pyramidal cell spiking that are provided by synchronizing the activity of fast-spiking inhibitory interneurons (see Whittington et al., this volume). This mechanism is clearly modulatory rather than driving, and relates dynamic coordination to synchrony, rhythms, and psychopathology.

To What Extent Does Dynamic Coordination Depend on Distributed Self-organization and to What Extent on Executive "Control"?

A pervasive idea holds that dynamic coordination is the result of interpretation and strategic planning by a central executive housed in prefrontal cortex (PFC). Conceptual difficulties with this idea and evidence from several disciplines suggest, however, that coordination is primarily distributed. It arises predominantly from many local coordinating interactions between the local processors themselves, rather than simply being imposed by commands from the PFC. This does not contradict the idea that PFC may function to formulate, initiate, and monitor strategic plans, but rather emphasizes the dynamic coordination that arises through self-organization in addition to any contribution from the PFC.

As it is now well established that different areas within the PFC have distinct roles, PFC activities may themselves need to be dynamically coordinated. Thus, it is possible that basic mechanisms of dynamic coordination also operate within the PFC. The classic concept of self-organization does not include goal orientation, which is a hallmark of strategic planning. The extent to which goal orientation can arise within a highly distributed self-organized system is therefore a major issue that remains to be addressed in future research.

These issues can be seen as analogous to those concerning the organization of society. The assumption that the PFC is necessary to coordinate activity is equivalent to the assumption that without commands from a central government, humankind is left only with anarchy. Originally, "anarchy" referred to the order that arises without a ruler. That it has now come to mean the absence of order clearly reveals the common but false assumption that without a ruler there cannot be order.

Themes and Variations

In seeking a better understanding of dynamic coordination, we begin with the assumption that it takes several different forms, all sharing some basic properties. Throughout, our search for common themes is not intended to imply that there are no variations. On the contrary, one of the central goals is to understand both the themes and the variations. For example, when we ask whether there is a canonical cortical microcircuit that includes mechanisms for dynamic coordination, we assume that both themes and variations will be important.

Computational Goals of Dynamic Coordination

Coordination is a major issue within distributed computing technologies (Singh and Huhns 2005) and must often be solved dynamically at run-time, rather than being hard-wired (Moyaux et al. 2006). The same applies to perceptual organization in computer vision, and the great amount of research devoted to this in many labs clearly shows that it is a major, and, as yet, unresolved issue. Within the brain, two basic goals can be distinguished: contextual disambiguation, which uses context to disambiguate local signals, and dynamic grouping, which organizes them into coherent subsets. These goals require knowledge of what predicts what, and thus some form of learning analogous to latent structure analysis (e.g., Becker and Hinton 1992; Kay et al. 1998). At least three other goals can also be distinguished: dynamic linking, dynamic routing, and dynamic embedding. All contribute to the flexibility and power of cognition, as well as to Gestalt perception, attention, working memory, and thought, in particular.

Contextual Disambiguation

Local activity can be ambiguous in several ways. How is it to be interpreted? Is it relevant to the current task? These ambiguities can be greatly reduced by using information from the broader context within which the signals occur. Neurophysiology, psychophysics, and neuropsychology all show context sensitivity to be very widespread, and examples will be given in later sections.

Long ago work on computer vision demonstrated a crucial role for contextual disambiguation at higher levels of object and scene perception (e.g., Clowes 1971). Many recent computational studies also show the effectiveness of context-sensitive disambiguation at the level of local feature processing (Kruger and Wörgötter 2002). Bayesian techniques or, more generally, probabilistic techniques such as factor graphs are highly relevant, as context, given in the form of conditional probabilities, can be used to reduce uncertainty of interpretation of local data.

Dynamic Grouping

At each level of processing, activity must be dynamically grouped into coherent subsets so that it can have appropriate effects at subsequent levels. Without such grouping, storing sensory patterns in their entirety in memory would be of little use, because they are unique and have no direct connection to each other. It is necessary to extract significant patterns from a given input, patterns that generalize from one situation to others. These significant patterns are to be found by grouping mechanisms. The principles of Gestalt psychology provide criteria by which this grouping may be achieved. They have been much studied and discussed in relation to what has been referred to as the "binding problem." Previous discussions of binding use inconsistent terminology, however, so that our intended meaning must be clarified: "Grouping" means organizing a set of data into subsets upon which some operations are to be performed. Dynamic grouping and prespecified grouping must be distinguished, as explained above, because they have different requirements and constraints. Dynamic grouping must then be divided into attentive and preattentive, because some forms of dynamic grouping occur rapidly and in parallel across the visual field, whereas others are slower, serial, and require attention (Watt and Phillips 2000).

Dynamic Routing

The evidence for functional localization and automatic processing up to high levels of analysis shows clearly that much inter-regional communication is prespecified. Many cognitive functions require flexibility in communication between brain regions (Fries 2005), however, so the fundamental capabilities and constraints of such flexible routing, its neural implementation, and cognitive consequences were major issues of discussion at this Forum.

Dynamic Embedding

Groups within groups can be described as being "embedded" or "nested." This embedding is dynamic when it is created as required by novel circumstances, rather than being given by a prespecified hierarchy. It is most obviously relevant to linguistic syntax and metacognition (e.g., thoughts about thoughts that incorporate novel syntactic structures within novel syntactic structures).

To demonstrate more precisely what we mean by dynamic embedding, let *ABCD* stand for variables and *abcd* represent their realized values in a particular instance. Prespecified grouping processes specify a subset of items on which to perform some operation: grouping *ABCD* as (*AB*) (*CD*) organizes four variables into two subsets in the same way for all values of the variables. Dynamic grouping processes also organize the items into subsets, but can do so only after the realized values are known: grouping *abcd* as (*ab*) (*cd*). Most models of hierarchical processing assume prespecified levels of embedding: grouping *ABCD* as ((*AB*) (*CD*)), i.e., as two groups at a lower level and as one group at a higher level. Such networks have considerable computational power, but cannot create novel hierarchies of embedding. An important issue is therefore whether meta-cognitive processes, such as analogies, syntax, and thoughts about thoughts, require dynamically embedded groupings (Hummel and Holyoak 2005). If so, some phylogenetic and ontogenetic differences may be due to the evolution and development of that specific capacity (Penn et al. 2008). The neuronal dynamics that could embody any such capability have yet to be discovered. Using a simple marker that signals whether two elements are members of the same group, such as whether the spike trains are synchronized or not, would not be adequate (Hummel and Holyoak 1993) because different markers would be required for different levels of grouping. One suggestion is that dynamic embedding may involve synchronizing activity within different frequency bands; however, this hypothesis now requires further examination.

Dynamic Coordination at Four Levels of Organization

Corresponding to the framework of this Forum, we outline the specific issues that were relevant to each discussion group.

Evolution of Dynamic Coordination

Working on the assumption that dynamic and prespecified coordination cannot be disentangled, Balaban et al. (this volume) focused extensively on comparative anatomy and coordination in general. Clearly, their discussions point out the need for a better understanding of the distinction between prespecified and dynamic coordination.

If processes of dynamic coordination make major contributions to Gestalt perception, attention, and working memory, then they should be found in many, if not all, species. If, in addition, processes of dynamic coordination play a central role in higher visuospatial cognition, relational reasoning, language, and fluid intelligence in general, then some of those processes may vary greatly across species, while some may be found in only a few. Important evolutionary issues arise, therefore, at the level of microcircuits, system architectures, cognitive capabilities, and subcortical structures.

Microcircuits

What microcircuit commonalities exist across species? What role do any such commonalities play in dynamic coordination? Is there evidence of differences in the microcircuitry across species that could be relevant to differences in their ability to coordinate activity? In particular, we need comparative studies of the local circuit mechanisms identified below as making major contributions to dynamic coordination.

System Architectures

System architectures vary greatly across species, but are any of these differences of particular relevance to dynamic coordination? For example, as long-range lateral and descending connections play a major role in dynamic coordination, comparative studies of their size or organization may reveal relevant evolutionary changes.

Cognitive Capabilities

To what extent does the evolution of cognitive capabilities arise from the evolution of new forms of dynamic coordination? Are there evolutionary quantum leaps in the ability to coordinate behavior? How do Gestalt perception, attention, working memory, and executive control evolve? Are some forms of thought or relational reasoning found in only a few species, such as humans? Do they depend upon dynamic mappings between novel relational structures? To what extent does human language depend upon capabilities already present in higher forms of visuospatial cognition, and to what extent upon underlying capabilities unique to language? Does consciousness involve dynamic coordination, and do new forms of consciousness evolve as new forms of dynamic coordination emerge? Do episodic memory and insightful problem solving require any special form of coordination? Though there is relevant comparative data on such cognitive capabilities (Grodzinski and Clayton, this volume), much more is needed.

Subcortical Structures

Do subcortical structures, such as the basal ganglia, play a major role in dynamic coordination, and if so, do their roles change in the course of evolutionary development? Such questions are of particular importance because there has long been evidence that tonically active neurons in the striatum play a central role in coordinating the distributed modular circuitry of corticobasal channels (Graybiel et al. 1994; Aosaki et al. 1995), thus providing an analogy in adaptive motor control to "binding" processes hypothesized to contribute to the coordination of perceptual processes.

Dynamic Coordination in Local Cortical Microcircuits

Much is known about the anatomical and physiological properties of local cortical microcircuits. Thus, Mehta et al. (this volume) focused on the following issues.

Canonical Cortical Microcircuits

Is there a canonical cortical microcircuit (e.g., as dicussed by Douglas and Martin 2007), and, if so, what are its major variants? Commenting on evidence for the basic nonuniformity of the cerebral cortex, Rakic (2008) concludes that although cortex is generally organized into vertical columns, the size, cellular composition, synaptic organization, and expression of signaling molecules varies dramatically across both regions and species. Our working hypothesis is that, although there is variation, it is not so great as to remove all functional commonalities.

Elementary Computational Operations and Mechanisms

What elementary computational operations are performed by cortical microcircuits, and, in particular, are there mechanisms that make a special contribution to dynamic coordination? Possible mechanisms include those underlying "modulatory" as contrasted with "driving" interactions (Sherman and Guillery 1998), "contextual fields" as contrasted with "receptive fields" (Kay et al. 1998; Phillips and Singer 1997), local circuit mechanisms for gain control (Tiesinga and Sejnowski 2004), and local circuit mechanisms for coordinating phase relations between rhythmic activities (Whittington and Traub 2003). Rhythms were thought to be highly relevant because they have a special role in dynamic grouping, attention, and other relevant cognitive processes (Buzsáki 2006).

Is Feedforward Transmission Driving?

How and why does feedforward transmission (e.g., from thalamus to primary sensory regions) have such a dominant effect on activity, when it constitutes less than 10% of synaptic input (e.g., to layer IV cells)? From a functional point of view, it seems likely that this occurs because feedforward transmission is primarily driving, whereas lateral and descending connections are primarily coordinating.

Glutamate Receptor Subtypes

Are some glutamate receptor subtypes of particular relevance to dynamic coordination? *Prima facie*, NMDA receptors seem to function as highly selective gain controllers and thus could help mediate coordination within the cognitive system (Phillips and Singer 1997). This hypothesis is supported by evidence that NMDA receptor malfunction or dysregulation constitutes a crucial part of the pathophysiology of cognitive disorganization in psychosis (Phillips and Silverstein 2003).

It is usually said that NMDA receptors are relevant to cognition because of their role in synaptic plasticity and learning. This view, however, neglects all of the evidence which shows that they have large and immediate effects on ongoing activity. NMDA receptors amplify activity that is relevant to the current context and suppress that which is irrelevant. Therefore, they have a major impact on current information processing—not simply on learning. Indeed, it can be argued that their role in learning is secondary to their role in current processing (i.e., synaptic plasticity tends to record those patterns of neural activity that were selectively amplified because of their coherence and relevance to the context within which they occurred). Differences between NMDA receptor subtypes may be of particular importance because they have different temporal dynamics. For example, the 2A subtype has faster deactivation kinetics than the 2B subtype, and thus may play a special role in rapid and precise cognitive processes.

Inhibitory Interneurons

What do inhibitory local circuit interneurons and the various subtypes of GABA receptors contribute? They have a central role in generating and coordinating rhythmic activities (Whittington and Traub 2003), in enhancing attended activities (Tiensinga and Sejnowski 2004), and in suppressing those that are irrelevant. Thus, in combination with NMDA receptors, they may play a central role in dynamic coordination. This hypothesis is supported by evidence implicating GABAergic neurotransmission in the pathophysiology of disorganized cognition (e.g., as in schizophrenia; Lewis et al. 2005).

Apical and Distal Dendritic Compartments

Are some coordinating interactions predominantly mediated by apical and distal dendritic compartments? *Prima facie*, it seems that basal and proximal synapses are well placed to have a central role in driving postsynaptic activity, and that apical and distal compartments are well placed to receive and integrate inputs from the modulatory context. Modeling studies support this hypothesis (e.g., Körding and König 2000; Spratling and Johnson 2006).

Windows of Opportunity Created by Rhythmically Synchronized Disinhibition

Pyramidal cells receive strong perisomatic inhibitory input from fast-spiking basket cells, which temporarily prohibits spiking. "Windows of opportunity" for pyramidal cell spiking are provided by the periods of recovery from this inhibition. The effects of excitatory inputs to principal cells can therefore be modulated by controlling these windows of opportunity. In part, this depends upon how synchronized the inhibitory inputs are, because when they are synchronized so are the periods of recovery from inhibition. Models of gain modulation through synchronized disinhibition show that it could play a major role in attention, coordinate transformation, the perceptual constancies, and many other cases of contextual disambiguation (Salinas and Sejnowski 2001; Tiesinga et al. 2008). These models show that such gain modulation is particularly effective at gamma frequencies. Furthermore, as synchronization of the phases of these windows of opportunity within and between cortical regions could contribute to dynamic grouping, linking, and routing, they could play a central role in all four main neurocomputational functions referred to jointly as dynamic coordination. Thus, this aspect of local circuit function may play a pivotal role in future studies of dynamic coordination. Furthermore, it is also closely related to NMDA and GABAergic function, gamma and beta rhythms, and the pathophysiology of cognitive disorganization in disorders such as schizophrenia (Roopun et al. 2008).

Synaptic Assemblies

The notion of "synaptic assemblies" is outlined by Mehta et al. (this volume). It potentially provides a far richer and more dynamic concept than does that of Hebbian "cell assemblies," and may be more centrally involved in dynamic coordination. Most importantly, it is inherently a relational concept, as it is concerned with connections between cells, rather than simply with the activity of the cells themselves. This resonates with the idea of the rapid formation, matching, and dissolution of effective network architectures as used within dynamic link architectures (Wolfrum et al. 2008).

Dynamic Coordination in Brain Systems

Lateral and Descending Connections

Is coordination in perceptual hierarchies predominately achieved through lateral and descending connections? Our working assumption is that this is so, at least to a first approximation. If feedforward drive is the primary determinant of receptive field selectivity, then lateral and descending connections may modulate the effects of that drive so as to increase overall coherence within and between different levels of the hierarchy (Lamme and Roelfsema 2000). An analogy with Bayesian techniques may be of relevance here because feedforward pathways can be seen as transmitting the data to be interpreted, whereas lateral and descending pathways carry information about conditional probabilities that are used to help disambiguate the perceptual decisions (Körding and Wolpert 2004). As some lateral and descending connections may actively fill in missing data, these connections will need to be distinguished from those that are purely modulatory.

Contextual Modulation

What does the neurophysiological evidence for contextual modulation tell us about dynamic coordination? Many such studies have been conducted and all show that activity in perceptual areas is modulated by contextual input from far beyond the classical receptive field (Gilbert 1992; Lamme and Spekreijse 2000). Contextual disambiguation is therefore a widespread and common process in perceptual systems. Similar forms of contextual modulation may also apply to the neural activities underlying working memory and thought.

Neural Bases of Gestalt Grouping

How does neural activity discover and signal perceptual Gestalts? It is clear that temporal proximity plays a central role in determining what events are grouped, but one version of this is of particular relevance here: the hypothesis that dynamically created groupings are signaled by synchronizing the spike trains of activities that are to be processed as a whole, and by desynchronizing those that are to be segregated. This hypothesis has led to many empirical studies and much debate.

It is possible that information about different properties of specified regions of space (e.g., color, texture, depth, and motion) is linked via their reference to common spatial locations. This cannot apply, however, to the linking of information that arises from different regions of space as required to distinguish figure from ground. Psychophysical studies of contour integration and of lateral interactions between distinct spatial channels suggest that, in addition to the inputs specifying their receptive field selectivity, cells in visual cortex have "association fields" that amplify coherent groupings. Neurophysiological studies of this possibility are therefore of central importance (Seriés et al. 2003), and a central goal for future research must be to assess the extent to which the Gestalt criteria for dynamic grouping are implemented by such connections.

Temporal Structure and Synchrony

What is the role of synchrony and high frequency rhythms in neuronal signaling? One possibility is that spike rate and spike synchronization operate in a complementary way such that salience can be enhanced by increasing either or both. Evidence for this is provided by physiological studies of brightness induction (Biederlack et al. 2006) and by psychophysical studies which suggest a central role for synchronized rate codes in figure–ground segregation (Hancock et al. 2008).

Both synchrony and high frequency (gamma and beta) rhythms have been implicated in attention, short-term memory, associative memory, sensorimotor coordination, consciousness, and learning (Melloni et al. 2007; Singer 2004; Wespatat et al. 2004). Both were major foci for discussion at this Forum (see Singer and Moser et al., both this volume). In conscious perception, for example, selective attention aids the formation of coherent object representations by coordinating the resolution of many widely distributed local competitions, and physiological studies (e.g., Fries et al. 2001) suggest that synchrony plays a role in attentional coordination. Furthermore, computational studies (e.g., Tiesinga et al. 2005) show that synchronized disinhibition could play a fundamental role in contextual gain modulation, thus extending the potential role of rhythmic synchrony to cover all four fundamental neurocomputational functions, referred to jointly as dynamic coordination.

Dynamic Linking

How is dynamic linking in object recognition and graph matching generally achieved? As mentioned earlier, pattern recognition is best based on assessment not only of the presence of figural elements (feature-based recognition) but also of the correctness of their spatial or temporal relations (correspondence-based recognition). Such recognition amounts to graph matching; that is, to the comparison of a stored model of the pattern and the potential instance to be recognized, where both have the form of a graph with feature-labeled nodes and links to represent neighborhood relationships. During graph matching or recognition, dynamic links between corresponding nodes must be established, as modeled in Wolfrum et al. (2008).

Neuronal Bases of Attention

What are the neuronal bases of attention? This issue is of central importance because selective enhancement of the relevant and suppression of the irrelevant are major forms of dynamic coordination. Many specific issues concerning attention arise: How is attention related to synchrony? Do stimulus-based and task-based influences on salience have essentially similar kinds of effect? What determines the balance between them? What is the current status of the neurobiological evidence for biased-competition models, and what further evidence is needed? Why do attention and working memory have such limited capacity? Did evolution get stuck in a dead end somehow, or do these limitations reflect fundamental underlying computational constraints? What can we learn from neurological disorders of attention? Do Bayesian models of salience and visuospatial attention clarify these issues?

One major conclusion was that biased competition, in the form of normalization models (Reynolds and Heeger 2009), is currently best able to explain the modulatory effects of attention at the local circuit level.

Dynamic Routing

How flexible is the routing of information flow between cortical areas, and how is that flexibility achieved? Communications between brain regions are largely prespecified by the cortical architecture. For example, V2 responds to activity in V1 and affects activity in V3 and V4, regardless of the current task. Some communication must be flexible, however, as some tasks and circumstances require novel routing of information (e.g., from sensory to motor areas). The central proposal discussed at the Forum was that such dynamic routing is achieved by coordinating the phases of transmission and receptivity across brain regions (Fries 2005; Womelsdorf et al. 2007), and much evidence was thought to support this proposal.

The Role of Prefrontal Cortex

In what ways does the PFC have a special role in coordination, and how well is activity coordinated by distributed processes of self-organization independently of any contribution from PFC? Does activity in the various components of PFC need to be coordinated, and, if so, how is that achieved?

Dynamic Coordination in Cognition and Behavior

The computational capabilities provided by dynamic coordination at the neuronal level have major consequences for Gestalt perception, attention, working memory and thought, and thus for learning as well. We assume that they are crucial to language, but have focused more on their relevance to visuospatial cognition.

Gestalt Perception

What do psychophysical studies tell us about Gestalt organization in perception? Does it operate within all perceptual pathways? How does it develop?

Evidence indicates a major role for Gestalt organization in dimensionality reduction and contextual modulation (Kovács, this volume). However, although knowledge of contextual relationships in image structure begins to be acquired in the first few months of life, evidence shows that it continues to develop over several years in human vision (Doherty et al. 2010; Kovács 2000).

Contextual Disambiguation

To what extent does cognition use context to reduce local uncertainty? A vast literature is available showing context sensitivity in a wide range of cognitive domains. Our focus was on consistency constraints in the perception of edges, surfaces and objects, because they provide paradigmatic examples of such coordinating interactions.

Structured descriptions are fundamental to cognition. It is obvious that language depends on descriptions generated by a grammar, which, though finite, enables us to produce and understand infinitely many novel sentences. Analogous "grammars" have been used to interpret line drawings of opaque polyhedra (solid objects with straight edges) in computer vision (e.g., Clowes 1971). Clowes demonstrated formally how local ambiguity can be reduced by using "co-occurrence restrictions," and showed their relevance to human vision using demonstrations of ambiguity, paraphrase, and anomaly, just as Chomsky did for language.

Figure 1.2a is ambiguous in several ways: it can be seen as a tent or as a book standing up (i.e., the central vertical edge can be seen as folding either outwards or inwards). Changes of viewing angle reduce this ambiguity (Figure 1.2a, b), thus demonstrating how context resolves local ambiguity; the arrow and fork junctions that are ambiguous in Figure 1.2a are disambiguated by the other junctions in Figure 1.2b and Figure 1.2c.

The equivalence of one interpretation of Figure 1.2a with Figure 1.2b and of another with Figure 1.2c shows that perception distinguishes between scenes and images, just as language comprehension distinguishes between deep and surface structures. Figure 1.2d is evidence that objects are seen as structures, and not merely as members of a prespecified category. Though you had never seen a tent with such a hole before, you have now, and that required you to create an internal description of a new distal structure and to map image fragments to parts of that structure.

The local ambiguities in such pictures and their resolution using contextual constraints are made explicit in Figure 1.3, which is based on the computer vision system developed by Clowes (1971). His analysis is deterministic, but would be more relevant to human perception if formulated in Bayesian or other statistical terms.

The arcs linking compatible local interpretations in Figure 1.3 are emphasized here because they clearly show essential properties of coordinating interactions. They resolve local ambiguity by applying co-occurrence constraints

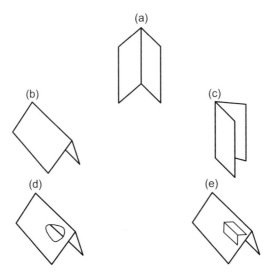

Figure 1.2 Local ambiguities disambiguated by context. The central edge of (a) can be seen as folding either inwards or outwards, because the arrow junction at the top and the fork junction at the bottom of that edge are ambiguous. This ambiguity is reduced in (b) and (c), where the context provided by the other junctions makes one interpretation more probable than the others. Drawing (d) shows that interpretation is not merely categorization, but relates structured descriptions in the picture and scene domains. Drawing (e) shows how context can reveal another possible (though improbable) interpretation of (a), i.e., as a chevron-shaped hole through which part of a more distant edge can be seen.

to select coherent subsets of the local interpretations that are made possible by the locally available data, thereby playing a major role in determining the Gestalt created. Assuming that different local interpretations are conveyed by different neuronal populations, this clearly demonstrates our assumption that macroscopic organization can have large effects on local microcircuit activity. Eleven vertices can be seen in the object in Figure 1.3. If each has at least three alternative interpretations, when considered independently, then there would be at least 3^{11} possible combinations. From that host of possibilities, coordinating interactions select the few that are coherent when taken as a whole. Our assumption is that this is done by selectively amplifying and grouping the neuronal activity representing local interpretations that are coherent.

These demonstrations do not imply that global coherence is necessary for perception. The importance of global coherence is often thought to be demonstrated by impossible figures. Take, for example, Figure 1.4, which is known as the devil's pitchfork because it has three round prongs at one end but two square prongs at the other. Such drawings do bring global inconsistencies to our attention, but they also show that contextual constraints are primarily local. We clearly see the various possible local interpretations, even though they are not globally coherent. Indeed, such drawings are only perceived as impossible

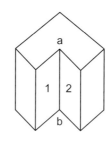
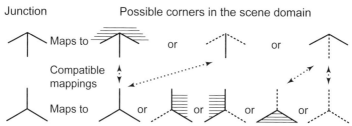

Figure 1.3 Example of how the Clowes' (1971) algorithm used contextual constraints to interpret line drawings of solid objects with plane surfaces and straight edges. The drawing at the top shows two further interpretations of junctions in Figure 1.2a. Here, regions 1 and 2 can be seen as being on two separate limbs of a block with a square column cut into one corner. The two upper edges of junction "a" would then be convex, and the edge joining "a" and "b" would be concave. Regions 1 and 2 can also be seen as the two sides of a square column rising into the top corner of a ceiling. The two upper edges would then be concave, and the edge joining "a" and "b" would be convex. Locally possible interpretations of the arrow junction at "a" and the fork junction at "b" as corners are shown below, with hatching indicating a surface that disappears behind an edge, solid lines indicating convex edges, and dotted lines concave edges. When junctions share an edge, as do "a" and "b," only certain combinations are compatible. Surfaces are regions surrounded by a set of compatible mappings from junctions to corners. Objects are a compatible set of surfaces.

because the local consistencies map so strongly on to partial three-dimensional structures. Without that, the drawing would simply look like an arbitrary collection of lines. Thus, such impossible figures can be thought of as showing the primacy of local over global coherence. An alternative perspective is to view this demonstration as showing that all elements of the whole figure may be bound if attended to as a whole, even though those elements are not seen as forming a single coherent structure.

Attention and Working Memory

What do cognitive studies of attention tell us about dynamic coordination? They seem to be of central importance, and show the conditions under which relevant signals can be enhanced and irrelevant suppressed. Many issues remain to be resolved, however, before we understand why and how this occurs. Cognitive studies support biased competition models of attention (Duncan

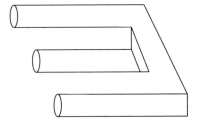

Figure 1.4 The devil's pitchfork. Although there is no globally coherent interpretation as a three-dimensional object, local consistencies are strong enough to produce partial perception of three-dimensional structure. This can be thought of as showing the primacy of local over global coherence.

2006), and their more recent formulation as normalization models (Reynolds and Heeger 2009), but how can that support be clarified and strengthened? Are competitions resolved by a single central executive controller or locally, as suggested by independence of their resolution across modalities (Hupé et al. 2008)? What is the role of attention in creating coherent percepts? How are conflicts between stimulus-driven effects on salience and task-driven effects on relevance resolved? Attention is needed to bind features that are not already bound pre-attentively (Treisman 1999). However, some dynamic grouping occurs pre-attentively (Watt and Phillips 2000). So, what determines whether or not attention is required? What do cognitive studies tell us about how and why the capacities of attention and working memory are so limited? Is working memory a major source of contextual influence on current processing, and how convincing are the neuronal models proposed as explanations of those influences (e.g., Kerns et al. 2004)? Some working memory studies have been interpreted as indicating the importance of dynamic grouping that is implemented by neuronal synchrony (e.g., Luck and Vogel 1999), but how strong is that evidence? Can theories of attentional selection based on dynamic link architectures (e.g., von der Malsburg and Schneider 1986) be tested psychophysically? Although these issues were discussed in depth at this Forum (see Engel et al., this volume), it is clear that they provide major goals for future research.

Reasoning

Are there any theories of relational thinking that are both psychologically and neurally plausible? Does relational reasoning require special forms of dynamic coordination? Though Hummel and Holyoak (2005) imply that it does, they propose a model in which the critical binding of roles to fillers is achieved using the same form of synchronization thought to underlie dynamic grouping in general.

The use of context to resolve local ambiguities, as shown in Figure 1.3, is an example of an approach to constraint satisfaction that can generally be applied to reasoning and problem solving. As the network of constraints grows, the possibility of checking for global consistency becomes less feasible; thus it is unlikely that fully global consistency can be established for rich knowledge structures, such as those in human cognition. The best that can then be done is to maximize local consistencies and minimize local inconsistencies (Dechter 1992).

Central Executive Functions

Do cognitive studies imply the existence of a central executive, and, if so, what are its specific responsibilities and capabilities? What do they tell us about its internal organization, and about the way in which the activities of its separate components are coordinated? Do these cognitive studies lead to the same conclusions as physiological investigations of PFC? Among the participants at this Forum, there was wide agreement that executive processes (e.g., cognitive control) play an important role in coordinating other activities, and are themselves highly dependent on the dynamic coordination of their various components.

The Distinction between Multimodal Coordination and Multimodal Sensor Fusion

How is multimodal coordination related to sensor fusion? Multimodal "integration" has been extensively studied, but the term "integration" is ambiguous. It often refers to sensor fusion, where inputs from different modalities are combined so as to produce an output signal within which the separate contributions are not distinguished. This is not the same as multimodal coordination, however, because in that case the separation between modalities is maintained even though the signals that they generate are coordinated.

Bayesian Analyses

Are Bayesian analyses relevant? Discussed in depth at this Forum (see Triesch et al., this volume), we think that they are useful because they account for much behavioral data. In addition, they compute posterior probabilities using data and priors in fundamentally different ways that match our distinction between driving and modulatory interactions, particularly if the Bayesian inference is implemented so as to amplify transmission of predicted signals (e.g., Kay and Phillips, submitted; Spratling 2008).

The Relation between Cognition and the Precise Temporal Structure of Neural Activity

What is the status of the evidence relating synchrony and brain rhythms to cognition? Addressed by both Tallon-Baudry and Engel et al. (this volume), it was agreed that such studies may be pivotal, potentially relating fundamental cognitive functions to neural activity at the population and local circuit levels. Furthermore, as noted next, they also provide a link to psychopathology.

Disorders of Dynamic Coordination

What do failures of dynamic coordination in various forms of psychopathology tell us about its role in normal brain function? What do studies of the pathophysiology of those conditions reveal about its neuronal mechanisms? Silverstein (this volume) shows that psychopathology can provide a rich source of evidence linking behavior and cognition to neural activity at population and synaptic levels. In support of this view, consider the findings of Roopun et al. (2008). They show how the windows of opportunity for pyramidal cell spiking at gamma and beta frequencies depend on GABAergic and NMDA receptor activity, and they relate those mechanisms to changes in cortical dynamics that occur in schizophrenia. Furthermore, they provide evidence that there are variations in those rhythms and mechanisms across cortical regions, so we now need to discover the computational, cognitive, and psychopathological consequences of such variations. Finally, it was agreed that the theories of Friston (1999) and of Phillips and Silverstein (2003) have much in common. Friston suggested that the "disconnection" theory of schizophrenia could be referred to as the "dysconnection" theory. This label would then apply to both theories if it is understood that it is not connections in general that are malfunctional, but coordinating connections in particular.

Conclusion

Dynamic coordination makes fundamental contributions to brain function and cognition. Much is already known about why and how it does so, but much more remains to be discovered. Many concretely specified issues and predictions have now been identified that can be investigated using well-established or recently developed techniques. Over the coming decades, we expect substantial advances in our understanding of dynamic coordination in the brain, thereby building reliable bridges between neurobiological and psychological perspectives on mental life.

2

Cortical Circuits

Consistency and Variability across Cortical Areas and Species

Jon H. Kaas

Abstract

Neurons in local circuits in the neocortex of mammals need both to extract reliably meaningful information from their dominant activating inputs and to modify their responses to these inputs in the context of inputs that are activating other local circuits. Neocortex has structural features to mediate both of these tasks. Neocortex is characterized by an arrangement of neurons with different types of inputs and outputs into six traditionally defined layers and a pattern of dense, vertical interconnections between neurons across these layers. This arrangement is consistent with the general conclusion that narrow, vertical columns of neurons interact intensely to form local-circuit processing units that reliably extract information from a small number of activating inputs that largely terminate in layer 4. In addition, neurons in these circuits are influenced by lateral connections that interconnect groups of columns, as well as by more widespread subcortical modulating inputs and feedback connections from other cortical areas. Some or all of these connections may provide contextual modifications within and dynamic coordination between the vertical arrays of cortical neurons. While basic features of columnar arrangements of cortical neurons and their connectional patterns with other columns are likely similar across cortical areas and mammalian taxa, they also clearly differ in ways that likely reflect areal and species requirements and specializations. Some of these differences are outlined here.

Introduction

The focus of this chapter is on the structural features of neocortex of mammals that allow information to be extracted from inputs to narrow vertical arrays of highly interactive neurons, the cortical columns or modules (Mountcastle 1997; DeFelipe 2005; Douglas and Martin 2007; Thomson and Lamy 2007), and on the features that allow neurons in different columns to interact dynamically.

Across mammals and within most cortical areas, neocortex is subdivided into six traditionally defined layers (Figure 2.1). Neurons within these layers have different functional roles based, in part, on having different inputs and outputs (Peters and Jones 1984). In brief, cortical activation is highly dependent on thalamic or cortical inputs into layer 4; layer 3 has lateral connections within the cortical area and projects to other cortical areas; layer 5 provides mainly subcortical projections; and layer 6 provides feedback connections to the thalamic or cortical source of layer 4 activation. The vertical connections between neurons of different layers are both very dense and very restricted in lateral spread (Figure 2.2). However, more sparse distributions of axons spread out laterally from neurons in layers 1, 3, and 5 to contact nearby neurons in other vertical columns of highly interconnected neurons. These lateral intrinsic connections provide the structural framework for interacting across cortical columns, possibly by inducing temporal synchronies between neurons in different columns (Singer and Gray 1995). In addition, widely distributed inputs from the brainstem and thalamus modulate the activity patterns across cortical columns and feedback connections from higher cortical areas, which are generally less specific in their terminations than the feedforward connections and are likely to have an integrating role.

While this brief depiction of the basic processing circuitry of cortex serves as a useful guide, it does not take into account the variability that exists in this circuitry across cortical areas and across mammalian species. As such

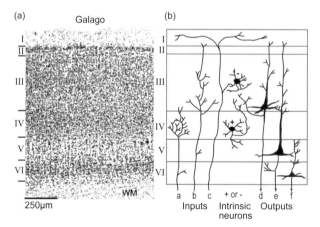

Figure 2.1 (a) The laminar arrangement of cells in the neocortex of a typical mammal (area 17 of a prosimian galago). Six layers are traditionally identified in most, but not all cortical areas. (b) The primary activating inputs from the thalamus or other areas of cortex, a, are to layer 4. Other thalamic and cortical inputs, b, are to other layers, while brainstem modulating inputs, c, are to layer 1. Intrinsic neurons are excitatory or inhibitory on other neurons, and they can be of several types. Output neurons are largely pyramidal neurons which project to other cortical areas, d, mainly subcortical targets, e, or provide feedback to thalamic nuclei or areas of cortex providing inputs.

Consistency and Variability across Cortical Areas and Species

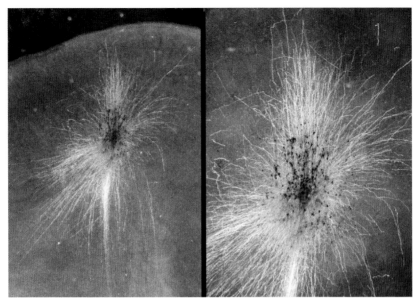

Figure 2.2 Neurons and axons labeled by an injection of a tracer into deeper cortical layers in a slice of neocortex of an owl monkey. Note the dense spread of axons to the more superficial layers immediately over the injection site (dense core of labeled neurons—dark spots), the sparseness of long lateral connections, and the bundles of output axons to the thalamus and brainstem, as well as to other areas of cortex.

variability in structure implies variability in function, some of the structural variability that occurs is reviewed here.

Cortical Layers Vary in Distinctiveness and Differentiation

Comparative studies suggest that early mammals had few cortical areas, on the order of 20–25, and that these areas were not very distinct architectonically from each other (Kaas 2007). In addition, cortical layers, although apparent, were not markedly different in cellular makeup or in histochemistry. No specialized agranular motor cortex was present (Beck et al. 1996). In sensory areas, layer 4 was populated mainly with intrinsic neurons rather than pyramidal neurons; however, because of their size, these layer 4 neurons could more appropriately be called stellate cells, rather than granule or powder (koniocellular) neurons, which are tiny cells found in the highly specialized sensory cortex of some mammals. The six main layers did not have very distinct sublayers. Overall, there were relatively few classes of neurons, the morphological and histochemical variations across neurons of a class were small, and the intrinsic connections in cortex were similar across areas in early mammalian species. Thus, except for functional differences imposed by the distinct activating inputs to areas of cortex, and the differing targets for outputs, the basic computations in the vertical cortical processing units across most cortical areas in early mammalian species were likely to be highly similar.

In many cortical areas of numerous mammals, this ancestral pattern of weak cortical lamination and limited cellular differentiation appears to have changed very little. There are, however, mammals where there clearly have been major modifications. In one example, the neocortex of whales and other Cetacea is so modified, and to some extent regressed, that some investigators have postulated that they had retained structural features of the "initial" mammalian brains (Glezer et al. 1988). More likely, their brains have been severely modified as an adaptation to their unusual marine lifestyle (Marino 2007). The cytoarchitecture of neocortex of Cetacea is unique in not having a detectable layer 4 (or possibly having a very meager layer 4), while layer 1 is very thick, layer 2 is pronounced, and other layers are very indistinct. Additionally, there is little difference from region to region that would suggest morphological specializations for different kinds of processing in different cortical areas (e.g., Hof and Van der Gucht 2007). Unlike other mammals, the majority of afferents from the thalamus, and presumably those that are feedforward from area to area, appear to go to layer 1 rather than layer 4. Thus, cortical circuits in these marine mammals appear to be quite different from those in most mammals.

The lamination pattern of primary visual cortex (V1) of tarsiers is perhaps at the other extreme of differentiation (Collins et al. 2005). In Nissl-stained sections, layer 3 has three distinct sublayers that differ in cell packing and cell sizes; layer 4 also has three distinct sublayers; layer 5 has two; and layer 6 has two (using the numbering of layers according to Hässler 1967, which places layers 4A and 4B of Brodmann in layer 3). The overall laminar appearance of area 17 is reminiscent of the distinct lamination of the optic tectum of predatory birds. Surely these morphological distinctions reflect functionally significant modifications of the basic circuitry of primary visual cortex in these highly visual predators.

These two extremes of cortical lamination patterns only hint at the great variability that exists in cortical lamination features across cortical areas and across mammals. Using histochemical, immunohistochemical, and receptor binding procedures, it is now possible to reveal laminar, areal, and species differences in a great number of factors that are likely to be important in neuronal circuit functions. For example, the monoclonal antibody Cat-301, which reacts with neurons associated with the magnocellular visual processing stream of primates, reveals different laminar patterns of antigen expression in primary visual cortex of cats, monkeys, and tree shrews (Jain et al. 1994). Layers across cortical areas vary in such features as the expression of synaptic zinc, cytochrome oxidase, parvalbumin, calbindin, vesicular glutamate transporters, and neurofilament markers. While layers in homologous cortical areas across species often have similar relative levels of expression of these markers, there is considerable variability (Hof et al. 1999; Wong and Kaas 2008, 2009a, b).

Layers and cortical areas differ in the patterns of expression of the various neurotransmitter receptors such that cortical areas can be recognized by a "fingerprint" of their neurotransmitter profile (Zilles et al. 2002). Although the functional significance of such structural and histochemical variability in cortical layers is not always clear, it suggests that the operations of cortical circuits likely vary with such features.

Neuron Densities and Proportions to Glia and Other Nonneural Cells Vary across Areas and Species

A frequent assumption among neuroscientists is that cortical modules contain approximately the same numbers of neurons (for a brief review, see Rakic 2008). In one study that is often cited in support of this assumption, Rockel et al. (1980) counted the number of neurons in a narrow strip (30 µm) of cortex through the depth of cortex for several areas and five species; he reported that these were a fairly constant number (about 110) as studied in all areas and species, except for the primary visual cortex of macaque and humans (about 270). The results of more recent studies do not support the conclusion that neuron density is constant, but instead indicate that there is considerable variation in neuronal density across cortical areas and across species. If all of neocortex is considered, one recent estimate is that the average number of neurons underneath 1 mm^2 of cortical surface varies by about three times across primate species (Herculano-Houzel et al. 2008). Furthermore, primate brains consistently have a larger number of neurons than rodent brains of a matching size (Herculano-Houzel et al. 2007). Consistent with the early observations of Rockel et al. (1980) on macaques and humans, all primates appear to have a much higher density of neurons in primary visual cortex than in other areas of cortex (Collins and Kaas, unpublished). Yet, this density varies with species, and other visual areas, especially V2, have higher values than most cortical areas. In macaques, primary somatosensory cortex (area 3b) also has an elevated density of neurons. If one makes the assumption that cortical processing columns are of the same width across areas and species, the numbers of neurons within such columns vary greatly. This would certainly impact the processing within a column.

Dendritic Arbors of Cortical Pyramidal Cells Vary in Extent across Cortical Areas and Species

Some of the best evidence for how pyramidal cell morphology varies across cortical areas and species comes from a series of studies by Elston and his coworkers. By injecting layer 3 pyramidal neurons with Lucifer Yellow and viewing labeled dendritic arbors in tangential cortical slices, these researchers have been able to demonstrate great variability in the sizes of the basal dendritic fields. In macaque monkeys, for example, basal arbors of layer 3 pyramidal cells were smallest in primary visual cortex and were progressively larger across visual areas V2, V4, TEO and TE; the largest arbors were for neurons in prefrontal cortex (Figure 2.3) (Elston 2003; Elston et al. 1999). In contrast, the layer 3 pyramidal cells in tree shrews had larger arbors in V1, while V2 and temporal visual cortex had neurons with progressively smaller arbors. Other features of dendrites are also variable (Elston et al. 2005). For example, the pyramidal cells in V1 of tree shrews had twice the number of dendritic spines as those of primates. In addition, Elston et al. (1999) reported that peak spine density, reflecting synaptic contacts, was over three times higher for layer 3 pyramidal cells in higher-order visual area TE than in primary visual cortex of macaque monkeys. Elston (2003:1134) proposed that such regional variations in pyramidal cell structure "are likely to underlie fundamental differences in cortical circuitry," leading to "different functional capacities." If the widths of cortical columns correspond to the widths of the fields of basal dendrites, columns with neurons that have widespread basal dendritic arbors would be larger than those having neurons with restricted arbors. Columns that are larger in diameter would typically have greater numbers of neurons, although this is not necessarily the case, as neuronal densities vary.

Pyramidal neuron sizes vary as well. The specialized Betz pyramidal neurons of primary motor cortex and Meynert pyramidal neurons of primary visual cortex are known for their extra large size, which appears to be a specialization for fast conduction of axon potentials over long distances. Meynert neurons project cortically to visual area MT (middle temporal) and subcortically to superior colliculus, whereas Betz cells project to motorneuron pools in the brainstem and spinal cord. Both Betz and Meynert neurons also have long widespread basal dendrites that summarize information over a larger expanse

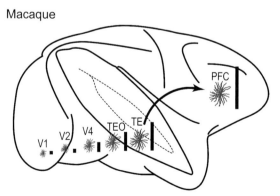

Figure 2.3 A dorsolateral view of the brain of a macaque monkey showing the relative sizes of the basal dendritic arbors of layer 3 pyramidal neurons in progressively higher-order visual areas (V1–TE) and in prefrontal cortex (PFC). Modified after Elston (2003).

of cortex than other pyramidal neurons. In a comparative study of Betz and Meynert neurons, Sherwood et al. (2003) found that terrestrial patas monkeys had larger Betz neurons than any of the great apes, even though patas monkey brains are about four times smaller. As Betz and Meynert neurons tend to be larger overall in larger brains, the authors suggested that the large Meynert neurons might be a terrestrial adaptation for visually detecting predators, a problem that could be more pronounced for patas monkeys, since they live in the open savannah. Also, several primates, including tarsiers, had larger Betz neurons than predicted from brain size. As small, nocturnal visual predators, tarsiers might benefit from rapid grasping as well as escape movements. Betz cells are smaller than expected in the relatively slow moving prosimian galagos (Otolemur). The large dendrite arbors of Betz and Meynert neurons suggest that they integrate information over several cortical columns to form a hypercolumn. Thus, cortical circuit processing would differ in cortical areas that have the large Betz or Meynert neurons.

As another modification of cortical pyramidal cells, humans, but not apes or monkeys, have a novel mesh of dendrites in layer 3 (Brodmann's layer IVA) of primary visual cortex (Preuss and Coleman 2002; Preuss et al. 1999). Because this meshwork is related to the magnocellular pathway, Preuss and Coleman (2002) suggest that this modification in cortical circuitry subserves the visual perception of rapid orofacial consequences of speech.

Other Neuron Types Also Vary with Cortical Area and Species

Perhaps the neuron type currently receiving the most attention by neuroscientists and other readers is the Von Economo neuron: a spindle-shaped neuron with a simplified dendrite arbor. These neurons were once thought to be found only in humans and certain great apes (Nimchinsky et al. 1999; Allman et al. 2002), but they now have been described in elephant and whale brains (Hof and Van der Gucht 2007). Von Economo neurons are unusual in that they are considerably larger than nearby pyramidal cells, while having a large apical dendrite extending toward the cortical surface and a single basal dendrite extending toward the white matter. They appear to be restricted to the anterior cingulate cortex and frontal insular cortex of great apes, humans, whales and elephants. One suggestion is that these spindle cells or von Economo neurons have a special role in neural mechanisms related to social and emotional functions (Seeley et al. 2006). Whatever the case may be, their presence, in only a few cortical areas of a few taxa with very large brains, indicates that all cortical circuits are not the same.

There is some suggestion that another rare type of pyramidal neuron—the inverted pyramidal cell—is more frequent in large-brained mammals (Qi et al. 1999). This neuron type, however, is sometimes thought to be a result of errors in development, rather than being a functionally distinct type. This possibility has not been suggested for the Von Economo neuron, although their appearance in the very large brains of members of distantly related taxa suggests that they could reflect developmental factors associated with such large brains.

Other studies have demonstrated variations across areas and species in the distributions of inhibitory interneurons (DeFelipe et al. 1999; Hof et al. 1999; Sherwood and Hof 2007). The inhibitory double bouquet cell is present in the cortex of primates but not in rodents, lagomorphs, or ungulates. The impact of this anatomical difference is uncertain, but this inhibitory neuron could be critical in constraining and forming receptive field response properties to sensory stimuli. Inhibitory neurons also vary in distribution across species and the thalamic nuclei that project to neocortex (Arcelli et al. 1997; Penny et al. 1984). For example, GABAergic neurons are typically found in the visual lateral geniculate nucleus, but often not in the somatosensory ventroposterior nucleus. Furthermore, in mammals with few GABAergic neurons in the thalamus, intrinsic and projecting neurons vary little in size, whereas intrinsic neurons are smaller than projecting neurons in species where thalamic intrinsic neurons are widespread.

Patterns of Intrinsic Horizontal Areal Connections Vary across Areas and Taxa

Horizontal cortical connections that link various vertical arrays of cells within cortical areas appear to exist in all mammals and in all cortical areas. Although those connections are most dense near their cells of origin, the sparser, longer horizontal connections seem well suited for the role of coordinating the activities of groups of columns of cortical neurons, as widely proposed (Gilbert 1992; Singer and Gray 1995). The surface-view patterns of the distributions of these horizontal connections, however, are quite variable, apparently in ways related to the functional organization of cortical areas. The differences in the patterns of intrinsic horizontal connections in primary visual cortex of tree shrews and squirrels are, perhaps, the most dramatic in this regard. Squirrels and tree shrews are highly similar, visually dominated, diurnal mammals with well-developed visual systems. They are also members of the same major branch of mammalian evolution (Euarchontoglires). Tree shrews, however, have a remarkably widespread, patchy distribution of intrinsic horizontal connections with any location in V1 (Rockland and Lund 1982; Sesma et al. 1984), whereas the distribution pattern in squirrels is diffuse and even, rather than patchy (Van Hooser et al. 2006). The reason for this difference appears to relate to how neurons selective for stimulus orientation are distributed in V1, as cells with similar preferences are adjacent in V1 of tree shrews but distributed in squirrels. Thus, the long-range horizontal connections in V1 of tree shrews (and some other mammals, including primates) are patchy as they interconnect distributed patches of neurons with matching orientation preferences, but they are not in squirrels (and most other mammals) where orientation selective cells are not grouped by preference similarity (for a review, see Van Hooser 2007).

The intrinsic connections in V1 of primates reflect another pattern that indicates that other functional classes of neurons are sometimes selectively interconnected. Primates have a distribution of cytochrome oxidase patches, called blobs, which are missing from V1 of most mammals. Neurons in the blob regions are thought to be especially involved in color processing, and not in mediating sensitivity to stimulus orientation. Overall, blob regions are connected over the long intrinsic connections with other blob regions, and interblob regions with the interblob regions (Yabuta and Callaway 1998); however, significant differences in these patterns exist such that in galagos, and likely other prosimian primates, the extra long intrinsic connections involve blobs (Cusick and Kaas 1988). Furthermore, in galagos even the callosal connections between blob regions of V1 are rather extensive, including blobs quite distant from the outer border of V1 representing the vertical meridian (Cusick et al. 1984). Thus, the intrinsic connection system involving blobs can be more widespread than that for interblobs, and species differ in the extents of these widespread connections between blobs.

As for V1, where intrinsic horizontal connections may or may not be patchy, motor cortex of cats has an even distribution of horizontal connections, leading to the conclusion that these connections "bind together" the representations of a variety of muscles (Capaday et al. 2009). In contrast, intrinsic horizontal connections are patchy in primary motor cortex of macaque monkeys (Lund et al. 1993), suggesting that motor cortex functions differently in monkeys than it does in cats. Finally, and for uncertain reasons, the intrinsic connections of prefrontal cortex in macaques terminate in stripes rather than patches (Levitt et al. 1993).

Distributions of intrinsic connections are influenced in other ways by the somatotopy of primary somatosensory cortex (S1 or area 3b). In the S1 representation of the whiskers of the face in rats, intrinsic horizontal connections are more extensive between the representations of anterior-posterior rows of whiskers than vertical arches of whiskers (Kim and Ebner 1999). In a similar manner, intrinsic connections in the hand representation in area 3b of monkeys are more extensive along the length of the representation of individual digits, than across these representations (Fang et al. 2002). In addition, although the face representation adjoins that of digit 1, there are few connections across the hand–face border.

Although there could be many more examples, these few illustrate the point that the universally present intrinsic connections are quite variable in extent and distribution pattern. This variability implies that cortical areas within and across species vary in the ways cortical columns interact with each other. Overall, it appears likely that patchy and stripe-like patterns of intrinsic connections in cortical areas signify a like-to-like pattern of connections between groups of neurons with similar response properties, while diffuse, evenly distributed (at comparable distances from origin) patterns suggest a lack of specificity in such connections.

Feedback Connections

Connections from higher to lower areas in hierarchies of cortical areas provide another source of widespread neuronal interactions, as feedback connections are generally thought to be more widespread and less specific than feedforward connections (e.g., Krubitzer and Kaas 1990). Thus, patchiness is less pronounced than in feedforward connections and can participate in the coordination of processing in different processing streams. Nevertheless, feedback connections are generally more dense in the matching than non-matching feedforward modules, and thus vary in ways that reflect the modular organization of target areas (Salin and Bullier 1995).

Conclusions

One of the great temptations for overworked neuroscientists is to ignore, deny, or oversimplify the complex variability within and across nervous systems. In initial stages of development, models of nervous systems need obviously to depend on a few simplifying assumptions, but ultimately realistic models must reflect the organizations of real nervous systems. If we focus on mammalian neocortex, it is useful to remember that mammals with neocortex emerged at least 250 million years ago, and since that time formed the many branches of the mammalian radiation. A cladistic analysis, together with evidence from the fossil record, suggests that neocortex of early mammals occupied proportionally little of the brain, and that it was divided into few cortical areas, perhaps 20–25, that were poorly differentiated in cellular structure and rather similar (Kaas 2007). No present-day mammals have completely retained their ancestral organization, although the brains of some extant mammals have clearly changed much more than others. Perhaps human brains have changed the most, with human cortex now having more neurons than any other mammal, and having perhaps 200 functionally and structurally distinct processing areas. In addition to variably increasing the numbers of cortical areas across species (and in some cases, reducing them), cortical areas variably become more different in laminar and cellular structure. Thus, it is now unreasonable to assume that all cortical local circuits are the same, and that cortex varies simply in numbers of such circuits and types of inputs and outputs. Instead, we should explore and document this variability further, and use this variability as experiments of nature to understand how local circuits function and interact.

3

Sequence Coding and Learning

Gilles Laurent

Abstract

The topic pertaining to "sequence coding and learning" is deceptively wide ranging. Sequences or activity patterned in time in the brain can be linked to the nature of the physical world, to the nature and needs of active motion in the world, and to the fact that brains are dynamic systems. Coding, as it pertains to brains, is by contrast a rather fuzzy, ill-defined, and plastic notion. In the context of this discussion, it refers to the idea that information is contained in patterned activity. This chapter discusses some of the assumptions that go into such a statement, and reasons why we should tread carefully when addressing the issue. For instance, circuit dynamics might exist and be useful, even if the spike patterns they generate are never explicitly decoded. Linking sequences to codes is thus far from trivial. While learning requires no definition here, it is useful to be reminded of the likely coevolution of learning and representation systems in the brain: perception and memory depend on one another; this implies that the rules underlying brain "codes" must be constrained, at least in part, by those underlying learning (and vice versa). Given this, how do we go about testing the idea that learning helps generate stable dynamic attractors for perception and action?

Sequences

I can think of at least three groups of reasons to introduce the importance of dynamics in brain activity. Some are a reflection of temporal structure in the world. Others are linked to the constraints of action in a physical world. The third set is internal and linked to the dynamic nature of neurons and circuits (Gerstner 1995; Theunissen and Miller 1995). This division is somewhat artificial and used here solely for the purpose of description. Most likely, all features, causes, and consequences of brain dynamics have been influenced by one another throughout the evolution of neural systems.

Temporal Structure in the World

Life, as seen from an individual's own perspective, could be described as a sequence of interactions with the world that occurs between birth and death.

This sequence is locally nonrandom, due mainly to the laws of physics—where this cat stands now strongly predicts where it will be some time from now; the earth-day has a certain periodicity, etc.—and therefore contains large amounts of information (e.g., correlations) which are of great value for survival. Such correlations occur over a variety of timescales (subseconds to years); not surprisingly, brains evolved to detect and take advantage of them. This can be seen, for example, in motion sensitivity, in the deciphering of speech/song sound patterns, in circadian entrainment, and in many other pattern-sensitive attributes of sensory systems (Carr 1993). This definition extends to sensory experience which is itself the result of sequences of motor action, as in the emergence of place fields in rodent hippocampus and in their sequential activation during behavior or certain phases of rest and sleep (O'Keefe 1993; Mehta et al. 2000).

Action as Ordered Sequences of Muscle Recruitment

Conversely, brains generate motor actions through the ordered recruitment of muscle groups. This, too, is greatly constrained by the physics of the world (air, ground, water), of the plant generating the action (skeleton—when there is one—tendons, muscles, etc.), and of the interactions between them. Not surprisingly, therefore, action is often the expression and consequence of highly optimized motor "programs" encoded as sequential muscle recruitments in specialized circuits (Georgopoulos et al. 1986; Ekeberg et al. 1991; Llinás 1991). The mapping between motor program and action is, however, not always simple. Insect flight, for instance, can be generated by the nervous system in two main ways. First, motoneuron firing frequency determines the wing-beat frequency in a one-to-one manner. In such species (most large flying insects), wing-beat frequencies are usually low. Second, motoneurons serve simply to bring the myo-skeletal system in a resonant regime; in this way, wing-beat frequency can far exceed neural driving frequencies (e.g., flies, mosquitoes). Neuronal patterning is thus adapted to the physics of the plant and to a number of complex features, such as inertial and viscous forces generated by motion in a viscous medium. Sound production by animals obeys the same basic rules and constraints. Most motor programs (speech/song production, locomotion, skilled action) require refinement through learning and thus must involve some comparison between a measure of execution and a kind of template or desired output. Some programs are executed in an open-loop configuration, but even those often require calibration through learning. Here, as well, the timescale of such programs of sequential activity is widely variable (subseconds to hours), implying learning rules with wide dynamic ranges, or many learning rules to cover this dynamic range well.

Brains as Dynamic Systems

A third source of dynamics in the brain is purely internal. Brains are groupings of neurons and ordered couplings between them. Because neurons and synapses are themselves endowed with dynamic properties (FitzHugh 1961; Connors and Gutnick 1990; Izhikevich 2001)—due, for example, to the capacitive and resistive (usually nonlinear) nature of membranes, to diffusion and depletion, to channels and receptors kinetics—neural circuits are complex dynamic systems, with preferred behaviors and collective emergent properties (Brunel 2000; Gerstner et al. 1996; Maass et al. 2002; Rabinovich et al. 2008; van Hemmen et al. 1990). Some of them are relatively easy to detect and describe: oscillations or limit cycles (von der Malsburg 1981/1994; Gray et al. 1989; Eckhorn et al. 1988; Singer and Gray 1995; Engel et al. 1991; Laurent 1996, 2002), propagating waves (Lubenov and Siapas 2009). Others are not: complex spatiotemporal patterns of activity lacking periodicity or obvious large-scale spatial order (Diesmann et al. 1999; Stopfer et al. 2003; Mazor and Laurent 2005). While the kinds of patterning described in the previous two sections are linked causally to interactions with the external world (e.g., think of both the sensation and the production of sound), the kinds of sequences and patterning alluded to here are linked to the external world in only an indirect way. The sequences of activity seen in the early olfactory system in response to odors, for example, are not a result of sequences in a time-varying input; they are a consequence of, and therefore reflect (possibly encode), a particular, time-invariant, odorant signal; they are solely generated by and located in the nervous system. Generally speaking, this describes cases where the dynamic attributes of the representation are not a simple reflection of the dynamics of the input that drives activity (Rabinovich et al. 2008). Oscillatory patterns of activity (in the retina, in visual cortex, in olfactory systems, in hippocampus) are one form of such temporally ordered activity. A looming question, therefore, is whether those dynamics are useful and if so, for what?

Coding

Linking Neural Dynamics to Coding Is Difficult

I would expect that most attendees at this Forum would agree that in our field of experimental systems neuroscience, there exist at least some incontrovertible examples of spatiotemporal patterning, indicative of some form of collective order (e.g., synchronization, periodicity, waves of activity, seemingly random yet deterministic sequences of firings). Clearly, more experimental work is needed to better assess the brain's dynamic landscape, to characterize the properties and behavior of specific systems, and to classify the observed diversity of patterning. I would argue, however, that single clear examples are sufficient to prove the existence of such ordered phenomena in neural circuits and justify the study of their causes, consequences, and significance. Where this is accomplished (i.e., with which circuits, brain regions, or animal species) has no real importance to me at this point, other than practical considerations.

What is less clear, in my opinion, is whether we can say with equal confidence that such brain dynamics have been clearly linked to coding in the above examples. I can see at least two reasons for this hesitation.

First, coding (as in, "neural codes") is a rather fuzzy notion in neuroscience. In sensory neuroscience, it is now conventional to assess coding (as in, these retinal ganglion cells code for X) on the basis of (for instance) a measure of mutual information between an input and a response set (or distributions) (Bialek and Rieke 1992; Berry et al. 1997). Yet, neural codes are not codes in the sense of Shannon's (1948) information theory: brains are not channels optimized for information propagation. A brain selects, prunes, "computes," and generates a highly selective, adapted, and specialized output—not a copy of its input. A brain is also rife with complex recurrent paths, impossible to map simply onto a source-receiver model of information flow. However, the information theory approach is undoubtedly valuable: it defines what can be extracted by an observer from a spike train. The problem, in the end, lies with defining the observer (can or should a neuron be considered to be equivalent to an ideal observer?) and its approach to read-out (e.g., spatiotemporal inputs, initial and time-varying states, nonlinearities). For practical reasons, I would thus define neural codes very locally and base those definitions on the properties of the "decoder."

Second, proving that temporal structure in neural activity has any functional consequence is an extremely difficult task. It is difficult experimentally because selective manipulation (in space and in time) of spiking has, until recently, been impossible on a large and distributed scale. (While still lacking high spatial specificity, recent optogenetic techniques now enable the temporal manipulation of cell populations.) Proving the relevance of temporal patterning is difficult as well because brain activity patterns are the expression of a collective behavior, itself prone to adaptive rearrangement: to take a crude analogy, the loss of a midfielder in a soccer team can be rapidly overcome when the other players assume new roles and behaviors. This compensation is not solely an expression of redundancy; it reveals an ability of the system to change adaptively. The criteria traditionally applied to experimental data (necessity and sufficiency) to assign functional significance may thus often be inadequate. We need to be open to different/additional assays and criteria.

Decoupling "Encoding" and "Decoding"

An argument often advanced by those who doubt the relevance of temporal codes is that they are difficult to read out. While this argument is disputable, I would like to introduce the notion that dynamics and patterning do not necessarily mean codes: more precisely, what we often describe as "encoding" and "decoding"—the two sides of the "coding" coin—may not always be coupled to one another. Stated differently, dynamics and temporal patterning may be useful even in cases in which the patterns themselves are not decoded.

Imagine that a particular sensory input, S (e.g., an visual scene), causes, in a particular part of the brain, a complex spatiotemporal pattern of activity, PS. In purely descriptive terms, one could say that PS represents or "encodes" S. This is technically true from our perspective as outside observers. It might be, however, that the PS is nothing more than the expression of the brain's physical response to S, as would be that of many other similarly stimulated dynamic systems. Furthermore, it is possible that the decoding of PS by downstream neurons or networks involves no sequence decoding per se. This does not mean that PS, as a dynamical pattern, is not relevant. It may, however, reveal that its value is only implicit in the code.

Analogy

If you have learned a racquet sport such as tennis at one point in your life, you will have learned the value of the "follow-through"—the seemingly stylistic exercise of keeping the racquet in motion toward some imaginary position after the racquet has hit the ball. This, of course, appears completely nonsensical to a beginner: "Why does it matter what I do after I have hit the ball"? Trivially, the answer is that what I do after the time of contact is at least in part a consequence of what I did before, including at the time of contact; what I do after the hit is a consequence of the unfolding of a sequence of actions that preceded it: I cannot have one without the other because it is part of the physics of the system (the outside world, the ball, the body and, possibly, the brain itself). As far as the outcome of the game is concerned, however, the decoding of this action (i.e., the quality and accuracy of the hit) is transitory: the action sequence is implicitly contained in the hit, but it is not decoded as a sequence of movements. To summarize, the action sequence ("encoding") is necessary, but the read-out ("decoding") is done only transiently, at some appropriate moment throughout the motion. Hence, the dynamics, as the expression of a system's physical properties, are indispensable but do not embody the code itself.

Example

The system that my lab has studied for 15 years or so, the insect olfactory system, is dynamically rich. When an odor is presented, a population of some 1,000 projection neurons (PNs) in the antennal lobe (the functional analogs of mitral cells in the vertebrate olfactory bulb) becomes transiently synchronized, forming a distributed oscillatory pattern (~20 Hz) in which different subsets of PNs fire at each oscillation cycle. An odor representation is thus a specific time series of PN-activity vectors, updated at each oscillation cycle; these sequences of PN-activity vectors can also be imagined as odor-specific trajectories in the phase space defined by this system. These trajectories are stimulus specific and, provided that the stimulus is sustained for a long enough duration (>1.5 s), they contain three epochs: (a) an on-transient that leaves the baseline state, (b) a fixed-point attractor, and (c) an off-transient that leads back to baseline along a path different from the on-transient. The transients and the fixed point are all odor specific, but analysis of pairwise distances between the trajectories that correspond to different stimuli shows that distances are greatest when the system is in motion (during both on- and off-transients), not between steady-state attractors corresponding to the different odors. Interestingly, the targets of these PNs, called mushroom-body Kenyon cells (KCs), are active mainly when PN activity is in motion—not when PN steady-state has been reached. In fact, KC responses are highest when separation between PN vectors is greatest (i.e., during the initial phase of the on- and the off- transients). This leads to two conclusions: First, while circuit dynamics seem to serve an important function (here, decorrelation of representations), they do not appear to serve a coding purpose in the traditional sense; that is, there is no evidence that downstream areas decode the sequences of PN output. Second, a significant portion of a trajectory, its fixed-point attractor, does not even appear to be read out by targets (Mazor and Laurent 2005). This begs the question of the use, if any, of fixed attractors in this system. Whatever this use may be, it does not seem to be in the embodiment of a representation.

In conclusion, the existence of dynamics in a representation does not, in and of itself, prove that the dynamics are part of the "code," or a message to be decoded. The dynamics may be the result of properties of the system, and they may even be useful (e.g., in optimizing representations). However, they are not necessarily a feature that requires explicit decoding.

Learning

Learning and Perception Are Not Separable

The practicalities and sociology of neuroscience as a human endeavor are such that the subareas of learning/memory and perception/coding overlap only occasionally. Attendance at any large neuroscience meeting will often confirm this impression. Yet, perception is difficult to imagine without learning, and vice versa. Perception and recognition would not be what they are without templates/representations stored for classification and comparison, or without some trace of recent history. Conversely, the ability to learn seems pointless if it did not serve future comparisons between immediate input and a bank of memory traces—comparisons that are needed for perception, recognition, and adaptive action. It follows then that the mechanisms underlying sensation and perception should, somewhere, express the requirements imposed by learning, storage, and recall; this is simply because circuits must have evolved with these coexisting constraints. When we talk about neural codes (ignoring for now the ambiguity of the term), therefore, we should not forget that their formats may be optimized not for coding per se, but for learning and recall as well. In other words, the attributes of biological learning rules, presumably adapted to the statistical structure of the world and to the intrinsic properties of the brain, should be interpreted as an added constraint on the formats of sensory and motor codes. This is particularly relevant to our thinking about "neural codes," especially if those (a) are at least partly dynamical in nature or (b) have substrates that express dynamical properties. I am reminded of this every time I use my bike's lock, a perfect example of procedural memory: while the lock's combination escapes me now, it springs back to me reliably every time I start spinning the rotating face of the lock. In conclusion, thinking about codes and representations can only benefit from including the constraints and necessities of learning and recall.

Learning Rules and Sequence Coding

Recent experimental results on the mechanisms of plasticity (long-term potentiation, asymmetric Hebbian rules, spike timing-dependent plasticity) have allowed realistic links to be drawn between learning and network activity (Hebb 1949; Herz et al. 1989; Bliss and Collingridge 1993; Markram et al. 1997; Bell et al. 1997; Bi and Poo 1998; Cassenaer and Laurent 2007). In a typical example, an externally imposed sequence of activity leads, through a learning phase, to the selective modification of synaptic weights via application of a learning rule with an appropriate time window; thus, the training sequence can, after learning is terminated, be recalled or replayed in the absence of the external drive, very much as in a feedforward "synfire" chain. The practical consequence can be the recall of a full sequence (as a path in physical space or as in procedural learning), a bias for recall of a particular pattern upon appropriate seeding (as in the recall of selective memories). In all cases, the main idea is that the pattern has become a stable attractor for activity. The theoretical questions that these ideas raise are many and complex:

- What are the best dynamical models to explain such behavior? What do they predict?
- How stable can such dynamic attractors be, given the known biophysical properties of neurons and synapses (and their non-fully deterministic behavior)?
- Similarly, what is the tolerance to external noise for such representation mode?
- What is the gain on memory capacity (relative to classical fixed-point attractor models) for such systems?

- What are the constraints that such rules impose on representation density/sparseness?
- Do such representation modes require (or work best with) a discrete timeframe?
- Do they allow time compression/warping?

As a dedicated experimentalist, I would argue that many of the critical answers (or hints to those answers) will be given by experiments and offer the following as examples of crucial issues to be addressed:

- What are promising experimental systems to study these issues?
- What are the criteria needed to establish firmly the existence of spatiotemporal patterning in circuits? In other words, if I were given a neuronal system with dynamical attractors, would I even detect its nature?
- What are the criteria needed to establish the functional relevance of spatiotemporal patterning?
- What are the different forms that this relevance could take?
- What constraints would the existence of such attractors impose on the learning rules we know?
- Assuming that many of the learning rules we know are used mainly for homeostatic regulation (e.g., on weight distributions, timing), how plastic must they be to underlie the formation of stable dynamical attractors?

4

What Can Studies of Comparative Cognition Teach Us about the Evolution of Dynamic Coordination?

Uri Grodzinski and Nicola S. Clayton

Abstract

The field of comparative cognition has provided examples of the cognitive abilities of many mammal and bird species, such as some "understanding" of physical properties and the use of episodic-like memory of past events to alter behavior flexibly. Although the seemingly complex behavior exhibited by an animal may be the outcome of cognitive mechanisms, it need not be: often, associative learning principles, or fixed action patterns that form without associative experience ("innate rules"), or a combination thereof are all that is required. Distinguishing experimentally between these accounts is one of the main concerns of comparative cognition. This chapter outlines what is necessary for connecting the field of comparative cognition with the subject of dynamic coordination. Because animal cognition studies are performed on many species from diverse taxa, analyzing the type of coordination required to pass the tasks they include may shed light on the evolution of (dynamic) coordination. Specifically, the convergent evolution of particular cognitive abilities in primates and corvids implies that the type of coordination necessary for these abilities has evolved at least twice. In this respect, one future challenge is to clarify which principles of coordination are shared (and which are different) between mammals, with their laminar cortex, and birds, with a forebrain that is not laminated. Several recent examples of animal cognition are explored and the sort of coordination they require is discussed.

Marrying Comparative Cognition and Dynamic Coordination

A common assumption in cognitive neuroscience is that cognition has evolved in a homologous manner, such that the cognitive abilities exhibited by different extant species have all developed from the same basic abilities present in their last common ancestor. Like other traits, however, cognitive ones can also evolve through a process of convergent evolution in which distantly related species independently evolve similar solutions to similar problems. When attempting to study cognitive abilities, there is much benefit to be gained from taking a comparative approach to the study of cognition, where many species from diverse phylogenetic taxa are compared for their cognitive abilities. This way we can assess the relative importance to the evolution of cognition of (a) phylogenetic homology (common ancestry) and (b) differences in selection pressures across species and taxa.

Comparing primates and corvids (the crow family) provides one clear case of convergence in cognitive abilities (Emery and Clayton 2004; Seed, Emery et al. 2009). Nonhuman primates and corvids show remarkably similar abilities to behave flexibly in ways suggesting that they understand features in their physical and social worlds. If the reptilian common ancestor of birds and mammals did not possess these abilities, then this supports the hypothesis that cognitive abilities have evolved convergently in these groups. The comparison between the neuroarchitecture of avian forebrains and the mammalian neocortex has suffered from mistakenly attributing a striatal developmental origin to most of the avian forebrain, while in fact most of it is derived from the embryo's pallium (Reiner et al. 2004; see also Balaban et al., this volume). Thus, areas in the avian forebrain are homologous to the mammalian neocortex, to which they also correspond functionally. Although these similarities are now well accepted, there are still major differences in the neuroarchitecure of mammalian and avian forebrains. Most notably, whereas the cortex of primates and other mammals has a laminar organization, the avian forebrain lacks such a laminated organization and is instead nuclear in its organization. This finding, together with the cognitive parallels, implies that it is feasible to evolve similar (computational?) solutions to complex problems, based on different underlying neural mechanisms.

The principles of coordination between different local brain functions may also have evolved convergently, along with cognition. However, in order to begin considering dynamic coordination in a comparative way, at least two routes are possible. First, we may investigate the brains of different species directly to determine how different brain functions are coordinated. This may be done for a small number of species, for which methodology has already been developed and whose brain is relatively well studied. Another, indirect route, is to use the extensive research of (behavioral) comparative cognition in order to draw a general picture of which taxa are likely to exhibit the capacity for dynamic coordination. Marrying comparative cognition and dynamic coordination may allow us to assess the sort of coordination in a very large number of species from diverse taxa. To do so, however, we must first clarify the relation between cognitive abilities exhibited by animals such as primates and corvids, on the one hand, and the (minimal) sort of coordination that is required in order to exhibit them, on the other hand. This is what we attempt to explore in this chapter. In the following sections we consider the type of coordination necessary in episodic-like memory and in several primate and corvid studies of physical cognition.

Concentrating on such studies requires justification because many, much simpler tasks require some sort of coordination between different local brain functions. In other words, different brain areas, in charge of different computational tasks, have to "talk to each other" to achieve many of the behaviors performed regularly by animals (e.g., grooming; see Balaban et al., this volume). Coordination might therefore be inherently dynamic if it involves the synchronization of oscillating groups of neurons, and many examples of behaviors and basic cognitive abilities (e.g., Gestalt perception, selective attention, and working memory) that involve this type of dynamic coordination have been discussed by both Moser et al. and Engel et al. (this volume). We may need to consider, however, additional or special forms of coordination when trying to understand what makes some species able to solve tasks that others cannot. In this chapter, therefore, we will distinguish between cases in which the coordination required is either prespecified in the brain itself (Phillips et al., this volume), or is specified directly by the outside world, and those tasks in which the solution must somehow come from within the brain. Performing a well-trained skill, such as walking, is an example of prespecified coordination (Phillips et al., this volume), and so is classical conditioning. Even if the stimuli to be associated are novel, precluding a prespecified connection between their representations in the brain, and even if the coordination involves synchronizing oscillations (and is therefore dynamic in that sense), the representations to be connected could be directly specified by external events. In contrast, we will discuss studies in comparative cognition where we suspect that solving the task requires the animal to come up with the solution itself, where the very identity of what should be coordinated, as well as the form of the coordination, needs to come from within the brain. Before we continue, however, we must first clarify how comparative cognition research distinguishes different accounts for flexible behavior.

Different Accounts in Comparative Cognition

Comparative cognition has devoted much empirical effort to distinguishing between different accounts to explain flexible behaviors. Consider the case in which an animal performs behavior *A* in context *a*, whereas it performs another type of behavior *B*, in another context *b*. Perhaps the simplest account for this flexibility in behavior is an "innate rule" that specifies that the type of behavior is dependent on a particular context. Such a rule is "innate" in the sense that its formation does not require the explicit ontogenetic experience with the different outcomes of behaviors *A* and *B* in contexts *a* and *b*, but instead develops even in an animal naïve to these outcomes, presumably as a result of selection pressures acting on the development of this rule in its ancestors. Another account of flexible behavior is associative learning, whereby the rule for performing A in context a (and B in context b), for example, may be created through reinforcing only behavior A in context a and only behavior B in context b (for a simple account of how this works, see Dickinson 1980). In comparative cognition research, innate rules and associative learning mechanisms are usually taken as the alternative accounts to a cognitive one. What a cognitive account of the above rule would imply is that the animal has some appreciation of the *reasons* for each behavior being appropriate in a different context. Although it is unclear how this can occur, such an appreciation would be beneficial as it allows much more flexibility in generalization to new contexts. For example, if in a new context c the reasons that made behavior A beneficial in context a remain the same, the animal will be able to transfer to this new context (or new task), whereas depending on associative learning or innate rules will not enable such flexibility.

We hope that we have made clearer the alternative accounts for flexible behavior that the field of comparative cognition addresses, and the empirical ways to distinguish between them. In many cases it is still unclear which account is more plausible in each case (for a simple associative-learning account of what may at first sight appear to be pigeons insightfully stacking boxes on top of one another to obtain food that would otherwise be out of reach, see Epstein et al. 1984), but this is not the focus of our chapter. Instead, we now attempt to consider what kinds of coordination between different brain functions are required to achieve flexible behavior according to these different behavioral accounts. Innate rules seem to require only the coordination that can be prespecified by natural selection. In the case of associative learning, previous experience may prespecify the coordination needed for behavior to be flexible so that it can be adequate in different contexts and tasks. In contrast, behavioral flexibility that is based on an appreciation of reasons that make such flexibility beneficial may require a dynamic coordination which is neither prespecified in the brain nor obvious from observing the outside world. In the following sections we offer a few examples of studies in comparative cognition and discuss their relation to dynamic coordination. In many cases there is still an ongoing empirical debate on whether or not simpler explanations can account for some of the findings and on whether human and animal abilities are comparable. Here, we will not go into these debates, but rather present the findings and their interpretations to make the connection between comparative cognition and particular forms of dynamic coordination. The particular forms to be considered are those involved in episodic memory and in problem solving, and which may require more than the basic forms of associative learning, selective attention, and working memory.

What Coordination Does Episodic-like Memory Require?

To recall a fact, such as "Frankfurt was first founded in the first century," we do not necessarily need to remember when, where, and how we learned this information, and neither do we have to have been there ourselves. In contrast with such factual or "semantic" knowledge about the past, episodic memory contains the specific "what, where, and when" information of a previous event that we have personally experienced (Tulving 1972). For example, UG recalls how, as part of a security check at the Frankfurt airport in the summer of 1994, he was required to exhibit his puny juggling skills to demonstrate that his juggling clubs were indeed what they were. At first sight, it seems that episodic memory would require new coordination in that the memory of a past event can be used to make a later decision. For example, imagine that you realize you have lost your wallet and try to trace your steps back to remember where you have left it. This is essentially different from making a mental note of putting your wallet in the drawer, but does it indeed require a novel coordination between different functions?

Episodic-like memory has recently become an area of research in comparative cognition, starting with experiments showing that western scrub-jays remember the what, where, and when of caching a food item (Clayton and Dickinson 1998, 1999; Clayton et al. 2001, 2003). The term "episodic-like" memory is used when considering animal memory (Clayton and Dickinson 1998) because current definitions of human episodic memory necessitate the subject being consciously aware that the past memories are his or her own and that he or she re-experiences the event when remembering (Tulving 2005), characteristics which cannot be assessed without language and ones that become part of our identity. With study species now including other birds, such as magpies and pigeons, rats, nonhuman primates, meadow voles, and honeybees (reviewed in Crystal 2009; Dere et al. 2006; Schwartz et al. 2005), episodic-like memory seems like a good candidate for exploring the relation between cognitive capacities and the sort of coordination they require. To do this, let us look more closely at some of the first experiments showing episodic-like memory in nonhuman animals.

To study episodic-like memory in western scrub-jays, Clayton, Dickinson and colleagues capitalized on the fact that these birds readily hide food in the lab as well as in the wild (Clayton and Dickinson 1998, 1999; Clayton et al. 2001, 2003). Indeed, western scrub-jays scatter-hoard many types of food in multiple locations and recover them hours, days, or months later (Curry et al. 2002). The types of food cached include various types of nuts as well as invertebrates, which perish much quicker. Thus, these jays might need to form a specific what–where–when memory for each caching event in order to recover the food on time. As mentioned above, distinguishing between cognitive and simpler accounts is a key feature in comparative cognition research. In a study that shows that scrub-jays remember the content of caches they made, as well as whether they have already recovered them, Clayton and Dickinson (1999) explicitly ruled out a simpler associative account, which is useful for our analysis of the type of coordination needed.

In the first experiment in that study, scrub-jays were first allowed to cache peanuts in one tray and then dog kibbles in another. Before they were allowed to recover from the two trays (simultaneously), they were pre-fed on one type of these food types. Consequently, they preferentially chose to recover from the tray where they had cached the other type of food. This suggests that they remember what they have cached where; however, there was a simpler explanation for these results, based on a classical (Pavlovian) conditioning between each tray and the food type that had been cached in it. According to this account, such an association was formed during the caching phases, such that the stimuli of each tray elicit the representation of a different food type. The motivation to approach one of the food types was decreased due to pre-feeding and therefore the sight of the tray associated with it elicited less approach than the sight of the tray associated with a food type that was still highly valued. For our purposes, it is obvious that the coordination between the representations of location, types of food, and motivation to approach could all be prespecified during the caching session or before then. There is no need, according to this account, for dynamic coordination; that is, for the type of coordination which is "created on a moment-by-moment basis so as to deal effectively with unpredictable aspects of the current situation" (Phillips et al., this volume).

However, a second experiment ruled out this simple account. In this experiment, both pine nuts and kibbles were cached in both trays, but only one food type was subsequently recovered by the jays from each of the trays. Then, the jays were pre-fed one of the food types and consequently chose to recover from the tray that still contained the other food type. This showed not only that jays remember what they had cached where (ruling out the simple conditioning account) but also that they register at recovery that the food they recover is no longer in place. The coordination needed here is obviously more complex, as we discuss below. Before we do that, however, it is worthwhile to consider briefly what else is known about scrub-jay memory for cached items.

Scrub-jays prefer wax moth larva ("wax worms") to peanuts, but after experiencing that worms take a short time to decay (and that peanuts do not decay), the jays searched preferentially for peanuts when allowed to recover both types of food at a time when worms would have already degraded. When allowed to recover after a shorter time had elapsed since caching, short enough for worms to stay fresh, they preferentially search for the worms (Clayton and Dickinson 1998), suggesting that jays remember when and where they cached each type of food. Another study (Clayton et al. 2001) showed that the what, where, and when components form an integrated memory which is retrieved as a whole, and also suggested that the memory of the caching location elicits the two other components; namely, time and content. In addition, the jays are able to incorporate flexibly new semantic information about decay rates, acquired after caching, and use this information to alter their recovery decisions (Clayton et al. 2003).

Taken together, the above studies depict a complex combination of functions, schematically illustrated in Figure 4.1. For each caching event, episodic information about food type, location, and time of caching is encoded to form an integrated event-specific memory. When deciding which caches to recover, jays take into account semantic information about the decay rates of different food types, e.g., "After 4 days, worms have already perished." This rule can be altered if new information is available, even after caching, providing further support that the jays do not encode their recovery decisions themselves during caching, but rather encode what–where–when information, which is later used when these decisions are made (see Clayton et al. 2003). As mentioned above, jays also remember whether they had already recovered each item (Clayton and Dickinson 1999). Finally, motivation for different food types can be changed through feeding. At recovery, motivation, episodic-like and semantic information are all used to make the correct decision: recover at the site which, according to the decay-rate rule previously experienced, contains a food type that is still edible, and for which there is current motivation.

Episodic-like memory in these jays is thus a complex decision-making system. Dynamic coordination is required between episodic, semantic, and motivational functions (Figure 4.1), and in each caching event a new, event-specific connection must be made between the functions for location, time, and content. However, all of the above does not necessarily require that the *fundamental way* in which these different functions are coordinated be dynamic and change according to the task or between caching and recovery. This is because the sort of coordination needed to make recovery decisions does not in itself need to change when it is time to decide. For example, the possible coordination

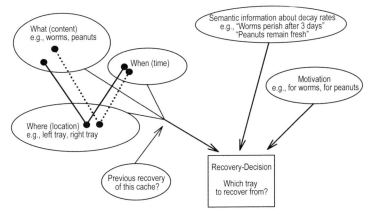

Figure 4.1 Schematic structure of coordination between different brain functions which may be required to solve the episodic-like memory tasks in Clayton and Dickinson (1998, 1999) and Clayton et al. (2001, 2003).

structure given in Figure 4.1 may allow the jays to solve all of the tasks in the above-mentioned experiments without changing anything but the data that is substituted in the different "brain functions." Information about the time that had elapsed since caching can be gained using the episodic-like memory of the caching event, whatever it is; then, this elapsed time could be compared with the semantic information about decay rates, whatever that information may be; motivation, despite changing frequently, will also affect the decision in a constant manner. All this does not change even if the test retention interval between caching and recovery has not yet been experienced by the jays (as in Clayton et al. 2003).

A somewhat different approach to studying episodic-like memory is to ask the subject unexpectedly to report their memory of past events. Zentall et al. (2001) investigated whether pigeons could report on whether they had just pecked a key or not. First they trained them to peck a red key, if they had just pecked, and green if they had not. In a second training phase, two different keys were used: yellow was followed by food whereas blue was not. Finally, the pigeons were given a test: either yellow or blue were presented, with only yellow eliciting a pecking response due to its previous association with food. Immediately thereafter, the bird was to choose between red and green, and they tended to choose the correct key according to their initial pecking or non-pecking action. Note that trials in the second training phase (presenting yellow or blue) did not include the question, "Did you just peck or not?" (i.e., choosing red or green, which was done during other "refreshing trials" in phase 2). In addition, a second test used novel stimuli which the pigeons either pecked or not, again followed by the red/green choice. Therefore, the authors suggest it is very unlikely that the pigeons would encode the correct response to this question (using semantic information) when they peck or do not peck during the tests. Rather, when given this question, they would need to retrieve their memory of what they had just done. While this provides more evidence for episodic-like memory in an animal model, even passing this test does not necessarily require dynamic coordination. The connections between pecking and choosing red (and between refraining and choosing green) were established in the first training phase; the connection between yellow (but not blue) and food was established in the second phase. All of these connections would have to be used during the test, but new ones, or the rearrangement of old ones, are not necessary. In that sense, again, the task does not require the coordination to be created moment-by-moment.

To conclude, passing the above tasks involves coordination of episodic and semantic information and, in some of the experiments, also motivation for different food types. However, as far as we can tell, it seems that episodic-like memory does not necessarily require the structure of coordination to change dynamically. There is, of course, a need for coordination throughout the task; however, the way in which the representations of the different keys relate to one another and to the task at hand do not change. Of course, there is always a possibility that the *actual* computational manner in which such tasks are solved involves special forms of coordination after all, at least in some species. All we can say at this stage is that they are not necessarily needed in this case. In a similar manner, we shall now try to analyze studies from another realm of cognition, the physical one.

What Coordination Does Physical Cognition Require?

When subjects are able to solve new tasks based solely on their underlying physical properties, this suggests that they appreciate or "understand" something about these physical properties, and that we may bestow the successful subject some "physical cognition" or "causal understanding." While the computational and neuroanatomical details of how such feats are accomplished are largely unclear, physical cognition seems, at first glance, to require the establishment of a novel coordination between the specific properties (affordances) of a new task and some physical "rules" which the subject has already learned. Many studies of physical cognition include an initial training period, where subjects learn to perform a task, and may or may not also learn the physical rules governing this task. Then, "transfer tests" are performed to determine what the subject understands about these physical rules. These tests are most relevant for our purposes, as they include a novel situation which is unsolvable using only the associative learning acquired in the training phase or generalizing from the stimuli in the training task. Rather, the subject has to somehow apply the physical reasons or rules that underlie the success of its behavior in the training phase. The question is whether passing these transfer tests really requires any rearranging of the coordination present before the test. To address this question, we shall explore a few examples of primate and corvid physical cognition and attempt to outline the properties of the coordination they require.

Understanding the Properties of a Trap

A classical example of a physical cognition experiment, first used with primates and recently also with birds, is the trap tube. In the first version of these experiments (Visalberghi and Limongelli 1994), the training phase included a transparent horizontal tube with a trap-hole at its center, next to which a food reward was placed. The idea is to test whether the subject will learn to push the food reward to the correct direction (i.e., away from the trap), using a stick tool. Some of the capuchins, chimpanzees, woodpecker finches, and others tested learned the first task (reviewed in Martin-Ordas et al. 2008). After a subject mastered this task, a transfer test was performed with the tube inverted so that the trap was no longer functional. All primates tested still continued to push the food away from the trap, even when the tube was inverted, whereas one woodpecker did not. It has been rightfully argued that continuing to push away from the nonfunctional trap bears no cost, and therefore this does not necessarily imply lack of physical understanding (Silva et al. 2005).

Since these studies were conducted, a modification of this task was introduced where two traps are placed in either side of the food reward, but only one is nonfunctional. This allows experimenters to assess what animals understand about the properties of traps, such as when an object will become lodged in a trap as opposed to when it will pass along the top of it or fall through the bottom. Versions of this experiment were conducted with rooks (Seed et al. 2006), and subsequently this design was adopted to test New Caledonian crows (Taylor et al. 2009) and chimpanzees (Seed, Call et al. 2009). The transfer tests suggest that some of the individuals tested have some understanding of the physical rules of the task in all three species. There is not yet a consensus among researchers as to the correct way to interpret success of only a small minority of the subjects. In addition, in some of these studies the transference was not made in the first trial but in the first block or blocks of trials; thus learning cannot be ruled out completely. We might need to wait for more experimental results in order to draw strong conclusions from our analysis. However, at this stage we are only trying to determine what form of coordination passing such transfer tests (on the first trial) seems to require.

In the rook and New Caledonian crow studies, a two-trap tube was used (Seed et al. 2006; Taylor et al. 2009), the configuration of which was altered by the experimenter to make only one trap functional. In the chimpanzee study (Seed, Call et al. 2009), a conceptually similar two-trap box design was used (Figure 4.2), which enabled testing physical cognition without the need for using tools. The subjects could reach the food with their fingers and push it to

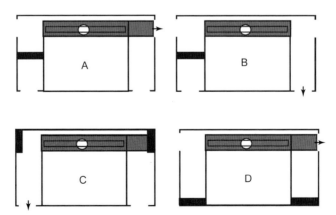

Figure 4.2 The two-trap box design used to test chimpanzees' understanding of physical properties (adapted with permission from Seed, Call et al. 2009). Configurations A and B were used in phase 1; configurations C and D in the transference tests (see text for details). The white circle marks the location of the food reward at the beginning of the trial, and the arrow marks the hole through which the food may be extracted if the subject pushes it in the correct direction.

either side. In configuration A, there was one functional trap (left-hand side of Figure 4.2A); the other trap was made nonfunctional by placing a shelf piece that prevented the food from falling, enabling its extraction through the side exit. In configuration B, the trap was nonfunctional due to the blocking piece being removed so that the food could drop through the bottom exit. Six out of the eight subjects successfully learned these two configurations and were then given the transfer tests (Figure 4.2, configurations C and D) to assess what they had encoded about the task. Note that in these transfer configurations, the blocking pieces which made the trap functional in A and B were removed to prevent a subject from acting according to the rule "push away from the blocking piece" to pass these tests (in D two different blockers were placed at the very bottom of the trap).

Applying such a rule, as well as rules such as "push toward the supporting shelf," does not seem to require any special form of coordination. However, to pass both transfer tests (C and D) the subject must apply some understanding of the fact that the shelf piece prevents the food from falling (e.g., "surface continuity prevents falling"). In C, the conclusion should be to avoid the shelf piece because the only way to get the food is by allowing it to fall through the bottom exit (the side exists are blocked), whereas in D the conclusion should be to push towards the shelf piece to prevent the food from falling and being trapped by the bottom blocker. Thus, an additional rule should have been encoded during the acquisition of A and B, or from previous experiences, to enable transference to both C and D. Namely, something in the form of "objects cannot go through barriers," which would be applied to the side and bottom blockers in configurations C and D, respectively. Such "abstraction from rules" may enable passing transfer tests without any understanding of unobservable forces such as gravity (for further discussion, see Seed et al. 2006; Seed, Call et al. 2009).

The relevant question for our purposes is whether applying such rules requires any special form of coordination between functions operating in these tasks, in the sense that the *type* of coordination between the representations of surfaces, solid objects, and the food item needs to be created moment-by-moment according to the attributes of each transference condition (see above). Our feeling is that there is no such requirement inherent to passing these transfer tests. If, say, the rules (a) "surface continuity prevents falling" and (b) "objects cannot go through barriers" are encoded before the transfer tests are given, then passing the transfer tests "only" requires applying them to the different surfaces and barriers. In other words, once such rules are encoded, a generalization of "barrier" is enough to enable the transference. How such abstract rules are encoded in the first place is another interesting question, and is probably far from being trivial. Evidently, only one rook (Seed et al. 2006) and one or two chimpanzees (Seed, Call et al. 2009) seemed to be able to do so.

"Insightfully" Bending a Wire to Make a Hook

New Caledonian crows are known to manufacture and use tools in the wild (Hunt 1996) and have recently been used in experiments to assess their understanding of physical properties. In one such study, a female crow (Betty) spontaneously shaped a hook tool out of a straight wire (Weir et al. 2002). This occurred during an experiment designed for different reasons, where subjects had to choose a hook tool over a straight one and use it to lift a bucket with a food reward, placed inside a vertical clear tube. In this instance, the provided hook tool was taken by the male, leading the female to shape her own hook from the remaining straight tool; in subsequent trials she (but not the male) continued shaping hooks when provided only with straight wire (Weir et al. 2002). In another study (Bird and Emery 2009), four rooks were first given experience using a hook tool to retrieve a food reward in a similar bucket-lifting task, and subsequently they succeeded in choosing a functional hook tool over a nonfunctional "hook" tool (with a backwards end). Then, they were provided with only a straight wire and spontaneously shaped it into a hook to retrieve the food bucket. It is very unlikely that hook shaping in the above studies could have been prespecified through associative learning or as an innate behavior, given that the rooks have not, in fact, been observed to use tools in the wild and that the birds showed spontaneous manufacture of hook-shaped tools.

In both studies, the subjects had presumably already encoded the connection between a hook tool and lifting the bucket, but not the possibility of bending a straight wire into a hook. Thus, the presentation of the apparatus may have elicited a representation of the appropriate hook tool, but it seems to us that the course of action leading from a straight wire to a hook required a new connection to be made. The fact that the connection is new to the animal means that it cannot be prespecified (Phillips et al., this volume). Perhaps, even more importantly, the type of coordination needed here (leading to the correct solution) is not prespecified either, nor imposed in its entirety by the observable properties of the task, as would be true for an associative learning task (see above). That is, the task affordances suggest a way of action, but do not explicitly specify it. Instead, the subjects must somehow creatively find the right answer. In other words, a form of dynamic coordination is required to solve the task "insightfully," apparently existing at least in these two corvid species. Interestingly, as Bird and Emery (2009) point out, one definition of "insight" states explicitly that it may involve a "sudden adaptive reorganization of experience" (Thorpe 1964:110). Claims for insightful behavior have been documented in primates and birds before, but in some cases simpler explanations involving the "chaining" of preexisting behaviors, which do not seem to require insight, were not excluded (e.g., see the associative account of box-stacking in pigeons by Epstein et al. 1984). Nevertheless, this field appears to hold a promise of providing a good behavioral tool for assessing coordinating abilities in different species, and it will thus be very interesting to explore what computational mechanisms are enough to enable the insight involved in each case.

Using a Stick to Do a Stone's Job

The four rooks in the above study were also trained to insert a stone into a clear vertical tube in order to collapse a platform at its bottom, so that a food reward placed on the platform will drop (Bird and Emery 2009). In a series of transfer tests, Bird and Emery show that the rooks can choose a stone of correct size for the tube diameter of the tube (both from provided stones and also from stones picked from the ground) and orient it correctly when necessary for insertion. These transfers suggest that during initial training the rooks encoded something about the stone being of correct size compared to the diameter of the tube. It is quite remarkable that they encode this property, especially given the fact that during initial training all stones were of correct size. However, once encoded, a rule such as "compare size of stone to tube diameter" can be applied to solve the above-mentioned transfer tests without requiring insight. Another transfer test, however, seems to require a novel connection to be made. The rooks were given either a heavy stick, which they could drop into the tube to get the reward, or a light stick, which would not collapse the platform if dropped. On the first trial, all four rooks retained their grip of the light stick instead of dropping it, inserted it into the tube and pushed it, providing the force necessary to collapse the platform and get the food reward. This immediate transfer shows that during previous training and testing with the stones, the rooks encoded something about the necessity of collapsing the platform, which they used. However, this representation of the collapse of the platform is not enough to plan how to achieve that with the stick (having only been given experience with dropping stones to do so). Success in this transfer task thus seems to require a novel, "insightful" connection between the properties of the light stick and the way in which it can be used to achieve the collapse of the platform, again suggesting a sophisticated form of dynamic coordination in this corvid.

Open Questions and Future Directions

In addition to the examples analyzed above, there are other areas of research in comparative cognition which may be interesting with respect to the type of coordination they require. "Transitive inference" is the ability to infer a previously unencountered relation between two objects from their relation to one or more other objects (e.g., if $A > B$ and $B > C$, it can be inferred that $A > C$). Does this inference of unencountered relations require any special coordination between their representations (through the representations of the relations which have been encountered)? Long thought to be uniquely human, evidence is accumulating that nonhumans have various degrees of transitive inference (reviewed in Vasconcelos 2008), making the coordination required of even greater interest.

The attribution of mental states (e.g., knowledge, ignorance, motivation, belief) to another individual may also require projecting oneself into another's perspective, but does this require any reorganization of information in a way which is not prespecified? Whether nonhumans are able to attribute mental states is still a matter of controversy (e.g., Penn and Povinelli 2007), with evidence showing that some nonhuman primates and corvids may have such abilities (e.g., Emery and Clayton 2001; Hare et al. 2000). In any case, it could be beneficial to analyze what computational work is needed for one individual to use representations of unobservable states of another, and whether this involved any special form of coordination.

Finally, the ability to "think about one's own thoughts," or metacognition, may require some reorganization of the existing data (i.e., to apply a cognitive function such as "knowing" on one's own knowledge). Recently, evidence has been reported that Rhesus monkeys choose to "opt-out" and not even be given the latter part of a match to sample a test when they do not remember the sample well, suggesting that they "know when they remember" (Hampton 2001).

Regarding the convergent evolution of primate and corvid cognition (and possibly of dynamic coordination), questions remain as to how the structurally different avian and mammalian brains allow comparable cognitive abilities, and whether these structural differences necessarily confer different pathways or constraints in the evolution of cognition and that of dynamic coordination.

Our analysis of the coordination between different cognitive functions in a few animal cognition studies was a very schematic description. This is mainly because little is known about the way in which abstract rules, for example, are encoded in an animal's mind, and which brain functions are involved in such computations. Future research into the mechanisms governing these recently discovered cognitive abilities may enable a more detailed description. However, perhaps even at this early stage there is a better way to represent the coordination required in such behavioral tasks of which we are unaware.

As mentioned above, studies in comparative cognition are usually concerned with distinguishing cognitive explanations from simpler explanations such as associative learning. In addition, many of the studies have been inspired by realms of human cognition and many of the current debates are concerned with whether or not a certain ability that animals exhibit is comparable with its human counterpart (for a discussion of the extent to which episodic-like memory in animals captures the core components of human episodic recall, see, e.g., Tulving 2005). As such, these studies might not easily lend themselves to an analysis of dynamic coordination of the kind we have attempted here. A more productive way to proceed may thus be to design behavioral experiments for the purpose of distinguishing between different types of coordination, and specifically for assessing an animal's capability to reorganize the information it already has in a novel way which can only be created when it is tested. If such experiments can be implemented in a diverse selection of species, it could enable drawing a phylogenetic tree of coordinating capabilities.

Acknowledgments

The authors are thankful to Shimon Edelman, Sten Grillner, and Bill Phillips for their constructive comments on previous versions of this manuscript and to Julia Lupp and the Ernst Strüngmann Forum for providing the intellectual infrastructure for this work to be carried out. UG is funded by a Human Frontier Science Program fellowship (LT000798/2009-L).

Overleaf (left to right, top to bottom):
Sten Grillner, informal discussion
Evan Balaban, Gordon Pipa, Erich Jarvis
Shimon Edelman, Plenary discussion
Jon Kaas, Uri Grodzinski
Group discussion, Gilles Laurent

5

Evolution of Dynamic Coordination

Evan Balaban, Shimon Edelman, Sten Grillner,
Uri Grodzinski, Erich D. Jarvis, Jon H. Kaas,
Gilles Laurent, and Gordon Pipa

Introduction

What insights does comparative biology provide for furthering scientific understanding of the evolution of dynamic coordination? Our discussions covered three major themes: (a) the fundamental unity in functional aspects of neurons, neural circuits, and neural computations across the animal kingdom; (b) brain organization–behavior relationships across animal taxa; and (c) the need for broadly comparative studies of the relationship of neural structures, neural functions, and behavioral coordination. Below we present an overview of neural machinery and computations that are shared by all nervous systems across the animal kingdom, and the related fact that there really are no "simple" relationships in coordination between nervous systems and the behavior they produce. The simplest relationships seen in living organisms are already fairly complex by computational standards. These realizations led us to think about ways that brain similarities and differences could be used to produce new insights into complex brain–behavior phenomena (including a critical appraisal of the roles of cortical and noncortical structures in mammalian behavior), and to think briefly about how future studies could best exploit comparative methods to elucidate better general principles underlying the neural mechanisms associated with behavioral coordination. In our view, it is unlikely that the intricacies interrelating neural and behavioral coordination are due to one particular manifestation (such as neural oscillation or the possession of a six-layered cortex). Instead of considering the human cortex to be the standard against which all things are measured (and thus something to crow about), both broad and focused comparative studies on behavioral similarities and differences will be necessary to elucidate the fundamental principles underlying dynamic coordination.

Comparative Approaches to Brain Structure and Function: Is Cortex Something to Crow About?

Conversations combining evolution and coordinative phenomena in behavior and the nervous system are intellectually bewildering, because they might progress along a diverse number of lines. For example, a common approach for evolutionary biologists is to attempt to pinpoint the ultimate reasons underlying selection for behavioral coordinating mechanisms. In contrast, neuroscientists often prefer to think about more mechanistic aspects of brain structure and function without the need for explicit specification of the evolutionary force(s) selecting for particular classes of functionality. For example, this volume is filled with attempts to elucidate meaningful connections between coordinative phenomena in the brain and behavioral/cognitive abilities. In many instances, the mammalian cerebral cortex is assumed to subserve many of the computations underlying complex behaviors, and it is also assumed that such computations are not possible with other neural organizations. In the following discussion, we emphasize an evolutionary perspective and the unique and powerful insights that can be achieved from a broadly comparative approach to the mechanistic neurobehavioral questions at hand.

A comparative approach can elucidate mechanisms that mediate behavioral and neural coordination by revealing broad classes of constraints that separate organisms. For example, a particular organism could simply have neural machinery that is incompatible with instantiating particular cellular or circuit functions, or that is unable to flexibly organize circuits into fleeting, larger-scale assemblies that are necessary to perform particular kinds of computations. We first consider whether there are any fundamental "phase transitions" seen across groups of organisms in the basic components that build neural circuits, and in the kind of computations that these can perform. Similarly, we consider whether there are any major transitions in kind in the types of basic neural building blocks that behaviors are assembled from, for example the often-heard distinction between "hardwired" and "flexible" behaviors.

The Fundamental Unity of the Functional Aspects of Neurons, Neural Circuits, and Neural Computations across the Animal Kingdom

It is a useful exercise to examine whether different animal groups, which people subjectively associate with different levels of behavioral complexity (e.g., roundworms, as compared to honeybees, as compared to sparrows, as compared to humans), have nervous systems that function in fundamentally different ways. Do there appear to be any major phylogenetic transitions in the basic building blocks of nervous systems that might limit the kinds of cell assemblies which can be realized, or the kinds of basic computations that can be accomplished? Such building blocks include the structural and functional components of cells, their molecular constituents, the types of substances they use to communicate, and the kinds of interactions they have. A quick way of obtaining an answer is to survey that part of the animal kingdom without backbones—the invertebrates—to see whether human brains contain some basic structural or functional feature that the brain of an insect or a mollusc lacks.

The few invertebrate species that have been studied to date do not do justice to the diversity of invertebrates, because there are a few dozen invertebrate phyla (e.g., molluscs, arthropods, flatworms, roundworms) compared to only one vertebrate phylum (chordates). Some of the invertebrate phyla have enormous numbers of species: there are about one million known (and between 5–10 million estimated) species of insects, thought to represent 90% of the differing life forms on Earth, compared to close to 60,000 species of vertebrates (The World Conservation Union's 2007 IUCN Red List of Threatened Species, based on summary statistics from 1996–2007). When discussing invertebrates, therefore, we talk about a tiny known sample in a pool with enormous diversity.

To perform evolutionary analyses, one needs to consider the phylogenetic relationships of the organisms under study. Based on current molecular and anatomical evidence, there are three major groups of metazoa (Figure 5.1): the deuterostomia (to which vertebrates belong), the lophorochrozoa (to which annelids, and molluscs, such as the octopus and squid, belong), and ecdysozoa (to which the nematodes, insects, and other arthropods belong) (Mitchell et al. 1988). Several representative subgroups from each of these major groups have been intensively studied in neuroscience, such as rodents and birds among vertebrates, annelid worms among lophotrochozoa, and fruitflies and roundworms among ecdysozoa. With this wealth of information it is now becoming possible to ask whether there are basic principles underlying nervous systems functions across phyla.

To facilitate this exercise, we have defined seven broad areas in which to compare invertebrate and vertebrate nervous systems, and we have examined how invertebrate nervous systems rate in each of these areas: molecular building blocks (e.g., structural, cell-signaling molecules, ion channels), neuronal geometry, nervous system size/scale, mapping/connectivity relations between neurons, local circuit motifs, local computation, and global emergent properties.

Molecular Building Blocks

The explosion of molecular data made available by gene-sequencing studies performed on vertebrates and invertebrates has clearly indicated that there are no known broad classes of molecules involved in mammalian or other vertebrate brains that are absent in invertebrate brains. This is true for ligand-gated channels, voltage-gated channels, gap junctions, the neurotransmitters, neuromodulators and their receptors used for intercellular communication, as well as for the second-messenger pathways used intracellularly to plastically change

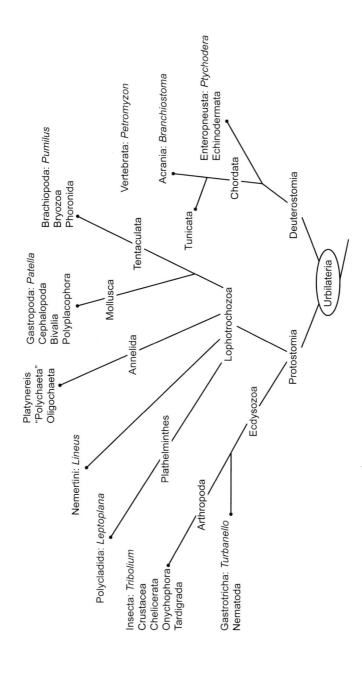

Figure 5.1 The phylogenetic tree of all organisms (after Tessmar-Raible and Arendt 2003).

operational characteristics of cells (some of these same molecules are also found in plants). For example, invertebrates have NMDA channels, inward rectifier current channels, and Ca-activated K channels (Bargmann 1998). Consequently, these neurons exhibit all of the complex phenomena shown by mammalian neurons, including dendritic nonlinearities and intrinsic resonance. Some neuromodulator systems involved in learning or "mood" regulation in vertebrates, such as dopamine and serotonin, are present in invertebrates and appear to be used in behavioral circuits in similar ways (Fiala 2007).

Some similarities are functional rather than sequence-based. Odorant receptors, for example, show cross-phylum similarities in the diversity of the three-dimensional structure of the sites on receptor cells that bind odorant molecules, even if vertebrates and invertebrates may not employ similar sequences in the parts of the proteins that define these binding regions (Benton 2006). In such cases, the functional similarity in vertebrate and invertebrate odorant-binding mechanisms may be the result of selection for similar functions in vertebrate and invertebrate olfactory receptors (convergent evolution), rather than shared ancestry. Vertebrates and insects also use different mechanisms to transduce molecular binding of specific odorants into neural impulses (secondary messenger systems vs. direct channel gating) (Wicher et al. 2008; Nakagawa and Vosshall 2009).

There are also other differences in the deployment of particular mechanisms that both vertebrates and invertebrates possess. For example, photoreceptor conductances are hyperpolarizing in insects but depolarizing in vertebrates. In vertebrates, glutamate is the main excitatory neurotransmitter in the CNS, while acetylcholine is the transmitter at the neuromuscular junction. In insects, this relationship is reversed. In both vertebrates and invertebrates, there are considerable species differences in the elaboration and functional specialization of classes of nervous system molecules. For example, the number of variant forms of particular neurotransmitter receptors may differ in the two groups. The NMDA receptor has two major forms found in both vertebrates and invertebrates, but vertebrates have a component of the receptor (the NR2B subunit) that is not found in insects (Ryan et al. 2008; Emes et al. 2008; Ryan and Grant 2009).

So, the basic molecular components of vertebrate and invertebrate neurons appear to be shared and, when this is not the case, similar functions appear to have evolved convergently. There are also examples of divergences from this general pattern in all groups to meet special circumstances. However, the places in the nervous system where common molecular components get deployed are not necessarily consistent across phyla.

Neuronal Geometry

Santiago Ramon y Cajal (1911) was of the opinion that insect brains are to vertebrate brains what fine watches are to grandfather clocks: "the quality of the psychic machine does not increase with the zoological hierarchy. It is as if we are attempting to equate the qualities of a great wall clock with those of a miniature watch." However, there exists as wide a diversity of dendritic/axonal geometries and shapes (linear, planar, three-dimensional, sparse to dense) among insect neurons as among vertebrate ones. Some invertebrate neurons are very strictly polarized with clearly separated input and output fields, linked via a neurite with a spike initiation zone. Others have intermingled pre- and postsynaptic sites forming local dendro-dendritic circuits. Some neurons, involved in motor coordination between different segments in arthropods can have multiple spike initiation zones (typically one per segment), a feature never described, to our knowledge, in vertebrates.

A feature characteristic (though not universal) of most vertebrate neurons—a soma interposed between dendritic and axonal segments—is absent in most invertebrate neurons, in which somata are often devoid of synapses. Invertebrate neuropil is thus typically devoid of cell bodies, and the spike initiation zone is located on a neuritic segment. Whether the incorporation of cell bodies within the neuropil seen in vertebrate brains is a requirement for their increased growth and/or lamination (or conversely, a constraint that precludes large size increases of invertebrate brains) is not known, but the correlation is suggestive.

Molecular fingerprinting studies within vertebrates (mammals, birds, reptiles, amphibians, and fish) and invertebrates (insects, nematodes, and annelids) have revealed that the centralized nervous systems in both groups are developmentally controlled by many of the same genes that are expressed in specific cell types of developing mammalian cortex, basal ganglia, and spinal cord. For example, the layer V cortex-specific transcription factor ER81 is also found in the forebrain projection neurons in the arcopallium of birds, and in the anterior part of the nervous systems of annelids (Laudet et al. 1999). Annelids have differentiating neurogenic zones that express the same molecules that have been used as cortical and basal ganglia markers in vertebrates. The Hox genes, which are involved in the control of body segmentation, divide the nervous systems of vertebrates and invertebrates into similar segments (Pearson et al. 2005). In spite of some differences in cellular structure between vertebrates and invertebrates, the same global structural and molecular principles appear to produce comparable cell types across the animal kingdom.

Brain Size and Scale

Some invertebrates are minute, such as the parasitic mites of insects, whereas others are gigantic, such as the giant squid (up to 14 meters in length). A sense of scale can be derived from the following numbers. The brain of Drosophila contains about 250,000 neurons, whereas that of a large insect contains about 1 million. The mushroom body of a cockroach—a structure containing odorant-processing cells and interneurons and which may be analogous to parts of the vertebrate forebrain—contains 300,000 Kenyon cells (greater than the number of pyramidal cells in a rat's hippocampus). A large arthropod, such as a ten-year-old horseshoe crab, has tens of millions of neurons in its brain, mostly in its mushroom body (Laurent 2002). Cephalopod brains are even larger; the brain size of a giant squid is thought to be the largest invertebrate brain currently on the planet, but is essentially unknown.

Synapse numbers are not very well characterized, though this is likely to change in the near future. Known ranges of convergence on single neurons in one species (locust) are between a few (order 10) in some early visual neurons, and many hundreds onto Kenyon cell dendrites (Jortner et al. 2007; Turner et al. 2008). Divergent synapses are even less well characterized, but known examples are between ~1 in the synapse between the lobular giant movement detector and the the descending contralateral movement detector of the locust (Rind 1984), and ~20,000 in locust antennal lobe projection neuron–Kenyon cell populations (Jortner et al. 2007).

Conduction velocity is one feature in which vertebrates were thought to have come up with an evolutionary novelty: the myelin sheaths that enclose axons. Whereas invertebrate groups do not appear to have a myelin basic protein—a building block often used in vertebrate preparations to detect the presence of myelin—crustaceans, shrimps, annelids, and copepods do have sheaths that enclose axons in similar ways and which function physiologically in the same way as myelin (Hartline and Colman 2007).

Mapping Relations between Neurons

Topographic neural maps exist in insects that are comparable to ones found in mammals. Some examples are the somatotopic map of the wind-sensitive cercal sensory system in crickets and cockroaches (Jacobs and Theunissen 1996), functionally similar to the dermatotopic somatosensory maps in the S1/layer IV of rodent somatosensory cortex; the tonotopic map in the auditory system of bush crickets (Imaizumi and Pollack 1999), which is similar to the tonotopic auditory maps of mammalian auditory pathways; and the nontopographic projections from the antennal lobe neurons to the mushroom bodies of the locust's "generalist" olfactory system, similar to the ones seen in the nonpheromonal portion of the olfactory system of mammals (Jortner et al. 2007). Finally, one finds regions in which there is clear connectional structure (e.g., the olfactory receptor cell projections to insect antennal lobes or to vertebrate olfactory bulbs) but in which the underlying rules of the mapping are equally unknown.

While it is often thought that the connectivity of invertebrate brains is rigidly specified genetically, earlier studies on genetically identical waterfleas and grasshoppers indicate that this was not always the case (Macagno et al. 1973; Goodman 1978). More recent data indicate that genetically identifiable Drosophila Kenyon cells cannot be identified on the basis of their tuning to odors, whereas the neurons presynaptic to them can (Murthy et al. 2008). Similarly, using genetic and developmental manipulations, it has been shown that the number of morphological local interneuron types in the antennal lobes of several thousands of individual flies exceeds the number of local interneuron types in any one antennal lobe (Liqun Luo, pers. comm.). These pieces of evidence suggest that interindividual variations of internal connectivity, similar in kind to those seen in vertebrate brains, also exist in the brains of insects. Experience-dependent modulation of the strength of local connections is well-known across a wide variety of invertebrate systems (Roberts and Glanzmann 2003; Cassenaer and Laurent 2007), so vertebrate and insect brains appear to share similar basic principles for establishing and changing connectivity.

However, there are three connectional features known in vertebrates that have not so far been found in invertebrates: (a) sensory and motor maps that are registered with each other in an interconnected way, such as those seen in the mammalian or avian superior colliculus; (b) massive feedback loops such as those seen between the primary cortices and thalamus in mammalian brains; and (c) nesting of sequences of modular local circuits, such as those seen in mammalian hippocampal circuits.

Local Circuit Motifs

There are no apparent differences between invertebrate and vertebrate circuits in local circuit motifs. Both vertebrate and invertebrate circuits can include: local or global–local inhibition, reciprocal inhibition, feedforward inhibition, lateral inhibition, lateral excitation, focal convergence (olfactory glomeruli), wide divergence (50%), and all-to-all negative feedback (Laurent 1999).

Local Computation or Operations

There are no major differences between vertebrate and invertebrate nervous systems, of which we are aware, for local computations and operations. Insect brain systems exhibit shunting inhibition, dendritic multiplication, infra- and supra-linear summation, plastic changes mediated via synaptic Hebbian rules, Elementary Motion Detection and directional selectivity, and gating by efference copies (Poulet and Hedwig 2002, 2006; Gabbiani et al. 2005; Cassenaer and Laurent 2007).

Global Emergent Properties

Invertebrate nervous systems exhibit the functional reconfiguration of network output (frequency and phase) in response to neuromodulators (e.g., crustacean stomatogastric system; Marder and Bucher 2007), the adaptive regularization of synchronized oscillatory output by synaptic, timing-dependent plasticity (e.g., locust olfactory system; Cassenaer and Laurent 2007), and various forms of oscillatory synchronization at frequencies from < 1 Hz (Limax) to 20–30 Hz (Schistocerca) (Laurent and Davidowitz 1994; Gelperin and Tank 1990).

The frequency range of oscillations discovered thus far is narrower than in vertebrates, and high frequency bouts nested within lower frequency ones have not yet been described.

The overall conclusion of these considerations is that there is a common mechanistic toolkit at multiple levels, from the molecules that participate in neural structure and function, to properties of single cells, to properties of cell assemblies, shared by all animals with nervous systems. While species differences can exist at many of these levels, the overwhelming impression is that the mechanisms underlying neural computations and the nature of those computations do not undergo dramatic phylogenetic shifts.

The reasons for such conservation of neural computational functions across phyla and ecological niches may be found in boundary constraints that apply to the evolution of neural computation. As noted by Herbert Simon (1973): "…nature is organized in levels, and the pattern at each level is most clearly discerned by abstracting from the detail of the levels far below.…nature is organized in levels because hierarchical structures—systems of Chinese boxes—provide the most viable form for any system of even moderate complexity" (cf. the discussion of hierarchical abstraction in Edelman 2008a:30–31).

The implication of Simon's insight for understanding neural systems is that homogeneously interconnected (i.e., unstructured) networks of basic units would not scale up well for all but the simplest tasks, placing them at a disadvantage relative to networks that embody hierarchical abstraction through the existence of multiple levels of organization and multiple functional units at each level (Edelman 2003). This computational constraint should be kept in mind as we attempt to understand the neural basis of complex coordinated behaviors that exhibit serial order (Lashley 1951). Indeed, computational considerations suggest that the most complex types of these behaviors (including language), which require dynamic coordination across many levels of abstraction, timescales, and individuals, would be unlearnable and unsustainable in the absence of a properly structured and presumably dynamically coordinated computational substrate (neural or artificial).

Computational considerations also offer a solution to the usual puzzle of explaining, without resorting to conceptual "skyhooks" (Dennett 1995), how complex functions, and the correspondingly complex neural architectures that support them, can evolve without disrupting the existing mechanisms. The solution arises from the concept of *subsumption architecture*: an approach to incremental and nondisruptive augmentation of function proposed by Brooks (1991) and developed by him and others in the context of evolutionary robotics (see Sloman and Chrisley 2005). In a subsumption architecture, modifications to an existing circuit are initially introduced as modulatory add-ons that do not disrupt its functioning; subsequent evolution may cause the original circuit to be eventually completely replaced by the novel components acting in concert, or its encapsulation and persistence as a fall-back mechanism that continues to provide basic functionality (e.g., the decorticate cat, which will be discussed later). Clearly, smooth functioning of subsumption-based neural systems requires dynamic coordination among their components at all levels. In this process, evolutionarily newer organizational entities may exert disproportionate amounts of control over preexisting entities; an example of the kind of downward causation (Thompson and Varela 2001; Edelman 2008b) that may arise in such cases has been aptly described by Shakespeare near the end of Hamlet's famous soliloquy:

> Thus conscience does make cowards of us all,
> And thus the native hue of resolution
> Is sicklied o'er with the pale cast of thought,
> And enterprise of great pitch and moment
> With this regard their currents turn awry
> And lose the name of action.

Brain Similarities and Differences Can Be Used to Gain Novel and Fundamental Insights into Complex Brain–Behavior Relationships

All Levels of Coordination between the Nervous System and Behavior Are "Complex" though Some May Be More "Complex" than Others

The idea of a common mechanistic and computational neuronal toolkit that is applicable across animal species is fundamentally antagonistic to notions that there is some kind of *scala naturae*, according to which species with "simpler" nervous systems exhibit "simple" forms of behavior, while it is only species that share particular brain features with humans (such as the possession of large swaths of six-layered cortex in their forebrains) that are capable of showing "complex" behavior.

With few exceptions, every perceptual/cognitive/motor act in which an organism engages involves problems of coordination and control which must be adapted to the immediate conditions and circumstances that exist when the act is performed. To do this, both vertebrates and invertebrates appear to have organized local circuits that are wired together at relatively early stages of development (typically in the embryo) called central pattern generators (CPGs). These circuits were originally discovered in the context of neural work relating the organization of circuits in the vertebrate spinal cord to locomotion, and such locomotory circuits have broad similarities in organization and function in species ranging from lampreys to mammals (Grillner 2003, 2006). Such circuits exist in both the motor and sensory domains, and may exist in more abstract domains of function not easily characterized as either sensory or motor. A good example of a fairly complex, nonspinal implementation of these circuits and the fundamental role that comparative work can play in elucidating them was shown by experiments using developmental manipulations of the species identity of brain cells in avian embryos.

Balaban and collaborators used early transplants of neural tube tissue between quail and chicken embryos to examine species differences in neural circuits associated with "crowing," a vocalization that male chickens and quails use in the context of mate attraction as well as agonistic interactions (the chicken form is the well-known rooster "cock-a-doodle-doo" vocalization). Unlike the functionally similar "song" vocalizations of songbirds, which will be discussed below, "crowing" is a vocal motor sequence that does not depend on imitative learning. A songbird (or a human) who is deaf from an early age will later sing a song (or produce speech) that is not generally recognizable, while a congenitally deaf chicken will sing a song that may have subtle deficiencies but is nevertheless unmistakably recognizable. By transplanting both large and small sets of adjacent cells along the entire brain primordium between chickens and quails at a time in development when brain regions have already been determined, but neurons have not yet started to differentiate, Balaban et al. (1988) found a midbrain region that transferred the acoustic characteristics of the vocalization between species but left the head movements used to deliver it unaltered. Balaban (1997) then found a brainstem region that transferred the head movements used to deliver the vocalization between species but not the acoustic characteristics of the sound. The latter transplants also revealed a rostrocaudal organization in the circuits mediating the sequence of head movements delivered with the crowing vocalization. Transplants that differed in their rostrocaudal extent reliably transferred different temporal portions of the head movement sequence of the donor species. These transplants had no effect on the kinematic characteristics of other head movements that chickens and quail perform identically (such as yawning), and transplanted animals showed perfect integration between head movement and vocal aspects of the behavior. In operations conducted between two chickens, animals with similar transplants showed normal chicken behavior.

Although the concept of a CPG may seem to belong exclusively in the motor domain, it does have a close counterpart in perception: the classical receptive field (RF). Computationally, the RF is simply a template: a filter that responds with a certain degree of selectivity to stimuli that appear within a region of the input space. In mammalian vision, for instance, the RFs of neurons in the lateral geniculate nucleus of the thalamus have a circular center-surround organization in the visual field whose response profile is well approximated by a difference of Gaussians, which is to say that they respond well to spots of light against a dark background or vice versa (depending on the neuron). Feedforward "recognizers" for progressively more complex shapes can be constructed from RF-like building blocks (Edelman 1999); these, however, need to be coordinated in some fashion if more sophisticated function, such as compositional treatment of shapes and scenes, is required (Edelman and Intrator 2003).

The existence of CPG-like circuits involved in perception was shown by Long et al. (2001), who found a region at the junction between the midbrain and thalamus that transferred preferential response to parental warning calls between the two species. Under normal circumstances, chicken and quail chicks have the ability to walk and feed themselves within a short time after hatching; in the presence of danger, there is an acoustically distinctive vocalization that parent birds give to call the young. This vocalization is acoustically different in chickens and quail, and young animals hatched in the absence of prior exposure to these calls, and given equal experience with them in a choice situation, show a statistically significant preference for approaching the calls of their own species (Park and Balaban 1991). Animals with the effective transplant produced individuals who had significant statistical preferences for approaching the call of the donor species, and the strength of their preference was significantly stronger than that of unoperated donor individuals (the quail donor species develops at a faster rate than the host chicken species). The vocalizations of transplanted animals were not affected by the transplants, precluding an indirect effect of changes to the individual's own vocalizations as an explanation for the effect on auditory-mediated approach preferences.

While many of the early scientists who discovered CPGs emphasized their stereotypy, CPGs must be modulated to adapt flexibly to the changing circumstances that are part of the everyday lives of all organisms. For example, in the case of chickens, young animals appear to learn the auditory characteristics of the danger calls of their particular parents rapidly (Kent 1987, 1989). If they suddenly find themselves in the care of new parents, they have the capability to "relearn" their responses to the calls of the new set of parents. In the case of locomotory circuits, they must flexibly regulate their parameters with feedback from the environment to compensate for path obstructions, changes in angle of the ground surface, and other behavioral acts that an organism may be engaging in during locomotion.

This interplay between circuit elements that attain their functional characteristics early in development (like CPGs) and other circuit elements that recruit them flexibly into higher-order sequences of behavior and mediate plastic changes within them is a feature shared by *all* behaviors in all organisms. There is no behavior that is truly "hardwired," and no behavior that is completely flexible. Locomotion in insects and fish, the detection of particular objects in visual scenes by cats and primates, and people's participation in conversations all utilize a complex interplay between CPGs and other circuits that are flexibly mediated. The neural difference between the singing behavior of a bird that learns its song by imitation (like a cardinal) and one that does not (like a chicken) resides in differences in the interaction of neural circuits that are recruited during these tasks (Jarvis 2004). These species differences depend on differences in connectivity patterns laid down during embryonic development. The net result is that songbird brains during the learning period bring together information about a desired song form, together with information on the vocalization that they have just produced, whereas chicken brains do not. Both species flexibly use a combination of sensory, motor, and other CPGs together with other circuits to perform their respective behaviors.

Learning General Principles from Different Brain Organizations That Solve Similar Problems

A major advantage of comparative studies using species differences is to analyze how different brain organizations solve similar problems. With regard to problems of behavioral coordination, social communication is perhaps the quintessential example of a widely spread complex, coordinated activity. Two major vertebrate taxa—birds and mammals—have achieved considerable sophistication in their use of vocal communication, and in the means by which vocal behavior is acquired.

Oscine song birds, like humans, need exposure to their species vocalizations during development to produce normal vocal social signals in adulthood (Doupe and Kuhl 1999). Both oscine songs and human speech are vocal examples of imitation, a complex type of learning which is considered important for human social cognitive abilities. Although vocal imitation ability is rare, evidence for nonvocal imitation is starting to accumulate in many taxa (reviewed in Huber et al. 2009; Zentall 2004). For a behavior to be considered an imitation, an organism needs to perform a motor output corresponding to the sensory input of another organism's behavior that it has observed prior to producing the behavior. This is known as "the correspondence problem"(reviewed in Heyes 2009). The discovery of so-called "mirror" neurons, which are active when an action is perceived as well as when it is performed (Ferrari et al. 2009; Gallese et al. 1996) or in songbirds when a song is heard versus sung (Prather et al. 2008), suggests a mechanism by which the brain solves this problem (Ferrari et al. 2009). However, the existence of such neurons in the adult brain does not explain how the "right connections" came to be made in the first place. For example, connections between the visual input of a grasping movement and the motor output of grasping require an ontogenetic explanation. Such connections can either be specified during brain development and preexist prior to their use, as in the CPG examples described above, or could be built during ontogeny by the co-occurrence of the corresponding output and input when the individual itself performs a behavior. In this latter case, the formation of the "right" connections between these two systems would depend on mechanisms of neural plasticity that would need to be deployed properly in these pathways to make such learning possible (the hypothesis of associated sequence learning, and the evidence for it, are reviewed in Catmur et al. 2009; Heyes 2001).

As we have discussed, the computational problem of correspondence needs to be solved to imitate. However, it is also important to distinguish between imitation and other types of social learning which do not require overcoming the correspondence problem. In many cases, the probability of performing a certain behavior, or the speed of learning a new behavior, is increased by observing another individual (a "demonstrator") perform the behavior. Instead of imitation, however, this can be attributed to observation increasing the probability that the subject will engage in similar behavior that it already has in its repertoire ("social facilitation," e.g., Fiorito and Scotto 1992), or will act to achieve *the same goal* achieved by the demonstrator ("emulation"). Note we are not at all dismissing the complexity of these other forms of social learning, but are simply pointing out the fact that they pose different computational challenges and that they need to be distinguished before anything can be said about what it is that the animal's nervous system needs to compute.

Mammals and Birds: Similar Levels of Behavioral Complexity Despite Major Differences in Brain Organization

Comparing Avian and Mammalian Brain Organization

One of the better understood comparative analyses involves brain comparisons between birds and mammals. This analysis has challenged and forced a revision to the classical view of brain evolution and the supremacy of the mammalian cortex (Reiner et al. 2004; Jarvis et al. 2005). The classical view is that the avian cerebrum, along with that of other vertebrates, evolved in progressive dorsal-to-ventral stages from so-called primitive to advanced species (Edinger 1908). The current view holds that the avian cerebrum, and those of other vertebrates, was inherited as a package consisting of pallial, striatal, and pallidal domains, which together function in perceiving and producing complex behaviors (Jarvis et al. 2005; see also Figure 5.2). This current view is associated with a new brain terminology for birds developed by an international consortium of neuroscientists.

According to the classical view, evolution was unilinear and progressive, from fish to amphibians, to reptiles, to birds and mammals, to primates, and finally to humans, ascending from "lower" to "higher" intelligence in a chronological series. Proponents of this view believed that the brains of extant vertebrates retained ancestral structures, and thus the origin of specific human brain subdivisions could be traced back in time by examining the brains of extant nonhuman vertebrates. They also believed that evolution occurred in progressive stages of increasing complexity and size, and culminated with the human cerebrum. Thus Edinger (1908) argued that there was first the old brain—the paleoencephalon (also called the basal ganglia or subpallium), which controlled instinctive behaviors—followed by the addition of a new brain—the neoencephalon (also called the pallium or mantle at the top), which controlled learned and intelligent behaviors. To support this view, he and his students named the telencephalic subdivisions with the prefixes "paleo" (oldest), "archi" (archaic), and "neo" (new) to designate the presumed relative order of evolutionary appearance of that subdivision. To these prefixes, the root word "striatum" was added for the presumed paleoencephalic subdivisions and "pallium" (meaning mantle), or "cortex," for the presumed neoencephalic subdivisions. Fish were thought to have only "paleostriatum" (old striatum) and paleocortex was said to be the antecedent of the human globus pallidus. Amphibians were said then to evolve an archistriatum (i.e., amygdala) above the paleostriatum and an archicortex, the antecedant of the human hippocampus. Reptiles were said to evolve a neostriatum, which they passed onto birds, who then evolved a hyperstriatum. Birds and reptiles were not thought to "advance" the paleocortex and archicortex. Instead, mammals were thought to have evolved from the paleocortex and/or archicortex, a "neocortex." The archicortex and paleocortex with their 2–3 cell layers were assumed to be primitive; the neocortex with its six layers was assumed to be more recently evolved and a substrate for more sophisticated behavior. The avian cerebrum was thought to consist primarily of basal ganglia territories, and these were thought to control mostly primitive behaviors. This classical view was codified in the major comparative neuroanatomy text by Ariëns-Kappers, Huber, and Crosby (1936) and became pervasive throughout neuroscience. However, this view is now known to be incorrect.

Based on molecular, cellular, anatomical, electrophysiological, developmental, lesion, and behavioral evidence, an international consortium of specialists in avian, mammalian, reptilian, and fish neurobiology published a new nomenclature that represents the current understanding of avian telencephalic organization and homologies with mammals and other vertebrates (Jarvis et al. 2005; Reiner et al. 2004). They concluded that the telencephalon is organized into three main, developmentally distinct domains that are homologous in fish, amphibians, reptiles, birds, and mammals: pallial, striatal, and pallidal domains. It is hypothesized that the telencephalon of early fishes possessed all three domains, which were then inherited as a package by later vertebrates, including birds, and independently modified by them. The consortium eliminated all phylogeny-based prefixes (paleo-, archi-, and neo-) that erroneously implied the relative age of each subdivision.

They also concluded that the organization of the true basal ganglia among vertebrates (i.e., distinct nuclear striatal and pallidal domains) is quite conserved. Some key similarities between vertebrates, best studied in birds and mammals, include a high enrichment of dopaminergic axon terminals in the striatum that originate from a homologous substantia nigra pars compacta and ventral tegmental area neurons of the midbrain. Both avian and mammalian striatum contain two major classes of spiny neuron types: those with the neuropeptide substance P (SP) and those with the neuropeptide enkephalin (ENK), which project to two different neuron populations in the pallidum. In both birds and mammals, the SP neurons seem to be involved in promoting planned movement, while the ENK neurons seem to be involved in inhibiting unwanted movements. Both the avian and mammalian striatum participate not only in instinctive behavior and movement, but also in motor learning. Developmental studies indicate that the avian and mammalian subpallium consists of two separate histogenetic zones that express different sets of transcription factors: a dorsal zone that corresponds to the lateral ganglionic eminence and that selectively expresses the transcription factors Dlx1 and Dlx2 but not Nkx2.1, and a ventral zone that corresponds to the medial ganglionic eminence and selectively expresses all three transcription factors. The lateral ganglionic eminence gives rise to the striatum; the medial ganglionic eminence gives rise to the pallidum. Similar striatal and pallidal territories have been found in reptiles.

In contrast, the organization of vertebrate pallial domains differs to a greater degree. Like the striatum, the avian and reptilian pallium has a nuclear type of organization. The avian hyperpallium, however, possesses a unique organization so far found only in birds; its dorsal surface consists of semilayered subdivisions and might have evolved more recently than the mammalian six-layered cortex, since birds evolved well after mammals (by ~50–100 million years) (Jarvis et al. 2005). The six-layered cortex is a pallial organization unique to mammals. As all major groups of living mammals (monotremes, marsupials, and placentals) have a six-layered cortex, it was presumably inherited from their common therapsid ancestor over 200 million years ago. As all nonmammalian therapsids are now extinct, it is difficult to trace the evolutionary history of mammalian telencephalic organization from stem amniotes to mammals: layered, nuclear, or otherwise. Thus, the reptilian nuclear pallial organization cannot be assumed to represent the ancestral condition for mammals, as it is for birds.

Comparing Avian and Mammalian Cognitive Behaviors

Based on the modern view, the adult avian pallium, as in mammals, comprises ~75% of the telencephalic volume. This realization of a relatively large and well-developed avian pallium that processes information in a similar manner to mammalian sensory and motor cortices may help to explain some of the cognitive abilities of birds. Recent studies show that some bird species may have behavioral complexity on a par with nonhuman primates. Some of the best examples come from studies of physical cognition, where the classical trap-tube test has been used both with primates and birds (reviewed in Martin-Ordas et al. 2008; see also Grodzinski and Clayton, this volume). Many primate and other species have been trained to use a tool to push a piece of food placed in a transparent tube away from a trap, but subsequently failed to show an understanding of the properties of the trap as they continued to do so when the tube was inverted; the trap became ineffective (Martin-Ordas et al. 2008). In fact, the first nonhuman species to demonstrate such an understanding in modified versions of the trap-tube design are two species of birds: rooks (Seed et al. 2006) and New Caledonian crows (Taylor et al. 2009), recently joined by chimpanzees (Seed, Emery et al. 2009; see also Grodzinski and Clayton, this volume). The striking abilities of tool-using New Caledonian crows (Weir et al. 2002) and nontool-using rooks (Bird and Emery 2009) to manufacture, manipulate, and use tools in novel ways suggests some understanding of the physical properties of the tasks at hand, the likes of which are yet to be shown in other nonhumans. In another field of comparative cognition research, social cognition, food-caching corvids have attracted recent attention. When these birds try to steal each others' caches, as well as when they apply a number of cache-protection strategies to avoid being pilfered, they are most sensitive to their competitors' location and previous knowledge (Bugnyar and Heinrich 2005; Emery and Clayton 2001). This suggests a sort of "theory of mind" comparable to that previously suggested in chimpanzees (Hare et al. 2001). The recent developments in the field of comparative cognition also include studies showing bird episodic-like memory (reviewed in Grodzinski and Clayton, this volume) and transitive inference (Vasconcelos 2008). Other work has also shown that pigeons can memorize up to 725 different visual patterns, learn to discriminate categorically objects as "human-made" versus "natural," discriminate cubistic and impressionistic styles of painting, communicate using visual symbols, rank patterns using transitive inferential logic, and occasionally "lie" (reviewed in Jarvis et al. 2005). Together, all of these studies point out that the behavioral complexity of some bird species is comparable with that of the most behaviorally advanced nonhuman primates.

Some bird species even possess traits found in humans and not in nonhuman primates. The most notable is the rare skill of vocal imitation, or more broadly vocal learning. Not only do oscine songbirds have this trait, but parrots and hummingbirds do as well. This trait is a critical substrate in humans for spoken language and with the exceptions of cetaceans, bats, elephants, and possibly sea lions, it has not been found in any other mammal (Jarvis 2004; Janik and Slater 2000). Parrots, in addition, can learn human words and use them to communicate reciprocally with humans. African gray parrots, in particular, can use human words in numerical and relational concepts, abilities once thought unique to humans (Pepperberg 2006).

In general, these cognitive functions include important contributions from the telencephalon, including the six-layered cortex in mammals and the nuclear pallial areas in birds. The mammalian six-layered cortical architecture does not appear, therefore, to be the only neuroarchitectural solution for the generation of complex cognitive behaviors. Pallial-cortical folding is also not required. Birds' brains do not exhibit the complex pattern of gyral and sulcal folds in their pallia that mammals do; among mammals, such folding is more related to absolute brain size than it is to behavioral complexity.

The best-studied comparative circuit example is the vocal learning/speech brain pathways in birds and humans (reviewed in Jarvis 2004). The major groups of vocal-learning birds are distantly related to each other and seem to have evolved similar solutions, although not identical solutions, as humans for the generation of imitative vocal learning behavior. Vocal learning and vocal nonlearning birds and mammals (i.e., nonhuman primates and chickens) have very similar auditory pathways to the telencephalon, used for complex auditory processing and auditory learning. Thus, this is not a rare trait. However, only vocal learners (songbirds, parrots, hummingbirds, and humans) have brain regions in their cerebrums (pallium and striatum with pallidal cells) that control the acoustic structure and syntax of their vocalizations. These systems in birds consist of seven comparable vocal brain nuclei segregated into two pathways: a posterior vocal motor pathway responsible for production of learned song and calls (determined only in songbirds and parrots) and anterior nuclei (connectivity examined only in songbirds and parrots), which are part of an anterior vocal pathway responsible for vocal imitation and modification (Figure 5.2).

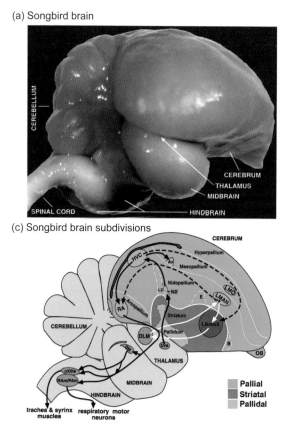

Figure 5.2 Avian and mammalian brain relationships. (a) Side view of a songbird (zebra finch) and (b) human brain to represent avian and mammalian species. The songbird cerebrum covers the thalamus, whereas the human cerebrum covers the thalamus and midbrain. Inset (left) next to the human brain is the zebra finch brain drawn to the same scale. Sagittal view of brain subdivisions according to the modern understanding of (c) avian and (d) mammalian brain relationships (Reiner et al. 2004; Jarvis et al. 2005). Solid white lines are lamina, which are cell-sparse zones separating brain

This motor pathway is similar to descending motor pathways in mammals, and the anterior pathway is similar to cortical-basal-ganglia-thalamic loops. These two pathways have nuclei that are functionally analogous to human cortical, striatal, and thalamic regions required for speech acquistion and production (Jarvis 2004).

Evolution of Brain Pathways for Complex Traits

How might a complex trait like vocal learning have independently evolved a similar circuit diagram in birds and mammals? Recent studies have suggested

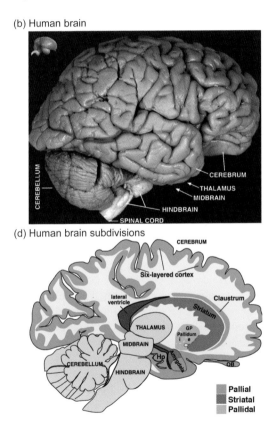

subdivisions. Large white areas in the human cerebrum are axon pathways called white matter. Dashed white lines separate primary sensory neuron populations from adjacent regions. The avian pallium consists of large nuclear regions, whereas the human is layered. Both are involved in vocal learning. The song learning system for the songbird brain is shown. Black arrows, the posterior vocal pathway; white arrows, the anterior vocal pathway; dashed arrows, connectivity between the two pathways. Figure based on (Jarvis et al. 2005; Reiner et al. 2004).

that the telencephalic vocal nuclei of vocal learning birds are embedded within a larger brain system that is active during the production of limb and body movements (Feenders et al. 2008). Likewise, in humans, the unique "spoken language" brain areas not found in nonhuman primates are either embedded within, or adjacent to, motor-learning brain areas found in nonhuman primates (reviewed in Jarvis et al. 2005; Feenders et al. 2008). These and related findings have led to a "motor" theory for the origin of vocal learning. The basic idea is that a preexisting motor system in a vocal nonlearner ancestor is organized as two sub-pathways: an anterior premotor pathway that forms a pallial-basal-ganglia-thalamic-pallial loop and a posterior motor pathway that sends descending projections to brainstem and spinal cord premotor and α-motor neurons. Subsequently, a mutational event or events caused projections of descending pallial-spinal/cortico-spinal neurons, which normally synapse onto nonvocal motor neurons, to synapse instead onto vocal motor neurons in vocal learners. Thereafter, cerebral vocal brain regions developed out of adjacent motor brain regions using the preexisting connectivity and genes. Such a mutational event would be expected to occur in genes that regulate synaptic connectivity.

According to this hypothesis, the vocal learning pathways in birds are analogous to those in humans, in that they are newly evolved neural systems performing complex computations for learned vocal communication. They are, however, homologous in that they share a deep homology with premotor and motor neural circuits that may have existed before the split of birds and mammals over 300 million years ago. This type of brain pathway evolution with shared mechanisms of a deep past is not only restricted to vocal learning, but can apply across multiple traits that require dynamic coordination.

Reconsidering How Different Forebrain Regions Apportion Their Labors

The six-layered mammalian neocortex is often assumed to control practically all aspects of behavior, from the simplest joint movement to the most complex aspects of cognition. Classic lesion studies from several decades ago teach us, however, that subcortical forebrain structures are able to handle many aspects of complex, goal-directed behaviors. For instance, Bjunsten et al. (1976) studied cats whose cortex was removed some weeks after birth (leaving all other parts of the brain intact), and who were able to move around in an exploratory way, become hungry, search for food, and eat. They could solve tasks in a T-maze and find their way out of a complex maze (Sten Grillner and Ulf Norrsell, per. comm.). They reacted emotionally, could successfully attack and drive normal cats away, and went through periods of sleep as well as displaying other aspects of relatively normal circadian rhythms. They were thus able to perform most, if not all, aspects of the standard goal-directed motor repertoire that cats typically show in a constrained laboratory environment, and their movements were well adapted to this environment. This clearly suggests that the subcortical infrastructure of the mammalian brain is capable of subserving a higher level of behavioral function than is generally assumed. These old observations were unfortunately not accompanied by precise quantitative studies of the "cognitive deficits" which inevitably will occur in animals without a cortex. We believe that specifying the unique contributions of cortex to behavior is an extremely important line of research using modern techniques for the quantitative study of behavior, as well as histochemical and imaging analyses to study the progression of the structural and biochemical effects of the lesions over time and their correlations with behavioral effects.

Subcortical forebrain structures, in particular the basal ganglia, are critical for maintaining the goal-directed aspect of motor behavior after the neocortical lesions. How could this come about? The striatum, the input level of the basal ganglia, receives a prominent topographic input from nearly all of cortex (making up about 55% of its inputs); it receives the other 45% of its inputs from the thalamus (Doig et al. 2009; J. P. Bolam, pers. comm.). Part of the thalamic input is sent to both cortex and striatum. Devoid of the cortical input to the striatum (after lesions), it will have to rely entirely on the direct input it receives from the thalamus. Although a fairly detailed knowledge is available on cell types, synaptic interaction, synaptic markers, dopamine innervation, and membrane properties, we do not yet understand the detailed mode of operation of the striatal microcircuitry that most likely plays a prominent role in determining which motor or cognitive programs are selected at any given instant. The striatum becomes severely incapacitated after dopamine denervation as in Parkinson's disease, which affects all aspects of action, motor and cognitive coordination. The output side of the basal ganglia is more well defined, and it contains subpopulations of spontaneously active GABAergic neurons which, at rest, are thought to keep the different brainstem motor centers under tonic inhibition (in addition to the thalamocortical projections). There are different subpopulations that control not only saccadic eye movements but also a variety of other motor centers (e.g., those that control locomotion, posture, chewing). The subcortical forebrain structures are required for the goal-directed aspect, whereas brainstem animals can be made to coordinate the different motor acts (e.g., walking, chewing, eye movements, or pecking), but not in the context of goal-directed adaptive behavior.

In summary, although a lot more work needs to done, the comparative work emphasized in this discussion shows that the opportunity to study analogous behavioral systems, which vary considerably in their complexity across taxa, is of great theoretical and practical importance. Such studies cannot only suggest which neural mechanisms and computations covary with behavioral complexity, but will also give us a better quantitative grounding for relating circuit complexity, computational complexity and behavior.

Functions That Appear to Be Solved in Similar Ways: Comparisons across Diverse Taxa

Although the emphasis in this section has been laid on comparative work examining differences, we thought it was important to point out the future promise of work examining how similar structures can be put to a variety of uses. One example could be provided by the glomerulus, a neural structure that is common to both invertebrate and vertebrate olfaction, and which also has a very high degree of correspondence in the first-order computations it carries out despite evidence that it convergently evolved in these separate taxa. We believe that much can be learned in the future from studies that elucidate the diversity of mechanisms and computations of functionally equivalent structures at a very fine scale of resolution

Need for a Broad Comparative Model System to Study the Relationship of Neural Structures, Neural Functions, and Behavioral Coordination, and for Universal Metrics for Quantifying the Complexity of Behavioral Tasks

There is a need to develop broad comparative neural and behavioral "model systems" to study analogous neural systems across diverse taxa. A promising area for such studies would be a behavior that is widely distributed among a variety of vertebrate and invertebrate species and which is organized into sequences. Grooming behavior is one ideal candidate, as it is widely distributed (Sachs 1988). Next, it would be important to identify instantiations in particular species that differ in complexity, with examples of several different species at each level of complexity studied, and to compare their neural correlates. This would have the joint advantages of providing people interested in dynamic coordination with an independent set of tools to bring to bear on questions relating neural oscillations, synchrony, and behavioral coordination, as well as providing scientists with an empirical method for sifting out the neural mechanisms that vary with particular aspects of behavioral complexity, which can then be targeted for more expensive and time-consuming mechanistic explorations.

Such studies could also bolster the adoption and improvement of common methodologies for quantifying behavioral complexity, a field that has great promise thanks to the introduction of new and powerful computational techniques. Traditionally, the description of coordinated behaviors, such as rodent grooming (Aldridge and Berridge 1998) or spoon-feeding a baby (Duncan 1997), have been primarily heuristic. A variety of computational tools are now available for conducting formal quantitative analysis of the complexity of animal behaviors. It is tempting to divide those a priori into continuous and discrete, but we stress that in many borderline cases, this decision itself should be left to a quantitative analysis with a clearly defined set of criteria.

For example, does the pre-shaping of the hand prior to grasping an object consist of a series of discrete steps? In the general case, the record of an instance of a behavior consists of a trajectory in the space defined by the measurement variables. A bundle of such trajectories can be processed to determine whether they are amenable to a low-dimensional description (i.e., in a manifold whose dimensionality is lower than the nominal dimensionality of the measurement space) or whether a discrete, sequentially modular representation in terms of "dynamical symbols" is in order (Dale and Spivey 2005; Edelman 2008b).

More complex behaviors exhibited by animals are likely to be sequentially modular and hierarchically structured, for reasons of computational tractability, and better fit to the structure of the environment. For such behaviors, a natural formal tool for representation and complexity analysis is grammar. For instance, if a behavior is described in terms of a finite set of states and the transitions among them, it can be represented concisely in the form of a so-called regular grammar. If the transitions are probabilistic, the grammar would have a corresponding annotation. Other classes of formal grammars, such as context-free or context-sensitive, as defined in computer science (Bod et al. 2003; Hopcroft and Ullman 1979) can be used to describe progressively more complex behaviors, including language. The computational methods of inferring a grammar from behavioral data and for using it for complexity analysis, which are akin to the problems of language acquisition and of parsing, may be highly nontrivial, but they are certainly worth the trouble. Behavioral science, and with it the neuroscience of behavior, cannot be considered sound unless it rests on a reasonable quantitative measurement methodology.

Finally, more comparative studies are needed which identify commonalities and differences in neuron structure, cell types, and the ways that brain regions are specified and connectivity develops across taxa. We believe that a more extensive across-taxa neural-comparative toolkit will enable better-informed conclusions about the neural architectures that support dynamic coordination and all other types of computational tasks accomplished by neural systems.

Conclusions

Broad comparative data are necessary to make stronger inferences about the relationship between structure and function. Vertebrates and invertebrates have a common basic neural toolkit that evolutionary processes build upon to generate diverse, but shared forms and principles. The tool kit consists of common genes, cell types, connections, and computations.

Within vertebrates, inroads have been made in understanding the relationships between birds and mammals. However, more of this type of work must be conducted with reptiles, amphibians, and fish.

There is, to date, no general understanding of the functional significance of having a layered (mammal) versus clustered (bird) pallium, since similar behavioral capabilities appear to be attained by both types of organization. Similarly, there is no rigorous evidence-based understanding of the division of labor between pallial (cortex) and nonpallial (basal ganglia) forebrain structures in mammals, despite widespread beliefs about this issue. Both the six-layered mammalian cortex and the nuclear pallial divisions of the avian brain are able to support vocal imitation and other complex cognitive behaviors once thought unique to humans.

Given that coordination occurs at the levels of neurons, circuits, and behavior within and among organisms across the whole animal kingdom, it is unlikely that coordination or oscillation as such is limited to particular neural architectures, such as a six-layered cortex. A cortex may turn out to be something to crow about for some as-yet-unidentified behavioral or computational traits, but it certainly does not work to work in isolation. Both broad and focused comparative studies on behavioral similarities and differences will be necessary to elucidate first principles underlying such phenomena.

6

Modeling Coordination in the Neocortex at the Microcircuit and Global Network Level

Anders Lansner and Mikael Lundqvist

Abstract

A key role for computational modeling in neuroscience is to connect cortical microscopic processes at the cellular and synaptic level with large-scale cortical dynamics and coordination underlying perceptual and cognitive functions. Data-driven and hypothesis-driven approaches have complementary roles in such modeling. The Hebbian cell assembly and attractor network paradigm has a potential to explain the holistic processing and global coordination characteristic of cortical information processing and dynamics. The pros and cons of such a view are described. A large-scale model of cortical layers 2/3 formulated along these lines exhibit the fundamental holistic perceptual and associative memory functions performed by the cortex. Such a model can provide important insights into the possible roles of oscillations and synchrony for processing and dynamic coordination and it highlights important issues related to cortical connectivity, modularization and layered structure.

Introduction

Today, massive amounts of data are available about the brain from many different sources as well as from molecular, subcellular, neuronal, and network levels, and more keeps accumulating due to increasingly advanced measurement techniques. Despite this, our mechanistic understanding of the normal and dysfunctional brain, in terms of how processes at these different levels interact dynamically to produce cognitive phenomena and overt behavior, is greatly lacking. Such an understanding would clearly open up new avenues in the search for more effective drugs and therapies for severe diseases and disorders affecting the brain. Perhaps the only tools that offer some hope of eventually reaching this understanding are mathematical modeling and computer simulation. A computational model can organize efficiently new data in a coherent fashion so as to generate new experimentally testable questions. This may eventually enable us to formulate quantitative theoretical models and thus achieve a more general understanding of the phenomena under study.

By necessity, the acquisition of data about the brain is largely determined by what is possible to measure, not by what we need to know to build better computational models. This tends to cause some detachment between experimentalists and modelers in the sense that the latter experience a patchiness of data where critical pieces are missing, whereas the former feel they produce data that is ignored by modelers. In the early stages of formalizing a scientific field, the approach has typically been one of starting from the fundamentals and gradually adding more detail and complexity. In some instances of less complex and more accessible neural systems, we have already seen a productive interaction between experimental and computational neuroscience (Kozlov et al. 2009), and we expect this to develop further in the near future.

As in other fields, multiscale modeling will be important in brain science, as it will allow us to relate detailed dynamical processes at the cellular, subcellular, and microcircuit level with cognitive phenomena at the level of brain-scale neuronal networks. Software tools, which allow large-scale network models comprising a mix of biophysically and biochemically detailed spiking neuron models, simplified integrate-and-fire neurons, and mesoscopic nonspiking population units to be defined and simulated, have been developed and are now available in the field of neuroinformatics. The most simplified mesoscopic models may represent some cortical regions with only one or two units and thus connect to, for example, dynamic causal modeling techniques used for analysis of brain imaging data.

The neocortex, which is the largest part of the human brain and the site of higher cognitive functions, has been a favorite subject for brain modelers. Computational models with some level of biological detail have been proposed and investigated for more than half a century. As more data has been acquired and as computers have become more powerful, models have increased in size and sophistication. Today, supercomputers are used to model large-scale neuronal networks with a high degree of detail in their component neurons and synapses. If current trends in computing continue, we should be able to simulate in real time detailed models of the entire human brain in about fifteen to twenty years.

Since there has been and still is quite some uncertainty with respect to the relevant experimental data, every neuronal network model to date is explorative and hypothetical. Nevertheless, several of even the early cortex models have been able to display interesting features of, for example, associative memory, dynamic activity, and coordination, replicating some of the key characteristics of the system modeled. Although abstract and without much biological detail, these models nevertheless provided a framework for studying important phenomena beyond what could be grasped intuitively and conceptually. Only by continuing in this same spirit and by bringing models closer to data can we maintain some hope of reaching a true mechanistic understanding of this very complex system.

In this chapter, we focus on attempts in computational neuroscience to connect cortical microscopic processes, at the cellular and synaptic level, with large-scale cortical dynamics and coordination more directly related to perceptual and cognitive functions. We discuss to what extent models might be data driven and the role of hypotheses in this research, as well as possible interpretations and functional implications of the dynamics observed in large-scale cortex simulations. We consider the role of oscillations and synchrony for processing and dynamic coordination. Finally, we argue for complementing the now dominating reductionist approach with more of a synthesis of components to achieve a coherent global picture. This will allow us to discuss the new directions in brain modeling, with more of model-driven experiments.

Data- and Hypotheses-driven Approaches to Modeling

Despite the enormous efforts in experimental brain science, one of the primary challenges of brain modeling is the relative patchiness of data. There is indeed a lot of data available, but this abundance has been driven largely by the capabilities of our recording equipment, rather than by what is the most important for building quantitative models. Heroic attempts to build "bottom-up" data-driven models of a part of the neocortical network without additional assumptions or hypotheses about function and mechanisms are therefore very high risk undertakings. It may be possible to gather high resolution data about single cells and synapses, as well as population data about types of cells and synapses and the distribution and intricate function of ion channels and neuromodulators acting on them. However, modeling work shows that, for instance, the dynamic function of and information processing performed by a cortical neuronal network is likely to be critically dependent on its synaptic connectivity at a local and global scale (Figure 6.1). The cortical networks are to a significant extent formed by activity-dependent processes individual for each brain. Therefore, with current technology it will be hard, if not impossible, to get the wiring right. Statistical approaches and pooling of data from different animals and regions is most likely to miss their target. Likewise, detailed reconstructions of local pieces of cortex are still very limited in their spatial extent. Nevertheless, such projects will provide valuable information to constrain and refine current and future quantitative and computational models.

An alternative to modeling is a hypothesis-driven or "top-down" approach. Starting from some conceptual theory of the fundamental functions of cortex, one builds a model gradually, by entering the basic elements first and adding additional relevant details one by one. This strategy has historically proven quite successful in many scientific and engineering fields. A useful theory should take into account data from many different sources and levels of description:

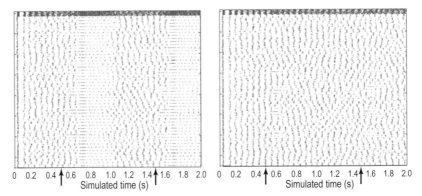

Figure 6.1 Dynamics of network with structured vs. permuted connection matrix. The figure shows the spike raster from runs of two networks. Patterned stimuli are given at the times indicated by the arrows. They differed only in that the left one had memory patterns stored in attractors; the right one had the same connections as the left, except that the pre- and postsynaptic cells were randomly permuted. Thus, both network connectivities obeyed the experimentally measured pair-wise connection statistics between different neuron types. As can be seen, this difference results in dramatically different responses to a patterned input.

from anatomical and physiological data on the neuronal and synaptic levels to experimental observations at the psychological and cognitive levels. Synaptic connectivity may be formed by using a Hebbian learning rule to store a set of memory patterns. Current hypothesis-driven models of cortex take only a fraction of the data available into account and thus are always at risk of leaving out relevant information. Typically, modelers need to fill in some critical data that is missing or conflicting. These parts of the model should be regarded as predictions. If later falsified by new information, models must be reexamined and, if no work-around is found, they should be revised or entirely discarded.

Hebbian Cell Assemblies and the Attractor Network Paradigm

Though the diversity of hypotheses about basic principles of operation of cortex has been pruned over the years, through exposure to increasing amounts of experimental data, a number of different ones still remain. Among some of the more vital and general theories of cortical function today, we find synfire chains (Abeles 1991), liquid state machines (Maass et al. 2002), and localist models of cognition (Bowers 2009). However, the oldest and currently most well-developed theory dates back to Hebb's cell assembly theory (1949). It has more recently been formalized in terms of attractor networks (Hopfield 1982; Amit 1989) and has been extensively studied, both mathematically and by means of computer simulations.

The most well-known component of Hebb's theory is that of coactivation-triggered (Hebbian) synaptic plasticity, and this part has been largely verified by later experiments, which have also extended our knowledge about synaptic plasticity far beyond the original proposal. The more controversial and less acknowledged aspects of Hebb's theory concern the network-level consequences of such processes. Hebb proposed that the wiring together of groups of neurons activated by the same stimulus would form mental objects in the cortex, which he called cell assemblies. They constitute the internal representations of the corresponding objects in the external world, and their existence changes the network dynamics, resulting in phenomena such as after-activity, figure–ground segmentation, and perceptual completion and rivalry effects. This view allows fundamental cognitive functions (e.g., content-addressable associative memory and association chains) to be conceptually understood in terms of processes at the neuronal and synaptic level. Several of these properties are prominent in the corresponding abstract computational models (e.g., a Hopfield network model of associative memory).

The term "attractor" stems from the analysis of the dynamical properties of such abstract recurrent artificial neural networks. Due to the symmetric connection matrix, their state always evolves from an initial state to the closest fix-point attractor and remains there. If a slow dynamics, corresponding to neuronal adaptation or synaptic depression, is introduced in such a model, the attractor state is destabilized and may shift its position in state space or even become transient; that is, it is not formally a fix-point attractor any longer. Such a network typically displays a more complex limit cycle dynamics, which can be described as the state jumping between the memorized states, visiting them for some time determined by the time constants of adaptation and depression, typically in the order of some hundreds of milliseconds. We refer to this here as the attractor states corresponding to memorized patterns becoming quasi-stable.

Adaptation and synaptic depression are common features of, for example, pyramidal cells and synapses between these in layers 2/3. Therefore, a similar dynamics as described above, but even more complex due to the spiking process itself, can be seen in networks of biophysically detailed model neurons and synapses. Here the state of a unit in the abstract network corresponds to, for example, the average instantaneous spiking frequency of the neurons in a local population of some tens or hundreds of neurons.

It needs to be emphasized that the basic attractor network paradigm primarily takes into account the recurrent cortical connectivity most prominent in layers 2/3 and 5. The holistic processing is assumed to be supported by this connectivity and the more refined, specific, and invariant neuronal response properties and representations likely found in higher-order sensory and association cortex. This view is thus incomplete as it disregards the important aspects of how such response properties and representations are formed by self-organization and learning. Such processes can be attributed to the feedforward processing stream of cortex with layer 4 as a key player in transforming the input from different sensor arrays to a sparse, decorrelated, and distributed code suitable for further attractor network processing. In the following, we will assume that the internal representations have already been transformed in this manner.

The cortical network, with its feedforward, lateral, and feedback projections, forms a brain-scale recurrent network structure which may support global attractor dynamics and top-down influences on earlier sensory areas. Attractors extending over large parts of cortex would exert a powerful dynamic coordination (e.g., in the form of multimodal integration). A current target of large-scale cortex simulations is to find out if the known strength and distribution of cortical long-range connectivity is sufficient to sustain such wide area coordinated and quasi-stable attractor states.

Criticisms of the Attractor Network Hypothesis

Though attractive as a working hypothesis for the primary holistic "Gestalt" processing functions of the cortex, this paradigm has been criticized on several points. First, it is obvious that the abstract models have very simplistic units and connectivity, and that they violate Dale's law of separate excitatory and inhibitory units. Further, their dense activity and full connectivity is quite different from the sparse activity and highly diluted connectivity typical for real cortex. In addition, the connection matrix needs to be symmetric to guarantee a well-behaved dynamics, and this seems highly unlikely to hold in reality. A spiking attractor network was also expected to have problems generating the low rate irregular discharge patterns of neurons *in vivo*, and a network with neurons firing at low rate was suspected to converge too slowly. Here we will discuss these criticisms and show how many of them have now been shown to be invalid. We will illustrate how attractor memory models have been brought closer to biological reality in terms of their constituent components and network structure as well as the kind of attractor dynamics and oscillatory activity they display.

There were further concerns about the storage capacity of the original Hopfield network being unreasonably low. However, later theoretical analysis and simulations have shown that an attractor network is actually an efficient associative memory in an information theoretic sense, and that its storage capacity scales linearly with the number of synapses (Amit et al. 1987). With realistically low activity levels, the number of distinct attractors possible to store is much larger than the number of units in the network.

Let us now turn to the biological plausibility of this paradigm and describe a plausible mapping of the attractor network paradigm to a simulation model of cortical layers 2/3. We have used this model to interpret experimental data

Cortical Local Subnetworks and Functional Microcircuits

When building a cortex model, one must confront, early on, the issue of possible repetitive elements other than neurons. It has repeatedly been suggested and discussed that the cortex actually comprises a mosaic of modules, such as functional columns or other types of subnetworks (Mountcastle 1978; Favorov and Kelly 1994; Rockland and Ichinohe 2004; Yoshimura et al. 2005). Such local networks would be more densely connected (like 10–25%) within themselves than to the outside and would likely be selectively targeted by incoming fibers from thalamus. They may be spatially segregated in, for instance, an anatomical minicolumn or may be anatomically diffuse (i.e., intermingled with other similar modules). Possibly in smaller brains (e.g., in the mouse), single or only a few neurons would be in each module; however, in larger cortices (e.g., in primates) each module may need to be more connected than a single pyramidal cell could reasonably sustain. This would make it necessary for several pyramidal cells to cooperate, such that the modules would be larger and perhaps even more anatomically distinct (DeFelipe 2006). Such a minicolumnar structure might also generate a patchy long-range connectivity, as is seen in cortex (Fitzpatrick 1996).

Other data suggests that these local subnetworks are organized in bundles to form larger modules (i.e., macrocolumns, hypercolumns, or barrels; Hubel and Wiesel 1977) and there are abstract models that incorporate such a modular structure (Kanter 1988; Sandberg et al. 2003). A hypercolumn may be assumed to represent, in a discretely coded fashion, some attribute of the external world. For instance, in the primary visual cortex the orientation or direction of an edge stimulus at a certain position on the retina is assumed to be represented by elevated activity in a corresponding orientation column in primary visual cortex. This also leads to sparse activity in the network, on the order of about 1–5%. Such an activity level is in accordance with overall activity densities of about 1% and an average spiking frequency of 0.16 Hz, estimated from metabolic constraints (Lennie 2003).

To investigate the extent to which the attractor network paradigm is compatible with this data, we have designed and simulated a biophysically detailed large-scale model of cortical layers 2/3 (Lundqvist et al. 2006). In this top-down model, the microcircuitry implements two types of modular structures, corresponding to minicolumns and hypercolumns. Each minicolumn is a network unit and comprises thirty pyramidal cells and a couple of dendritic targeting and vertically projecting interneurons, for example, double bouquet (RSNP) cells that inhibit cells in the minicolumn to which they belong. Their role is described in the next section. Each hypercolumn is a bundle of up to a hundred minicolumns together with about an equal number of basket cells.

The latter are driven by nearby pyramidal cells and their axonal arborizations extend over the entire hypercolumn, providing negative feedback and turning the hypercolumn into a kind of winner-take-all module.

Global Recurrent Connectivity Supporting Holistic Processing

Once this model of the local modular structure of the cortex is in place, it is quite obvious to see how the global, long-range connectivity could implement the memory-encoding connection matrix of the attractor network. The largest-scale model that we have simulated comprised several thousand of such hypercolumn modules, where memory patterns were stored by connecting sets of coactive minicolumns, one per hypercolumn. Hebbian plasticity rules typically give a connection matrix with a positive and a negative component. Our network represented the positive elements of the connection matrix by direct intracortical or corticocortical pyramidal-pyramidal synapses, whereas the negative elements of the connection matrix were disynaptic via double bouquet cells in the target minicolumn. This suggests a specific role and synaptic input for this type of cells, which is compatible with known data but where the details are not yet confirmed. Given some spatial modularization of the target area, in terms of functional minicolumns, this will give rise to the patchy long-range connectivity experimentally observed in several cortical areas.

In our model, the long-range connectivity is very sparse at the cell-to-cell level. Typically, in our largest network, out of the 900 possible long-range excitatory synapses from a source to a target minicolumn, only five are instantiated randomly. This gives a connection matrix which is sparse and nonsymmetric at the microscopic level, but symmetric at the macroscopic level of minicolumn units. As mentioned above, in theory, attractor networks need to have a symmetric connection matrix to have a guaranteed convergence to an attractor. In practice, we find that the network state does converge despite the sparse connectivity and this microscopic deviation from symmetry.

In the largest-scale model, comprising 22 million model neurons and 11 billion synapses, the connection densities and PSP amplitudes have been set to reflect experimental data from, for example, Thompson et al. (2002) and Binzegger et al. (2007). The average number of synapses onto single cells of different kinds are then held fixed as the network is scaled (Djurfeldt et al. 2008), whereby a single cell receives approximately the same synaptic input current regardless of network size. The network is driven by input from a few layer 4 pyramidal cells that are activated directly via noise injection and connects with a probability of 50% to the above pyramidal cells in the same minicolumn.

The sparse and low rate activity in the layer 2/3 network has some important implications for the balance between thalamic input, local connectivity, and long-range contextual influence. Though the latter connections are more numerous, only a few percent will be active due to the sparse activity in the presynaptic population. Despite their relatively low numbers, thalamic input influences the network activity significantly in the model and can by itself switch the activity from one active state to another.

Attractor Memory Dynamics and Holistic Perceptual Processing

What can we say about the dynamics of memory retrieval in a biologically detailed cortical network model, such as the one described above? The most important questions concern holistic processing in terms of attractor convergence, after-activity, associative memory, and perceptual completion and rivalry. The two latter are clear examples of how long-range "contextual" and coordinating interactions can modify and even override (rivalry) local processing. This aspect has been examined in numerous simulations; all of these operations were found to be prominent, occurring on the same timescale as observed psychophysically, in less than a hundred milliseconds (Lansner and Fransén 1992; Fransén and Lansner 1998; Sommer and Wennekers 2001). This is true even if a fragmented and noisy pattern is given as input or if two patterns are stimulated simultaneously, as in a pattern rivalry situation. The capacity to store and reliably retrieve patterns parallels that of the isomorphic abstract associative memory network.

Due to adaptation and synaptic depression in the network, after-activity will typically be restricted to a few hundred milliseconds, following which the activity terminates. If the input is still on, the second-best matching pattern may activate, and there may be a slow oscillating activity with the two memory states alternately in the foreground. This is reminiscent of the alternating perception of ambiguous stimuli as in the Necker Cube illusion. In other situations (e.g., hippocampal place fields), attractors may be overlapping and temporally chained, thus forming more of a line attractor structure (Lisman and Buzsáki 2008).

Notably, the network dynamics is highly sensitive to the higher-order connectivity structure. The statistics provided by pair-wise recordings or from morphological reconstruction is not enough to determine uniquely the network structure or its dynamics. For instance, the prominent attractor dynamics seen in our model, which results from a long-range connectivity set up according to a Hebbian learning rule to store a number of distributed patterns, is not seen in the same network where the long-range connectivity is permuted randomly (Figure 6.1). In both cases, pair-wise connectivity statistics is constrained in the same way according to experimental data; however, in one, an attractor state is entered whereas in the other only a weak disturbance of the ground state results.

With sufficient unspecific background activation, the state of our simulated network will under certain conditions spontaneously jump between attractors in a complex, presumably chaotic fashion and at about theta frequency (~5 Hz). This may relate to the dynamics of different kinds of ongoing activity recorded in cortex *in vitro* and in slices *in vivo* (Lehmann et al. 1998; Grinvald et al. 2003; Ikegaya et al. 2004). During such activity, the internal state of a cell may show plateaus and hyperpolarized periods. If the network is stimulated during ongoing spontaneous activity, it typically switches quickly to the stimulated pattern. This switch is, however, somewhat less likely to occur during the early part of a newly triggered attractor state, which has been suggested to provide a mechanistic explanation for the cognitive bottleneck phenomenon of attentional blink (Lundqvist et al. 2006).

The duration and intensity of an attractor state is quite sensitive to changes in model parameters relating to Ca–KCa channels and dynamics (Lundqvist et al. 2006). Several of the monoamine neuromodulators (e.g., serotonin and acetylcholine) act on these processes. This suggests a possible connection between these parameters of the model and neuromodulation of ongoing and stimulus triggered cortical dynamics.

Coordination in the form of burst synchronization at about theta frequency occurs over long distances (with conduction delays of tens of milliseconds) in the model, and it could be assumed to extend over large parts of the cortex. Such coordination would be important, for example, for cross-modal integration and top-down expectancy phenomena and could be observed in local field potential (LFP) and EEG recordings (Engel et al. 2001). Top-down excitation via back-projections from attractors formed in higher cortical areas, as well as facilitatory influences from subcortical structures involved in motivational and drive regulation, may guide attention (Ardid et al. 2007). At the global cortical level, a single coherent activity mode seems to dominate the entire network most of the time due to lateral inhibition. Such competition may ultimately be a mechanism responsible for the apparent unity of our consciousness and the serial nature of thought processes at the macroscopic level, in contrast to the massively parallel processing at the microscopic level.

Microcircuits and Fast Oscillations

Network dynamics, with its quasi-stable attractor states, has the potential to generate a slow rhythmic activity in the theta frequency range. This can be seen even in the mesoscopic models with nonspiking population units. While implementing this kind of network with spiking units, faster rhythmic dynamics emerges in the higher frequency ranges of alpha, beta, and gamma oscillations. The latter has gained special interest due to its correlation with attention, perceptual processing, and consciousness (Engel et al. 2001). Alpha and beta oscillations characteristic of resting conditions are replaced by faster oscillations during active processing. The general trend in experimental data seems to be that faster oscillations are more localized to layers or columns and more short-lived (Sirota et al. 2008). This suggests that the origin of fast (e.g., gamma) oscillations may be the microcircuit whereas slower oscillations arise from interactions over longer ranges.

Oscillations in the gamma frequency range emerge readily in simulated microcircuits, as a result of strong feedback inhibition, and can even do so in a network of recurrently connected inhibitory neurons when subject to some background excitation. Notably, gap junctions between inhibitory interneurons are not necessary to generate prominent fast oscillations. The main factors that decide the frequency in excitatory-inhibitory (E-I) networks is the type of connectivity, the relation between excitation and inhibition, and the synaptic time constants (Brunel and Wang 2003).

In our own large-scale simulations (Djurfeldt et al. 2008), global alpha and beta frequency oscillations during inactivity were replaced by gamma frequencies and a more focal pattern of activation (Figure 6.2). Under certain conditions with low background activation, the network model displayed a nonspecific, noncoding ground state in which beta frequency oscillations were prominent (Figure 6.1), and the pyramidal cells fired at around 1 Hz. Stimulus-triggered activation of a coding state increased significantly the power in the gamma band (Figure 6.3).

Gamma oscillations emerged due to the interplay between pyramidal cells of an active minicolumn and feedback inhibition from nearby basket cells. This regime also allowed for an approximate balance between excitatory and

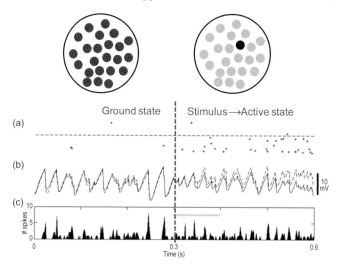

Figure 6.2 Oscillatory activity in ground and active states. Distribution of spiking within a hypercolumn during ground state (left) and active state (right). Average spiking frequency is coded by gray scale for each minicolumn: dark gray < 1 Hz; light gray ~ 2 Hz; black > 5 Hz. (a) Spike raster for the background and foreground minicolumns separated by at dashed line. (b) Average V_m for a foreground (dashed) and background (solid) minicolumn during ground state (0–0.3 s) and active state (0.3–0.6 s). (c) Time histogram of n:o spikes in the entire hypercolumn.

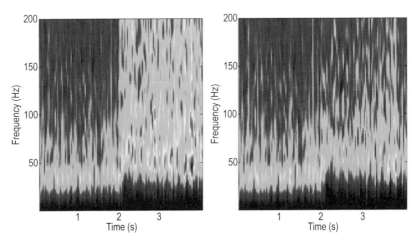

Figure 6.3 Synthetic LFP spectrogram. The network started out in the ground state and entered an active state after two seconds due to stimulation. The signal was produced from 30 local pyramidal cells entering foreground (a) and background (b). The average signal from five runs is plotted.

inhibitory currents in both ground and active states, and as a result cells fired irregularly as seen *in vivo*. In contrast to gamma oscillations, those in the alpha and beta frequency range were generated by activity and feedback inhibition of a much more diffuse and low-rate nature. In the ground state of the network, single cell-firing rates were very low, and the increase in oscillatory frequency thus signaled an increase in selectivity during attractor states.

Furthermore, the increase of cortical gamma oscillations can be entrained by the slower theta rhythm in hippocampus (Sirota et al. 2008). An interpretation of this from the attractor paradigm standpoint would be that (a) theta rhythm results from wandering between task relevant attractors and (b) gamma oscillations can be generated in each attractor state. Thus, according to our simulations, oscillations in several different frequency bands emerge from the underlying attractor dynamics during spontaneous as well as stimulus-driven activity. The slower oscillations seem to reflect the very important mechanism of attractor destabilization which paces spontaneous network activity and the "flow of thought," but what about the functional role of gamma oscillations?

Does Synchrony Hold Information or Have Computational Advantages?

Due to the local feedback inhibition via basket cells, adjacent pyramidal cells tend to be synchronized in simulated E-I networks. Most network models simulate only this local part of a neural network so that the same inhibitory network of basket cells controls the activity in the entire network. However, if larger networks that extend beyond the lateral extension of basket cells were investigated, they would involve several hypercolumns. If each of these code for a distinct feature, with minicolumns discretely coding for specific values, would the phase of the gamma oscillations signal which feature values belong together in a group, binding the different parts of an object together globally? If so, synchronization over long distance might help, for instance, to discriminate figure–background, to bind several segments of an object as a single perceived entity, or to form the perception of a novel object by temporary synchronization of previously learned features. Experimental observations support the idea that synchronization may be involved in such binding; for example, if attention is shifted toward a stimulus that is processed by the recorded cells, there is an increase in local coherence in gamma oscillations (Fries, Reynolds et al. 2001). True phase coding should, however, imply synchronization on the millisecond level spanning over several hypercolumns.

One model that displayed feature binding through phase locking was that of Ritz et al. (1994). They showed that a semi-realistic network model could segment a complex scene into previously learned objects by such a mechanism. The representations of different objects oscillated out of phase relative to each other so that temporal synchronization on a fine timescale signaled how features (pixels in this case) were grouped into objects. When the network oscillated at gamma frequencies, up to four objects could be separated. Each population, however, was coupled to distinct interneurons; thus inhibition did not induce competition. Furthermore, the network was not supporting persistent activity and was strictly input driven. Such network architecture is hard to reconcile with what is currently known about cortical connectivity.

Simulations highlight two problems that must be solved for phase coding to work in a realistic cortical network model. First, how can several local populations of pyramidal cells be active simultaneously within about 25 ms if they share the same local inhibitory network? In oscillatory networks pyramidal cell firing is a self-terminating process since it activates strong feedback inhibition. Therefore each oscillatory cycle will set up a racing condition or winner-take-all process, where the phase-leading population will terminate the activity of the other populations (Fries et al. 2007). This is presumably an important mechanism behind the rivalry observed in perceptual processing. According to our experience, it is very hard, if not impossible, to tune a network model with recurrent inhibition so that more than one population fires in each gamma cycle. To resolve this issue it is important to investigate if nearby cortical pyramidal cells receive feedback inhibition from distinct inhibitory networks, or if selective inhibition, via dendritic targeting inhibitory interneurons, is predominantly of a feedforward kind, activated from distant pyramidal cells as in our network model described above.

Second, we need to understand how populations of pyramidal cells that do not share the same local inhibition (i.e., belong to distinct functional hypercolumns) are able to fire phase-locked with each other. Since inhibitory interneurons are key players in generating locally synchronous oscillatory activity, how is the pace and phase kept over distances longer than those directly accessible to them? One solution might be long-range excitation; however, in our own model, this induces only a weak synchrony between minicolumns in different hypercolumns (Lundqvist et al. 2006). Recent modeling has shown further that synchrony between populations can only be achieved with strong inter-regional coupling (Honey et al. 2007) and that increase in synchrony cannot be separated from an increase in firing rates. This suggests that phase locking does not occur in the absence of rate modulation and that oscillations and synchronization are emergent features of spiking attractor dynamics in networks controlled by strong feedback inhibition.

Thus, it appears to us unlikely that several gamma oscillation modes could robustly exist independently in the same cortical location, or that assemblies are bound together by phase locking over longer distances. Still, oscillations in the gamma range are obviously prominent features of active cortical processing, and they readily emerge in network models. They might still convey computational advantages since the saliency of synchronized signals is heavily increased, allowing for self-sustained activity with very low (metabolically favorable) firing rates. Furthermore, transitions between coding states are very fast in the oscillatory regime (Fries et al. 2007), since there are only small differences in net excitation between foreground and background states, and relatively few thalamic inputs can strongly influence the network activity as described above.

Plasticity in the Microcircuit and Global Network

Activity-dependent plasticity has an obvious role in the attractor network paradigm; namely, in the formation of global cortical connectivity that establishes the attractor structure of the network. A standard form of Hebbian learning seems to be sufficient to support the holistic processing of such networks. However, several interesting issues open up when viewed from a broader perspective.

For example, an STDP-type learning rule seems well suited for a gamma oscillating network and might even impose a temporal structure reminiscent of synfire chains to the attractor dynamics, where subgroups of the assembly are sequentially activated. True sequence learning, as already suggested by Hebb, could readily be achieved in an attractor network provided that the pre–post timing window is somewhat widened and possibly made asymmetric. This would suggest that the synapses involved in global cortical connectivity could display a spectrum of temporal selectivity for LTP induction.

In addition, we note that our depiction is to a significant degree incompatible with attractor models of working memory, based on persistent activity (Camperi and Wang 1998). These models rely on attractor states for actual storage of memory items over time. Since an attractor network model supports only reluctantly the simultaneous activation of more than one state, multiple-item working memory would pose a problem for a realistic model. Alternatively, we suggest that working memory is based instead on fast-inducing, volatile Hebbian synaptic plasticity (Sandberg et al. 2003), with possible spontaneous cyclic activation of a limited number of attractor states representing items in working memories. Such a mechanism would unify long- and short-term forms of memory, and data on fast synaptic plasticity suggest that this is a realistic possibility (Hoffman et al. 2002).

Furthermore, the type of learning rules expected to operate should be quite different, depending on what type of processing and cortical layer we consider. From the perspective of the attractor network paradigm, the feedforward processing stream via layer 4 is likely to be involved in transforming internal representations to make them suitable for processing in higher-order attractor networks. Decorrelation and sparsification of representations might rely on learning rules that combine Hebbian synaptic plasticity in incoming feedforward synapses with competitive learning of some anti-Hebbian type in the recurrent (inhibitory) connectivity. This may be complemented by some form of intrinsic excitability that acts to avoid unused units (e.g., depolarization of neurons that respond infrequently relative to those that respond frequently). Different approaches to generating such distributed, sparse, and overcomplete neural representations are currently under investigation.

Discussion and Conclusions

We have surveyed and summarized work related to the attractor network paradigm of cortical function, based on simulation results from a large-scale biophysically detailed network model with a modular structure of minicolumns and hypercolumns. We have concluded that such a network model readily performs associative memory and attractor network functions likely to underlie holistic cortical processing. Attractor dynamics may come in the form of spontaneous ongoing activity at about theta frequency, or the network may display a stable ground state that can be replaced by some active coding state triggered by an incoming stimulus. Neuronal adaptation and synaptic depression in the model makes the coding attractor states quasi-stable. The precise sequence of spontaneous recall depends on the overlap structure of the connection matrix in this model. However, in real cortex, additional mechanisms are likely to influence attractor state duration and transitions (e.g., sequence storing connections and loops involving various subcortical structures related to goals, drives, and motivations).

In this context we proposed that working memory is based on fast Hebbian synaptic plasticity rather than persistent activity, which we see as a memory readout process rather than storage per se. Attended and task-relevant objects are loaded into working memory in a palimpsest fashion such that older memories are written over, and the items thus encoded may be cyclically activated. This view unifies mechanisms behind short- and long-term forms of memory in a biologically plausible manner and makes it quite straightforward to envision interactions between working memory and long-term memory, as well as memory consolidation based on repeated reinstatement.

Fast oscillations in the gamma frequency range emerge readily and influence potently the dynamics of the network model studied. Spiking statistics during different states show prominent spike synchronization between nearby cells that share the same inhibitory subnetwork; there is also a weak tendency for spike synchrony and phase locking over longer distances. Oscillations and spike synchrony in the gamma range may provide a functional advantage (e.g., promoting low rate firing and quicker and easier transitions between different attractor states). One could expect that a network with continuous STDP learning could induce further fine structure in spiking patterns, like synfire chains. Random networks with STDP learning have been simulated, but this needs to be further examined in a network with attractors stored.

We addressed the important question of whether more than one attractor can be active in the same gamma cycle of about 25 milliseconds, independently and out of phase with each other. We conclude that in our simulation model, lateral inhibition induced rivalry and competition is prominent, thus preventing such uncoupled ongoing activities. However, coordination in the form of theta burst synchronization occurs globally over the entire model network and could be assumed to occur over large parts of real cortex. This might be related to generation of low frequency brain rhythms in the theta–alpha range and phenomena like EEG microstates; it also could be important for dynamic coordination in the form of cross-modal integration, for bottlenecks in perception (e.g., attentional blink), and ultimately for the sequential nature and unity of consciousness.

Future Perspectives

The cortical model of holistic processing discussed here is still incomplete, and work is ongoing to incorporate the missing cortical layers, most importantly layer 4, which presumably provides the circuitry for transforming iconic sensory representations to a sparse and decorrelated format better suited for subsequent attractor network processing. Applying such mechanisms recursively in the cortical feedforward processing stream from one cortical area to the next could be expected to form units in higher-order areas with specific response properties that support some degree of invariant recognition. More large-scale simulations of network models comprising several interacting cortical areas, based on density data and distribution of synaptic connections as well as conduction delays of long-range inter-areal connectivity, are required to understand dynamic coordination and synchronization phenomena better at this macroscopic scale. Another component function lacking so far in the model, yet crucial for perception and memory, is temporal sequence processing and generation. A plausible addition to the model is Hebbian synapses with a time-delayed and possibly asymmetric pre- and postsynaptic induction window.

We are now beginning to see a constructive partitioning of the main functions of the brain and have had initial success in constructing computational models of its different parts. Phenomena like perceptual grouping, multimodal integration, rivalry, and associative recall from fragmentary information can already be well understood from the perspective of attractor network dynamics, and understanding the interaction between working and long-term memory as well as the mechanisms behind memory consolidation by reinstatement processes seems to be within reach. In addition to perception and memory, there are obviously other equally important major functions to consider: motor control, motor learning, sensorimotor integration, and behavior selection, as well as the emotional and motivational systems so important for goal-directed behavior and learning.

Though it is often convenient to study the different functions separately, efforts must be made to reassemble the modeled results. Integration is necessary to ensure that partitioning is plausible and that the different models can really cooperate efficiently so as to provide sufficient composite functionality. Multiscale modeling at the systems level, using high-performance computers, is necessary to address all of the relevant aspects simultaneously, with a maintained degree of biological detail.

As a final verification, such a whole brain model, or constructed brain, needs to be embodied in the form of an agent and immersed in a complex environment that provides dynamic input and feedback, depending on the output actions generated. The new challenges include, for instance, the detection of task-relevant data as well as the processing involved in the timely selection of the best action, while allowing on-line reinforcement learning to enhance internal representations, motor programs, and stimulus-response mappings to improve performance in future similar situations. Such embodied models will link the computational branch of brain science directly to core problems in information science and technology, including cognitive computing and robotics. An increased mechanistic understanding of brain function will not only have a major impact on improving health in our society; it may also catalyze breakthroughs in information technology, thus paving the way for important applications of novel brain-inspired technology in the service of humankind. Today, this may appear as fantasy and speculation. However, if our knowledge of the brain, together with the capacity and parallelism of computers, continues to develop at the current pace, we may very well reach this stage in a few decades.

7

Oscillation-supported Information Processing and Transfer at the Hippocampus–Entorhinal–Neocortical Interface

György Buzsáki and Kamran Diba

Abstract

As information is propelled along the multisynaptic feedforward loops of the entorhinal–hippocampal system, each stage adds unique features to the incoming information (Figure 7.1). Such local operations require time, and are generally reflected by macroscopic oscillations. In each oscillatory cycle, recruitment of principal neurons is temporally protracted and terminated by the buildup of inhibition. In addition to providing a temporal framework in which information can be packaged, oscillatory coupling across networks can facilitate the exchange of information and determine the direction of activity flow. Potential mechanisms in the entorhinal–hippocampal system supporting these hypotheses are described.

Oscillations Provide the Structure for Information Processing in the Hippocampus

Two major network patterns dominate the hippocampal system: theta oscillations (4–10 Hz) and sharp waves with their associated ripples (140–200 Hz). Theta and sharp-wave patterns also define states of the hippocampus: the theta state is associated with exploratory ("preparatory") movement and REM sleep, whereas intermittent sharp waves mark immobility, consummatory behaviors, and slow-wave sleep. These two competing states bias the direction of information flow to a great extent, with neocortical–hippocampal transfer taking

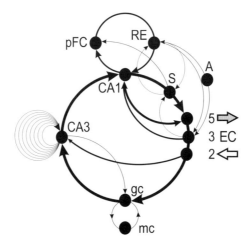

Figure 7.1 Multiple loops of the hippocampal–entorhinal circuits. The long loop connecting the layer 2 entorhinal cortex (EC), granule cells (gc), CA3, CA1, and subiculum (S) back to the layer 5 EC is supplemented by multiple shortcuts and superimposed loops. The shortest loop between the EC and hippocampus is the path from the layer 3 EC to CA1 and back to the layer 5 EC. Excitatory traffic in the multiple loops is controlled by a large family of interneurons, whose connections are not loop-like (Freund and Buzsáki 1996). mc: mossy cells of the hilus; A: amygdala; RE: nucleus reuniens of thalamus; pFC: prefrontal, anterior cingulate cortex.

place mainly during theta oscillations and hippocampal–neocortical transfer during sharp waves (Isomura et al. 2006).

The extracellularly recorded theta oscillation is the result of coherent membrane potential oscillations across neurons in all hippocampal subregions (Buzsáki 2002). Theta currents derive from multiple sources, including synaptic currents, intrinsic currents of neurons, dendritic Ca^{2+} spikes, and other voltage-dependent membrane oscillations. Theta frequency modulation of perisomatic interneurons provides an outward current in somatic layers and phase-biases the power of ongoing gamma frequency oscillations (30–100 Hz), the result of which is a theta-nested gamma burst. Excitatory afferents form active sinks (inward current) at confined dendritic domains within cytoarchitecturally organized layers in every region. Each layer-specific excitatory input is complemented by one or more families of interneurons with similar axonal projections (Freund and Buzsáki 1996; Klausberger and Somogyi 2008), forming layer-specific "theta" dipoles. The resulting rich consortium of theta generators in the hippocampal and parahippocampal regions is coordinated by the medial septum and a network of long-range interneurons. Furthermore, the power, coherence, and phase of theta oscillators can fluctuate significantly in a layer-specific manner as a function of overt behavior and/or the memory "load" to support task performance (Montgomery et al. 2009).

Theta-nested gamma oscillations are generated primarily by the interaction between interneurons and/or between principal cells and interneurons. In both scenarios, the frequency of oscillations is largely determined by the time course of $GABA_A$ receptor-mediated inhibition. Neurons that discharge within the time period of the gamma cycle (8–30 ms) define a cell assembly (Harris et al. 2003). Given that the membrane time constant of pyramidal neurons *in vivo* is also within this temporal range, recruiting neurons into this assembly time window is the most effective mechanism for discharging the downstream postsynaptic neuron(s) on which the assembly members converge. Although gamma oscillations can emerge in each hippocampal region, they can be coordinated across regions by either excitatory connections or long-range interneurons. The CA3–CA1 regions appear to form a large coherent gamma oscillator, due to the interaction between the recurrently excited CA3 pyramidal cells and their interneuron targets in both CA3 and CA1 regions. This "CA3 gamma generator" is normally under the suppressive control of the entorhinal–dentate gamma generator, and its power is enhanced severalfold when the entorhinal–dentate input is attenuated (Bragin et al. 1995). Entorhinal circuits generate their own gamma oscillations by largely similar rules, and these (generally faster) rhythms can be transferred and detected in the hippocampus.

When the subcortical modulatory inputs decrease in tone, theta oscillations are replaced by large amplitude field potentials called sharp waves (SPW). SPWs are initiated by the self-organized population bursts of the CA3 pyramidal cells (Buzsáki et al. 1983). The CA3-induced depolarization of CA1 pyramidal cell dendrites results in a prominent extracellular negative wave, from which the SPW derives its name, in the stratum radiatum. The CA1 SPWs are associated with fast-field oscillations (140–200 Hz), or "ripples" confined to the CA1 pyramidal cell layer (O'Keefe and Nadel 1978; Buzsáki et al. 1992). At least two factors contribute to the field ripples. First, the synchronous discharge of pyramidal neurons generates repetitive "mini populations spikes" that are responsible for the spike-like appearance of the troughs of ripples in the pyramidal cell layer. Second, the rhythmic positive "wave" components reflect synchronously occurring oscillating inhibitory postsynaptic potentials (IPSPs) in the pyramidal cells because the CA3–CA1 pyramidal cells strongly drive perisomatic interneurons during the SPW. In the time window of SPWs, 50,000–100,000 neurons discharge synchronously in the CA3–CA1–subicular complex–entorhinal axis. The population burst is characterized by a three- to fivefold gain of network excitability in the CA1 region, preparing the circuit for synaptic plasticity (Csicsvari et al. 1999a). SPWs have been hypothesized to play a critical role in transferring transient memories from the hippocampus to the neocortex for permanent storage (Buzsáki 1989), and this hypothesis is supported by numerous experiments demonstrating that the neuronal content of the SPW ripple is largely determined by recent waking experiences (Wilson and McNaughton 1994; Foster and Wilson 2006; Csicsvari et al. 2007).

Reciprocal Information Transfer by Oscillations

Oscillations and neuronal synchrony create effective mechanisms for the storage, readout, and transfer of information between different structures. Oscillations impose a spatiotemporal structure on neural ensemble activity within and across different brain areas, and allow for the packaging of information in quanta of different durations. Furthermore, oscillations support the bidirectional flow of information across different structures through the changing of the temporal offset in the oscillation-related firing (Amzica and Steriade 1998; Chrobak and Buzsáki 1998b; Sirota et al. 2003; Buzsáki 2005; Siapas et al. 2005). Most importantly, exchanging information across structures by oscillations involves mechanisms different from what is usually meant by the term "information transfer."

In the usual sense, transfer of information involves two structures, or systems, which can be designated as the "source" (sender) and "target" (receiver). Typically, the information transfer process is assumed to be unidirectional and passive: the source sends the information to an ever-ready recipient. In systems coupled by oscillations, however, the method appears to be different; we refer to this process as "reciprocal information transfer" (Sirota and Buzsáki 2005; Sirota et al. 2008). The reciprocal process implies that a target structure takes the initiative by temporally biasing the activity in the sender (information source) structure (Sirota et al. 2003; Fries 2005; Sirota and Buzsáki 2005; Isomura et al. 2006; Womelsdorf et al. 2007). Biasing is achieved by the strong output ("duty cycle") of the receiver so that the information, contained in finer timescale gamma-structured spike trains, reaches the recipient structure in its most sensitive state (the "perturbation" cycle), ideal for reception. Below, we illustrate this principle using the state-dependent communication between the hippocampus and neocortex.

In the waking state, transfer of neocortical information to the hippocampus can be initiated by the hippocampus via theta-phase biasing of neocortical network dynamics, as reflected in the local field potential (LFP) by transient gamma oscillations in widespread and relatively isolated neocortical areas (Sirota et al. 2008). As a result, locally generated gamma oscillations from multiple neocortical locations are time biased so that the information contained in their gamma bursts arrive back at the hippocampus at a phase of the theta cycle optimal for maximal perturbation of hippocampal networks and plasticity (Huerta and Lisman 1996; Holscher et al. 1997). In the CA1 region, this corresponds to the positive (least active, i.e., recipient) phase of the theta cycle (Csicsvari et al. 1999b). This is also the phase at which a hippocampal neuron discharges when the rat enters its place field (O'Keefe and Recce 1993). In short, through theta-phase biasing, the hippocampus can affect multiple neocortical sites so that it can effectively receive information from the resulting neocortical assemblies, by way of the entorhinal cortex (EC), at the optimal time frame.

The direction of information transfer during slow-wave sleep at the hippocampal–neocortex axis is largely opposite to that in the waking theta state. As discussed earlier, in the absence of theta oscillations, the CA3–CA1 region gives rise to SPWs. The synchronous discharge of CA1 neurons and downstream subicular and EC neurons provides a most effective output to the neocortex (Chrobak and Buzsáki 1994). During sleep the information source (sender) is the hippocampus but, again, the transfer of information is initiated by the receiver (now the neocortex). This latter process has been hypothesized to be critical in consolidating the learned information acquired in the waking state (Buzsáki 1989).

A caveat in this two-stage model of information transfer and memory consolidation is the absence of a mechanism to coordinate and guide the hippocampal output temporally to the ongoing activity in the neocortical circuits that gave rise to the experience-dependent hippocampal input. In other words, a mechanism must exist to allow the hippocampal output during sharp wave ripples (SWRs) to address the relevant neocortical circuits. Below, we outline one such potential mechanism.

While the hippocampus is generating SPWs during slow-wave sleep, large areas of the neocortex oscillate coherently at a slow frequency (0.5–1.5 Hz) (Steriade et al. 1993a, b; Destexhe et al. 1999). During these slow oscillations, large areas of the neocortex and paleocortex (Isomura et al. 2006) toggle coherently between active (UP) and silent (DOWN) states, although isolated cortical modules can also shift between states, independent of surrounding areas (Figure 7.2). The DOWN–UP transitions can trigger K complexes and

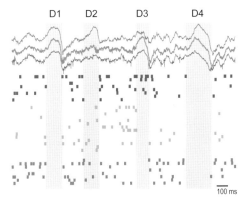

Figure 7.2 Global and local neocortical DOWN states (delta waves) during slow-wave sleep. Example of simultaneously recorded local field potentials (LFPs) and unit activity at three intracortical locations in the somatosensory area of the rat (~1 mm spaced). Note that DOWN states (shaded area) can be synchronous and global (D1, D4) or localized only to a small area (D2, D3). While the effects of global and synchronous neocortical patterns can be detected in the hippocampus by the macroscopic LFP, more localized firing patterns may exert only a subtle effect on the activity of the hippocampus. After Sirota and Buzsáki (2005).

associated sleep spindles in the thalamocortical system (Amzica and Steriade 1997; Molle et al. 2002; Massimini et al. 2004). These same shifts also affect the timing and presumably cellular composition of the hippocampal SPWs (Siapas and Wilson 1998; Battaglia et al. 2004). The synchronous cortical unit discharges, associated with the thalamocortical spindles, can lead to an increased firing of hippocampal neurons within 30–50 ms, and this increase in activity is often coincident with SWRs. The impact of neocortical oscillations on hippocampal circuits can be demonstrated by the DOWN–UP transition-induced sinks in the hippocampus, mediated by the entorhinal input (Figure 7.3).

Thus, the temporal coordination of thalamocortical sleep spindles and hippocampal SPWs (Siapas and Wilson 1998) by the slow oscillations offer a reasonable framework for hippocampal–neocortical information transfer. The DOWN–UP transitions and associated thalamocortical spindles trigger organized firing patterns of neocortical neurons which, in turn, lead to the

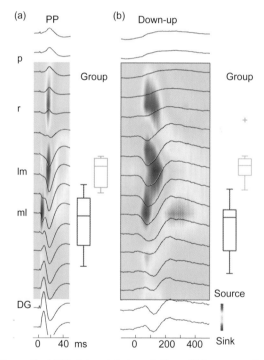

Figure 7.3 Neocortical UP state-related excitation of hippocampal neurons. (a) Perforant path (PP) stimulation-evoked LFP responses (black traces) and the derived current-source density (CSD) map. Note sinks (blue) in the dentate molecular layer (ml) and CA1 str. lacunosum-moleculare (lm). Box plots, group data of sink maxima positions (green, lm; blue, ml). (b) Averaged hippocampal LFP traces and CSD triggered by DOWN–UP transitions in an intracellularly recorded layer 3 entorhinal neuron. Box plots, group data. Similar observations were also made in the naturally sleeping rat. Modified after Isomura et al. (2006).

activation of specific subpopulations of hippocampal neurons. These activated hippocampal neurons then give rise to SPW-related synchronous outputs and readdress the neocortex. Because the SPW is a punctuated event (~100 ms), whereas the UP state and sleep spindle are temporally protracted (~0.5–1 s), the hippocampal output can be directed to the still active neocortical assemblies. The temporal coordination of these events facilitates conditions in which unique neocortical inputs to the hippocampus and, in turn, hippocampal outputs to the neocortex might be modified selectively (Buzsáki 1989; Wilson and McNaughton 1994; Sirota et al. 2003; Steriade and Timofeev 2003). In this information transfer process, the neocortex serves as the target ("receiver") of the information from the "source" (sender), hippocampus; nevertheless, the initiator of the events is the neocortical slow oscillation.

Propagation of Activity through Multiple Stages of the Hippocampus Is State-dependent

Propagation of neuronal signals across multiple anatomical regions are frequently explained by "box-and-arrow" illustrations, where large populations of neurons in each layer or region are replaced by a single "mean neuron," representing a homogeneously behaving population (Figure 7.4). While it is tempting to designate circumscribed and specific computations for each layer or region, such a simplified view may not adequately describe information processing and propagation. Much computation can take place at the interface between layers with control being exerted on local circuit computations by the global hippocampal states. Furthermore, representation of an initiating event is not merely transferred from one layer to the next but changes progressively. Depending on the previous history of the brain and the event, each layer may add unique information to the representation.

Timing is critical to the propagation of novel information. For example, a strongly synchronous input, such as an artificial electrical pulse or an epileptic interictal spike, may propagate through multiple layers at a high speed, limited primarily by axon conduction and synaptic delays. However, physiological information rarely advances at such high speed. The fastest physiological speed of spike transmission in hippocampal networks occurs during SPW ripples. During SPWs, the CA3-initiated population burst propagates through the CA1, subiculum, entorhinal layer 5, and layers 2/3 in just 15–20 ms. While the pattern is propelled through these feedforward layers, the large SPW-related increase in excitation in the hippocampus is balanced by the progressive buildup of inhibition in successive layers. In layer 5, inhibition balances the SPW-induced excitation and inhibition in layers 2/3 overcomes the excitation. Because of the increasing inhibition in successive layers, SPW activity rarely reverberates in the hippocampal–entorhinal cortex loop, although multiple reverberations can occur in epilepsy.

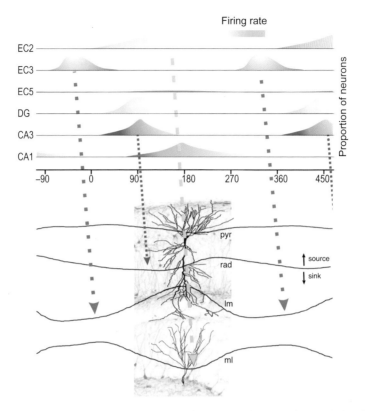

Figure 7.4 Temporal relationship between layer- or region-specific population firing patterns and theta current sinks in the hippocampus. In each region and layer, most neurons are silent or fire at low rates, with only a minority of neurons discharging at high frequency. The preferred theta phase of low and high firing neurons is different (advancing to earlier phase in EC2, DG, CA3, and CA1 neurons). Firing rate is illustrated by color intensity. The height of the histograms reflects the proportion of neurons with different discharge rates. Below: CSD theta traces are superimposed on a histological section in the CA1–dentate gyrus axis, with highlighted pyramidal cell and granule cell. Note phase-reversed sinks in CA1 str. lacunosum–moleculare (lm) and dentate molecular layers (ml) and phase-shifted sink (relative to lm sink) in str. radiatum (pyr, pyramidal layer). Tilted arrows indicate the temporal (phase) offsets between the peak of population firing in an upstream layer and the theta sinks in the target layers with the expected delays (based on axonal conduction velocity; 30° or ~ 10 ms). Note that whereas the population patterns correctly predict the timing of the dendritic sinks in their respective target layers, the propagation of spiking activity between upstream and downstream neuronal populations cannot be deduced from a simple integration of the inputs (after Mizuseki et al. 2009 and Montgomery et al. 2009).

The situation is dramatically different from SPWs in the theta state. The delay between the peak population firing rates in the entorhinal input layers (layers 2 and 3) and that of their respective target populations in dentate/CA3 and CA1 is severalfold longer during the theta state than during SPWs. Typically, the delays correspond to approximately half of one theta cycle (50–70 ms). Importantly, the current sinks in dentate/CA3 and CA1 pyramidal cell dendrites occur within 10–15 ms after the peak of the population activity in entorhinal layers 2 and 3, as expected by the conduction velocities of the entorhinal afferents. However, the buildup to maximum population activity in these hippocampal regions actually takes another 50 ms or so (Figure 7.4).

Addressing the potential causes of the delayed spiking activity during theta oscillations requires a thorough understanding of the temporal evolution of spike patterns of principal cells. As described above, the hippocampal theta oscillation is not a single entity but a consortium of multiple oscillators. Hippocampal principal cells can be activated by either environmental landmarks ("place cells"; O'Keefe and Nadel 1978) or internal memory cues (Pastalkova et al. 2008). During its active state, the spike train of a principal cell oscillates faster than the LFP, and the frequency difference between the neuron and the LFP gives rise to phase interference (known as "phase precession"; O'Keefe and Recce 1993). As an example, entering the place field of a typical CA1 place cell by the rat is marked by a single spike on the peak of the locally derived theta LFP. As the animal moves into the field, the spikes occur at progressively earlier phases. The "lifetime" (i.e., the duration of activity) of pyramidal neurons in the septal part of the hippocampus corresponds to 7–9 theta cycles, during which a full wave phase advancement (360°) may take place. In addition to spike phase advancement, the number of spikes emitted by the neuron increases and decreases as well, with the maximum probability of spiking at the trough of theta, coinciding with the center of the place field (Dragoi and Buzsáki 2006). In short, spikes can occur at all phases of the theta cycle but with the highest probability at the trough. Neither the phase advancement of spikes nor the increased probability of spiking in the firing field of the CA1 pyramidal cell can be explained by simple integration of the direct entorhinal layer 3 inputs, because spikes of layer 3 pyramidal cells are phase-locked to the positive peak of the CA1 pyramidal cell layer theta (as reflected by the sink in the str. lacunosum moleculare; see Figure 7.4), which is at the theta phase with the least probability of spiking for CA1 neurons. Therefore, the entorhinal input cannot be the sole initiator of each spike, especially for those occurring in the earlier phases of the theta cycle. The situation is similar in the entorhinal layer 2–dentate granule cell/CA3 cell network because peak firing of these neuronal populations is also delayed by approximately half of one theta cycle (Mizuseki et al. 2009). It is important to emphasize that neither the evolution of spike discharge activity nor the associated theta phase precession of spikes are necessarily controlled by environmental inputs; identical patterns can also occur during memory recall, route planning, and even REM sleep (Pastalkova et al. 2008).

Although the exact source of the additional spikes at unseeded phases of the theta cycle is not known, they may derive from local circuit mechanisms, according to the following hypothesis: Hippocampal neurons, initially discharged by the entorhinal input, begin to interact with each other to form transient assemblies, which individually oscillate faster than the ongoing population theta, reflected in the LFP. The oscillation frequency of the active cell assembly determines the magnitude of spike phase advancement and the "life time" of the assembly (i.e., the size of the firing field in spatial metric). From this perspective, the role of the entorhinal input is to add new members to the perpetually shifting and oscillating cell assemblies rather than to "drive" each spike directly in the hippocampus. The selected assembly members then begin to interact with each other for a limited time period, which is the theta cycle, and excitation is spread to connected neurons: within each theta cycle, multiple (7 to 9) assemblies interact with each other.

What is the functional significance of these interactions? For hippocampal cell pairs with overlapping place fields, the temporal structure of spike trains within a theta cycle reflects the distances between the place field centers and their sequential activation during the run (Skaggs et al. 1996; Dragoi and Buzsáki 2006). Within the theta cycle, the relative timing of neuronal spikes reflects the upcoming sequence of locations in the path of the rat, with larger time lags representing larger distances. These cross-neuronal time delays are independent of the running speed of the rat and are not affected by other environmental manipulations (Diba and Buzsáki 2008). The cross-neuronal temporal lags are specific to theta dynamics because the same sequences can be observed during SPWs but with shorter interneuron time delays (Diba and Buzsáki 2007).

The "fixed" temporal delays driven by theta dynamics have consequences for mechanisms of hippocampal coding. The first is a sigmoid relationship between within-theta time lags and distance representations, because the natural upper limit of distance coding by theta-scale time lags is the duration of the theta cycle (120–150 ms). As a result, upcoming locations that are more proximal are given better representation, with poorer resolution of locations in the distant future; distances larger than 50 cm are poorly resolved by neurons in the dorsal hippocampus because their expected temporal lags would otherwise exceed the duration of the theta cycle. Therefore, they fall on the plateau part of the sigmoid. Another consequence is that temporal resolution scales with the size of the place field; smaller place fields provide temporal lags, which represent very fine spatial resolution, whereas larger place fields that encompass the enclosure simultaneously provide a much coarser distance representations (Figure 7.5). These multiscale representations take place simultaneously, and possibly scale along the septotemporal axis of the hippocampus (Maurer et al. 2005; Kjelstrup et al. 2008). Assuming that locations can be regarded as analogous to individual items in a memory buffer (Lisman and Idiart 1995; Dragoi and Buzsáki 2006), this temporal compression mechanism limits the "register capacity" for the number of items that can be stored within a single theta cycle "memory buffer." By the same analogy, the sigmoid relationship suggests that episodic recall is high for the spatiotemporal conditions that surround a

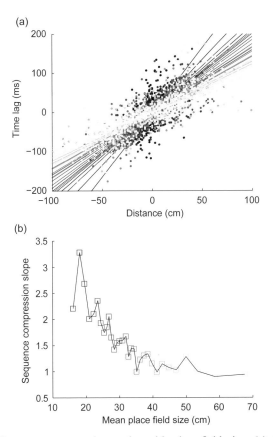

Figure 7.5 Sequence compression scales with place-field size. (a) Cell pairs are shown in groups of 100 (3200 pairs in total) with colors linked to the mean place-field size for the pairs; cf. (b) for corresponding values. Best linear fit is also depicted, illustrating that a smaller place field shows finer resolution but lower distance compression. Similarly, larger place fields provide the majority of points representing larger distances. (b) The slope of the best-fit line ("sequence compression slope") decreases with increasing field-size.

recalled event, whereas the relationships among items representing the far past or far future, relative to the recalled event, are progressively less resolved.

How can the mechanism responsible for maintaining theta-scale time delays be protected from firing rate changes, environmental modifications, and other factors that constantly affect hippocampal neurons? A working model is illustrated in Figure 7.6. The simple hypothesis is that interneuron-mediated inhibition provides a "window of opportunity" during which a postsynaptic neuron may spike, according to its excitatory inputs. The timing of this window may be established by the combined effect of presynaptic excitatory activity and inhibition. Through recurrent and feedforward connections, the spiking

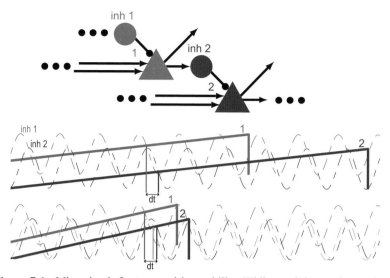

Figure 7.6 Microcircuit for temporal lag stability. While receiving excitatory input from unspecified sources, pyramidal cells (e.g., cell 1, red) are effectively inhibited by circuit interneurons (inh 1, red) and, in turn, drive other interneurons (inh 2, blue) that inhibit additional cells (e.g., cell 2), etc. In this figure, the synaptic efficacy of the individual components is inversely proportional to the length of the connection arrow. For example, to produce an action potential, pyramidal cells integrate excitatory input over many sources, whereas a small amount of excitatory input is sufficient to trigger a spike in an interneuron. Cells fire when excitation exceeds inhibition. The middle panel depicts the excitatory drives for the two interdependent place cells 1 and 2 (colors correspond) on a 2-m long track, with inhibition for each superimposed with a dashed line. Excitatory and inhibitory input to cell 2 are delayed relative to excitatory and inhibitory input to cell 1, resulting in net time lag dt. When the track length is shortened (1 m-long; bottom), the rise in excitatory drives occurs over a shorter duration (i.e., fewer theta cycles), and the place fields are shifted relative to each other. Inhibition to cell 2 (e.g., inh 2) is strongly coupled to the spiking of the earlier firing place-cell (1), and in our model, oscillates at this cell's frequency (and shows phase-precession). Hence the time lag of the trailing cell (2) is maintained relative to that of the leading cell (1), with the consequence that distance representations of the two neurons scale with the size of the apparatus (Diba and Buzsáki 2008). A similar mechanism may be responsible for the stable time lags across neurons at different travel velocities.

of interneurons (e.g., inh 2 in Figure 7.6) is tightly coupled to changes in the drive from the leading assembly (e.g., by neuron 1 in Figure 7.6), thus effectively determining timing for the trailing assembly (represented by neuron 2), which is in turn coupled to other inhibitory partners, and so on. In short, the stability of time lags between neurons arises from the theta network dynamics.

Using this hypothetical mechanism, let us consider the following paradox: Since nearly every place cell in the hippocampus oscillates faster than the ongoing LFP, how does the combined population generate a slower frequency output than its constituents? The answer lies in the strict temporal delays between active neurons. Consider 100 identical, partially overlapping, place cell assemblies that evolve while the rat navigates. With zero time delays between the neurons, the population frequency would have to be identical to the frequency of place cells. However, the insertion of temporal lags between cell pairs, in proportion to their distance representations of the environment, can slow the momentary population firing frequency. In this scenario, it turns out that the mean population frequency, also reflected by the LFP, is equal to the mean of the oscillation frequencies of the individual neurons plus the mean time lag (Geisler and Buzsáki, unpublished). In summary, the period of theta oscillations is largely determined by the time lags between active neuron pairs. Conversely, the ensuing theta dynamics constrains the propagation of activity across neurons. Such "bidirectional causation" is the essence of emerging dynamics of interacting neurons, and these constraints determine the speed of state-dependent computations in hippocampal circuits.

Conclusion

Our discussion on the temporal dynamics of networks was largely confined to hippocampal networks, which reside in the dorsal (septal) part of the structure. Although recent findings point to quantitative differences in place representations of more ventral hippocampal neurons (Maurer et al. 2005; Kjelstrup et al. 2008), the mechanisms discussed above may apply to the entire hippocampus. Since the hippocampal theta oscillations are coherent along the entire septotemporal axis of the hippocampus, they may serve as a temporal integration mechanism for combining local computations taking place at all segments and representing both spatial and nonspatial information. Furthermore, the computational principles discussed for the operations of the hippocampal circuits likely apply to other systems with similar forms of oscillatory dynamics.

8

What Are the Local Circuit Design Features Concerned with Coordinating Rhythms?

Miles A. Whittington, Nancy J. Kopell,
and Roger D. Traub

Abstract

This chapter outlines some of the basic features of neuronal circuits that underlie the rich diversity of population rhythms that can be generated in very small regions of cortex. Areas of cortex less than a millimeter square can generate rhythms from slow waves up to very fast oscillations using combinations of intrinsic neuronal properties combined with chemical and electrical synaptic connectivity profiles. Multiple concurrently generated rhythms can display many different forms of coordination: While mechanisms underlying coordination within individual frequency bands may play a role across frequencies, it is becoming clear that novel modes of coordination, such as concatenation, may also take place. The number of different neocortical rhythms capable of being generated so far shows a fixed relationship in the spectral domain. Building lower frequencies through concatenation of coexistent higher frequencies, across the EEG frequency range, provides a putative way to reconcile the existence of discrete frequency bands with the power law continuum observed in long-term EEG recordings.

Rhythm Generation in Cortex

The mammalian cortex *in situ* generates rhythmic activity over a very broad range of frequencies. The majority of these frequencies are capable of being replicated *in vitro* in small slices of tissue containing all layers of cortex with dimensions down to less than 1 mm. Rhythmic bistability in neuronal membrane potential, corresponding to repetitive periods of population activity and quiescence, occurs at frequencies around and below 0.1 Hz spontaneously in the absence of external excitation or neuromodulation. At the other end of the spectrum, neocortical tissue excited with glutamatergic or cholinergic receptor activation generates transient population frequencies up to 400 Hz. Analysis of long-term local field potential, far-field potential, and extracranial EEG recordings suggest a continuum of spectral activity between these extremes, typically with a power law relationship between spectral energy and frequency. However, carefully controlled cognitive and motor-behavioral tasks, and specific patterns of cortical activation and neuromodulation, reveal a range of discrete frequency bands *in vivo* and *in vitro* respectively. In older cortex, with one main layer of principal cells, these discrete rhythms exhibit an interrelationship such that modal frequencies have a ratio approximating to the natural log (e, c.2.7; Buzsáki and Draguhn 2004). In polymodal areas of neocortex, with at least two main principal cell layers (crudely, deep and superficial pyramidal cells), approximately twice as many peaks per spectral band are seen with ratio near $e^{0.5}$ (c.1.6; Roopun, Kramer et al. 2008). In the latter case, combination of such sets of frequencies can produce spectra that approximate very well to the power law relationships seen in long-term *in vivo* recordings (Figure 8.1).

The apparent correlation between the number of different populations of principal cells and the number of discrete frequency bands seen, and the ability to reproduce a huge range of these frequency bands in very small sections of cortex, suggests that the majority of basic rhythm-generating properties of neuronal populations may be held within local circuits. This counters the notion that slower and slower frequencies are generated as emergent properties of larger and larger networks. Instead, it supports the idea proposed by Mountcastle (1978) that the neocortex is functionally and anatomically modular, down to the scale of individual columns. Is this, however, feasible? Columns, of which there are many functional subtypes, each contain neurons of many different types based on morphology, spiking patterns, synaptic targets, and immunocytochemical signature. There are over seven different types of principal cells and many more different subtypes of inhibitory interneurons (the primary coordinators of local circuit behavior). Thus, if specific rhythms emerge as a property of specific local circuits containing different, interconnected principal cells and interneurons, then a great many different rhythms are possible in such a small region. Next we discuss some of the better-understood properties of local circuits and their member neurons that influence network rhythm generation and coordination.

Which Features of Local Circuits Generate Rhythms?

Even the simplest local circuits have many properties that favor rhythm generation. While many of these are often found to operate synergistically, it is worth considering them separately to get a sense of why rhythms are so ubiquitous a feature of electrical activity in cortical resting state and response to input. In addition, the local circuit mechanisms that generate rhythms are often

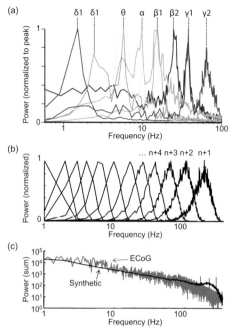

Figure 8.1 Multiple discrete frequencies of local circuit rhythm: relation to long-term EEG spectra. (a) Multiple modal peak frequencies of persistent rhythms generated in isolated neocortex *in vitro*. All rhythms were generated in secondary somatosensory (parietal) cortical slices maintained in artificial cerebrospinal fluid (aCSF). Rhythms were recorded as local field potentials (LFPs); resulting spectra (from 60 s epochs of data) are plotted with powers normalized to modal peak. In control slices, $\delta 1$ (~1.5 Hz) rhythms were generated spontaneously after > 1 h incubation in normal aCSF; $\delta 2$ (2–3 Hz) rhythms were generated by bath application of cholinergic agonist carbachol (2 μM). Both delta rhythms had maximal amplitudes in layer 5. In the presence of the glutamatergic receptor agonist kainate (10 nM), θ (6–8 Hz) rhythms were recorded in layers 2/3 and occurred concurrently with $\delta 2$ rhythms in layer 5. Following transient activation of cortex by pressure ejection of glutamate, α (~10 Hz) rhythms were generated. Peak amplitude was in layer 5 and was present concurrently with θ and $\beta 1$ rhythms in layers 2/3 and layer 4, respectively. Following tonic activation by kainate (400 nM), $\beta 1$ (13–17 Hz) rhythms were generated alone by partial blockade of AMPA/kainate receptors; $\beta 2$ (22–27 Hz) rhythms were generated in layer 5 by kainate (400 nM) and always occurred concurrently with $\gamma 1$ (30–50 Hz) rhythms in layers 2/3 in this brain region. Also generated by kainate (400 nM) were $\gamma 2$ (50–80 Hz) rhythms, but these occurred in layer 5 in aCSF with reduced chloride ion concentration. Additional peak frequencies at ≥ 100 Hz are generated by brief, intense periods of excitation but rarely met the criteria for persistence and thus are not considered here. (b) Similar ratios of adjacent frequency bands can be generated by concatenation of an initial Gaussian white noise source with mean frequency 200 Hz and standard deviation 10% of mean. Lower frequencies were generated from this source by iterating period n = period n–1 + period n–2. (c) By transforming the initial noisy signal into a set of periods (1000 consecutive period widths) iterative concatenation of this set, keeping all previous iteration sums, a power law spectrum ("synthetic," black line) results which closely resembles that from 10 minutes of human temporal cortical ECoG data ("ECoG," red line).

also those mechanisms manipulated in larger-scale networks to coordinate rhythms spatially and spectrally.

Intrinsic Properties

Individual neurons often show subthreshold membrane potential oscillations. In some cases, these oscillations occur over a narrow range of frequencies, such as for inferior olivary neurons, but more often the frequency is determined, to some extent, by background mean membrane potential (Hutcheon and Yarom 2000). Subthreshold oscillations occur mainly at the lower end of the EEG spectrum. Frequencies around delta, theta, and alpha are most common but oscillations within the beta band are also seen in neocortical pyramidal cells. While these rhythms occur in the absence of action potential generation, they can precipitate spiking when neurons are close to sodium spike threshold. The resulting interaction between a subthreshold rhythm and all-or-none spike generation results in mixed-mode oscillations that may lead to very robust, regular outputs from neurons in the absence of patterned input, or more complex oscillations on multiple timescales with exquisite sensitivity to individual intrinsic conductances (Krupa et al. 2008). They are also closely related to the phenomenon of resonance in neurons (Hutcheon and Yarom 2000). Here, intrinsic properties produce a highly selective frequency filter for neuronal inputs, effectively dictating in which local circuit rhythm a neuron may actively participate.

The intrinsic conductances that give rise to subthreshold oscillations are manyfold. In general at least two conductances are required, with at least partially overlapping, but opposing membrane voltage sensitivities. If such pairs of conductances were instantaneously active, or both constantly active, then a stable equilibrium for membrane potential would be reached. However, with a temporal component—essentially the activation and inactivation/deactivation kinetics of the channels involved—the system "hunts" constantly for, but never reaches, a stable equilibrium state. The resulting oscillation, therefore, has a frequency related to the kinetics of the component conductances. Most commonly involved conductances include those generated by persistent sodium channels, low-threshold activated calcium channels, HCN channels, and a wide range of potassium channels. Of the latter, it is worth noting the m-current, which can "tune" frequencies of axonally generated rhythms (Roopun et al. 2006) and potassium channels, which are sensitive to the ATP content of neurons, allowing network rhythm generation to be linked to the metabolic state of the cortex (Cunningham et al. 2006).

Synaptic Excitation

In cortex, a proportion of principal neurons are locally coupled by excitatory synapses. While it is theoretically possible for such networks to generate local rhythms alone, the occurrence of these excitatory connections locally is rather sparse. Estimates from paired principal cell recordings put connectivity probabilities between ca. 1:400 and ca. 1:40 per randomly sampled local cell pair. In addition, the pre- and postsynaptic properties of local excitatory synapses indicate this form of local communication is, for the most part, rather weak. Action potential-initiated presynaptic glutamate release probability is very low for single action potentials (but grows markedly for rapid trains), and postsynaptic unitary events generate target neuron membrane potential changes of a fraction of a millivolt. Excitatory synapses are usually compartmentally localized on dendrites, with sometimes considerable electrotonic distances between synapse and cell soma, again making such activity a rather weak driver of local network rhythms. However, in certain situations the properties above can combine to produce strong local activity. In models of epiletiform activity, a preponderance for burst discharge generation favors glutamate release and temporal summation of dendritic excitatory postsynaptic events. This, coupled with lowered levels of inhibition to reduce postsynaptic voltage shunt can lead to overt, locally synchronous network bursting rhythms in the theta to low beta frequency ranges (Traub et al. 1987). Combinations of recurrent excitation and intrinsic conductances (above) are seen to generate very low (>1 Hz) frequency rhythms in more physiological conditions. For example, temporal summation of background kainate receptor-mediated recurrent excitatory events can combine with intrinsic conductances to generate slow-wave oscillations in some cortical areas (Cunningham et al. 2006), a phenomenon recently termed "group pacemaking" when seen in central pattern-generating circuits.

Synaptic Inhibition

Synaptic inhibition is a critical, causal feature of rhythm generation in local networks for frequencies ranging from the theta to gamma range (~ 4–80 Hz). Local circuit interneurons are readily induced to fire by even extremely low levels of excitatory neuronal activity within a network. Differences between excitatory inputs to interneurons and to other principal cells may underlie this. Presynaptically, glutamate release occurs with a high probability, even for single action potentials reaching the terminal. The resulting unitary postsynaptic responses are considerably larger than counterparts in principal cells and have much faster kinetics, which affords a high degree of temporal precision to interneuron activation. Coordinated recruitment of local interneurons is also common, with a high degree of recurrent synaptic inhibition and gap junction-mediated excitation between interneurons of many different types observed. This, coupled with the enormous convergence of excitatory inputs to interneurons and divergence of inhibitory outputs back to principal cells, provides an ideal substrate for the generation of locally synchronous population rhythms.

With synaptic inhibition-based rhythms, the frequency is set predominantly by the kinetics of the inhibitory postsynaptic potentials onto participating neurons. For gamma rhythms generated solely by tonic depolarization of interneurons, it is the inhibitory postsynaptic potential between interneurons that sets the frequency. The size and kinetics of these events can support local population rhythms from a frequency of ~30 Hz up to ~80 Hz, but not higher (Traub et al. 1996). Recent reports of cortical activity patterns with frequencies above this have labeled the rhythm "high gamma." This may be slightly misleading as such rhythms cannot be supported by synaptic inhibition and clearly involve different network mechanisms (see below). Inhibition-based rhythms dependent on phasic-synaptic excitation of interneurons are more dependent on the kinetics and amplitude of inhibitory events onto principal cell perisomatic compartments. As such, their frequency range tends to be lower (only going up to ~60 Hz), reflecting the slower kinetics of this form of inhibition compared to recurrent interneuronal inhibition. Lower frequencies than gamma can also be produced by inhibition in a manner that is also dependent on postsynaptic inhibitory event kinetics. Theta rhythms depend on GABAergic inhibition to distal dendrites of principal cells. While the resulting inhibitory postsynaptic event at somata is broadened and slowed by cable properties of dendrites, it has also been shown that the dendritic synaptic current itself has much slower kinetics than those generated by perisomatic targeting interneurons (Banks et al. 2000).

Gap Junctions

Local circuit rhythms occurring at frequencies faster than the conventional gamma band ("high gamma" VFO, "ripples") may occur transiently associated with high frequency discharges in interneurons (Buzsáki et al. 1992), but appear to be generated primarily by nonchemical synaptic communication between neighboring principal cells when studied *in vitro*. In particular, gap junctions between axonal compartments have been shown via dye coupling and electron microscopy (see Hamzei-Sichani et al. 2007) and can support very rapid transmission of action potentials from one axon to another. Sparse (>2 gap junctions per axon) local connectivity of this type generates rhythms up to several hundreds of Hertz, with period duration determined as a statistical property of the "random" network of coupled axons: the mean "path length." These very fast oscillations (VFO) are often seen nested within slower rhythms: In layer 5 neocortical neurons, gap-junctionally coupled axons generate very transient epochs of VFO which are organized into a beta2 frequency population rhythm by the intrinsic properties of individual axons (mainly the m-current). During persistent gamma rhythms, VFO is also seen nested within each gamma period and is the fundamental driving force behind such rhythms (Traub et al. 2003). VFO is also seen accompanying neocortical theta rhythms, alpha rhythms, and even up states, suggesting that gap-junctionally coupled axonal networks may represent the primary source of "noise" used by local cortical circuits to generate rhythms.

Which Features Are Involved in Dynamic Coordination?

Coordination within a Frequency Band

Synchronization, with near-zero millisecond delay, is a feature of spike generation in populations of neurons responding to patterned sensory input (Gray and Singer 1989). The occurrence of a set of neurons that, at least transiently, generate spikes at the same time is used as a definition of a cell assembly—a subset of all active neurons in a population whose coordinated spike timing represents a central code for features in the sensorium. Zero-phase synchrony among cortical neurons can be readily achieved if each neuron receives a precisely identical pattern of ascending input, but it is difficult to see how such an input can occur, at least for thalamocortical inputs, when the duration and onset kinetics of activation are relatively slow. In such a case the assumption is that synchrony is generated from properties of the cortical neuronal network itself, yet how can this occur when direct connectivity between principal cells is far from 1:1 (see above), kinetics of such connectivity are rather slow, and conduction delays between neurons are finite and distributed over a broad range relative to the temporal precision observed? The key appears to be in the dominance of feedforward inhibitory neuronal activation in corticocortical connections. Here, both local and distal principal cell outputs can converge on single interneuron populations to generate temporally precise postsynaptic events which, in turn, can tightly control the timing of local principal cells. For these reasons it is perhaps not surprising that cortical rhythms during which synchrony is most readily observable are often inhibition-based (e.g., gamma and beta oscillations).

For gamma rhythms, the mechanisms that are perhaps best understood are those involving the coordination of activity in multiple, spatially separated neuronal populations. Consider a local gamma-generating circuit that comprise principal neurons and perisomatic targeting interneurons and provide a source of phasic, $GABA_A$ receptor-mediated synaptic inhibition. Such a circuit controls the timing of principal cell spiking via simple, local feedback inhibition, effectively providing a window of opportunity for principal cell spike generation a few milliseconds wide every gamma period. Many such local circuits may exist independently in cortex in the absence of functional connectivity between them, but such independence implies no fixed phase relationship between spike timings for neurons in different local circuits (Figure 8.2a). However, with connectivity between separate local circuits, coordination of each gamma rhythm is readily observed. The key feature of such coupled networks that provides the mechanism for this coordination is feedforward inhibition. Interneurons in a local circuit receive temporally precise excitatory synaptic input from both local and distal principal cells in such a manner that the phase difference, and conduction delay, between local circuits is effectively coded in the time difference between spike doublets generated in one gamma period—the first spike arising from local synaptic excitation, the second coming from the distal feedforward input (Figure 8.2b; Traub et al. 1996).

The consequence of spike doublet generation is a compound inhibitory input to principal cells, which effectively prolongs the gamma period during which spike doublets occur. For such a system to provide a means for principal

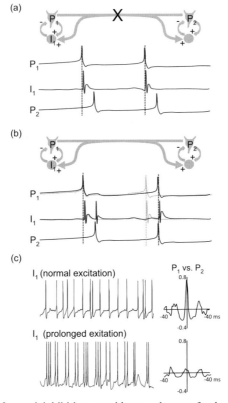

Figure 8.2 Feedforward inhibition provides a substrate for long-range synchrony during gamma rhythms. (a) At its most basic level, the local circuit gamma generator consists of reciprocal synaptic connections between principal neurons and interneurons. Principal neurons also have long-range axon collaterals that preferentially target distal interneurons. Without these, principal cell spiking in two such local circuits is not coordinated. Scheme shows the pattern of interneuron and principal spiking for the left-hand local circuit compared to the principal spiking in the unconnected right hand circuit. (b) With functional long-range feedforward inhibition, interneurons receive excitatory inputs from both local and distal principal cell populations (defined by a common mean delay, illustrated as P_1 and P_2, respectively). This generates spike doublets and prolongs the period of inhibition projected to local principal cells. This prolonged period of inhibition causes phase delay and brings the two principal cell spikes into synchrony. (c) The above mechanism is exquisitely sensitive to the amplitude and kinetics of excitatory input onto fast spiking interneurons. Disruption of this with genetic manipulation of AMPA receptor subunits causes excessive spike generation in interneurons and consequent disruption in long-range synchrony (Fuchs et al. 2001).

cell spike times to converge on synchrony, the relationship between spike doublet interval and period duration change has to be nonlinear. This nonlinearity is observed both experimentally and in computational models: The shorter the doublet interval, the smaller the second component of the principal cell–inhibitory synaptic event through basic paired pulse depression. If the time difference between distal and local inputs to an interneuron is very short, then the probability of generating the second spike in a doublet decreases owing to the kinetics of spike after hyperpolarizations in interneurons. This latter mechanism is sufficient, alone, to induce synchrony with spike doublets in computational models. In addition, this method of "correction" for spike timing differences can only work when spike delays are less than ca. 20% of the period length. This implies that coordination of rhythms over longer conduction delays requires longer period widths (slower frequencies). A comparison of the synchronizing properties of inhibition-based gamma and beta rhythms has shown that beta rhythms are indeed more effective coordinators of activity across longer conduction delays for this very reason (Kopell et al. 2000). A prediction from this mechanism is that disruption of interneuronal spike doublet generation should have a detrimental effect on coordination of gamma rhythms over distance. Such an effect is observed if the kinetics of excitatory inputs to interneurons is altered genetically (Figure 8.2c; Fuchs et al. 2001).

The mechanism described above links principal cell spike generation causally to the generation of coordinated gamma rhythms themselves. However, more recent evidence using powerful and elegant analysis of relative spike times has revealed that stable sequences of spikes can occur within the duration of the window of activity afforded by gamma rhythms (about 5 ms maximum). In the visual cortex, stable phase differences between spike times and the ongoing gamma rhythm correspond to the "goodness of fit" between stimulus presented and the orientation preference for individual cells (Fries et al. 2007). It is difficult to consider how such small but robust interspike intervals can be maintained using conventional models of rhythm generation which involve orthodromic neuronal spike generation: Combinations of a stable population of rhythm and spike rate differences around the population frequency would lead to phase precession rather than stable phase differences, and heterogeneity in neuronal excitation amplitude and timing do not generate stable, non-zero phase differences—at least during gamma rhythms in hippocampus.

To generate robust phase differences in the order of a few milliseconds, a system is needed with network time constant of equivalent order or less. Such a system exists when considering gap junctional connectivity between principal cell axons (see above). Networks of interconnected axons generate rhythms with periods dictated by the network structure itself. Activity percolating through a randomly connected network has a mean path length: the average number of intermediate axons an action potential needs to "jump" across to travel from any given axon to another. Each "jump" takes a finite time, ca. 0.25 ms. The period of oscillation in such a system is therefore the mean path length multiplied by this time constant. Types of slower population rhythm exist in which such activity underlies the pattern of synaptic events observed (e.g., persistent gamma rhythms and beta2 rhythms in association cortex). During persistent gamma rhythms, principal cell spikes are predominantly antidromic, with activity originating directly in axons. The temporal structure generated by such axonal network activity is directly observable in the compound excitatory synaptic events that recruit interneurons (Figure 8.3; Traub et al. 2003). An

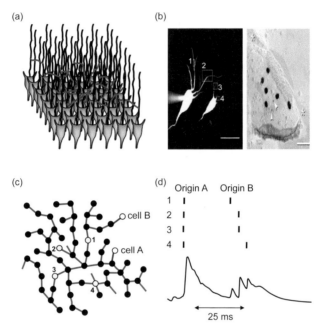

Figure 8.3 Evidence for nonsynaptic axo-axonic connectivity underlying persistent gamma rhythms. (a) The underlying feature of local circuits that supports persistent oscillations in the gamma and also beta2 band appears to involve direct, gap-junctional connectivity between axonal compartments of principal cells. Local axonal collaterals form a sparsely interconnected network that allows action potentials to pass from one principal cell to another. (b) Anatomical evidence for this is far from equivocal. Time-lapsed dye-coupling images, however, show passage of dye from one axonal compartment to another in a nearby cell (left), and FRIL electronmicroscopy shows small-diameter gap junctions between hippocampal principal cell axons (right). At least some of the connections involved in these gap junctions are of the type cx36, shown by the anti-Cx36 immunogold beads (Hamzei-Sichani et al. 2007). (c) Computational modeling of random, locally connected axons from regular spiking pyramidal cells reproduces many of the features of nonsynaptic communication underlying VFO and gamma rhythms. Scheme shows a small part of a large network of axons demonstrating the profile of connectivity needed. (d) Such connectivity allows activity to percolate through the network, generating sequences of action potentials with short, but finite, phase delays when seeded from the same source. Changing sources of initial action potentials also changes the pattern of phase delays, as observed in compound synaptic potentials to local circuit interneurons during gamma rhythms.

additional consequence of such a mechanism is that stable, though brief, relative spike delays are readily generated, with the sequence of spike generation dictated by the network structure and the origin of the first spike in a sequence (Figure 8.3)—a simple, nonsynaptic analog of synfire chains. While such a mechanism provides an attractive working hypothesis for very rapid spike sequences, during a slower rhythm it is not without problems. Spike propagation through axonal plexi is exquisitely sensitive to well-characterized synaptic excitatory and inhibitory events, but requires that somatic action potentials be predominantly antidromically generated—something that runs counter to the conventional neuronal doctrine.

Coordination between Frequency Bands

In addition to coordination of rhythms of the same frequency, generated in spatially separate cortical regions, coordination of rhythms of different frequencies is also seen. This can occur in a single brain region or in multiple, interconnected regions, and takes a variety of forms. One of the most commonly observed interactions between frequencies is the amplitude modulation of a higher frequency coordinated with the phase of a lower frequency—nesting. This is seen for multiple rhythms nested within neocortical delta activity (Lakatos et al. 2005; Figure 8.4) but is currently best studied in hippocampus, where VFO is seen nested within gamma rhythms which are, in turn, nested within theta rhythms.

The local circuit features required for nesting of gamma within theta rhythms are understood at a basic level. Gamma rhythms are generated by reciprocal interaction between principal cells and fast-spiking, perisomatic-targeting interneurons, whereas theta rhythms may occur from the coordinated output of a different subset of interneurons providing slower inhibition to distal dendritic compartments of principal cells. This compartmental separation of two inhibitory inputs allows a division of labor such that theta rhythms organize dendritic excitatory inputs in "packets" during a broad window of opportunity around 50–100 ms wide. Perisomatic gamma frequency inhibition then coordinates the output from these excitatory epochs at a finer timescale. Evidence suggests that appropriate nesting of gamma and theta inputs, and thus timing bursts of gamma frequency input with excitatory input, is done by mutual interaction between the two interneuron subtypes: Theta frequency outputs from oriens-lacunosum molecular interneurons organize basket cell outputs into packets of spikes coincident with the period of maximal dendritic principal cell excitation. In turn, output from basket cells provides trains of inhibition to oriens-lacunosum molecular interneurons to activate an intrinsic h-current and thus time rebound spiking in these cells (Rotstein et al. 2005).

Other forms of interaction can be seen in phase synchrony measures across frequency bands in cortex. Here synchrony is seen when the period length of one locally generated rhythm is an integer multiple of another locally generated rhythm (Figure 8.4). Such a situation requires that outputs from at least one neuron be critical to both rhythm-generating circuits. Paired neuronal recordings in human cortex suggest that this is entirely feasible, with multiple delay times evident for a single neuron activation of a number of different, interconnected local neurons (Molnar et al. 2008). When period lengths for two or more local rhythms are closely—but not exactly—matched, variable phase relationships may occur. Rhythmic changes in phase relationship are seen between superficial and deep neocortical laminae-generating gamma and beta2 frequency rhythms, respectively. Here, the pattern of phase change is determined by the trigonometric identity relating the two coexistent frequencies (Roopun, Cunningham et al. 2008; Figure 8.4).

This latter example of coordination of rhythms between different local cortical laminae reveals some aspects of the dynamic complexity inherent in a system with discrete sequences of rhythm, illustrated in Figure 8.1. In particular, with the ratio of adjacent frequencies seen in neocortex (ca. 1.6), the rate of phase change seen between two adjacent rhythms in the sequence corresponds to the frequency of the next rhythm in the sequence. For example, superficial gamma rhythms and deep beta2 rhythms in parietal cortex generate a phase relationship that cycles at beta1 frequencies (Figure 8.5a). What is also apparent

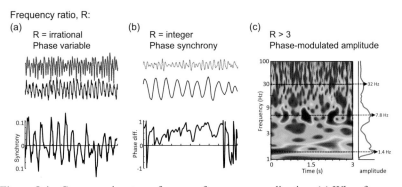

Figure 8.4 Common signatures for cross-frequency coordination. (a) When frequency ratio (R) is irrational, the phase relationship between two concurrently generated frequencies is nonstationary. Traces show LFP activity in superficial and deep cortical laminae during concurrent $\gamma 1$ and $\beta 2$ rhythm expression (Roopun, Cunningham et al. 2008). In this case, average synchrony between oscillators is near zero, with a periodic change in instantaneous phase corresponding to the sum of the periods of the two rhythms shown. (b) When R is an integer, a stable phase difference can be seen between the two rhythms recorded from single or pairs of brain regions. The figure shows bandpass filtered activity within the $\gamma 1$ and α bands in MEG recordings, and the corresponding plot of phase difference between the bands (modified and reproduced with permission from Palva et al. 2005). (c) When R is relatively large, "nesting" of coexpressed frequencies may occur. The phenomenon is observed as an amplitude modulation of one frequency relative to the phase of a lower frequency. The figure shows the nesting of spontaneous $\gamma 1$ rhythms within a concurrent θ rhythm, and the concurrent amplitude modulation of the θ rhythm by a coexistent $\delta 1$ rhythm in macaque supragranular cortex. (Figure modified and reproduced with permission from Lakatos et al. 2005).

is that, following a period of concurrent gamma and beta2 rhythm generation, reduction in excitation to the cortex stabilizes the phase relationship between deep and superficial layers, with the beta1 frequency now manifest directly in

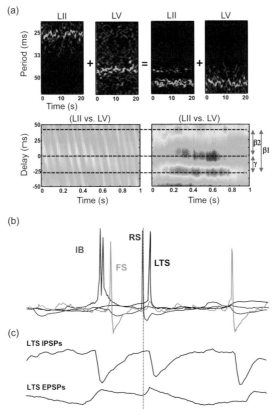

Figure 8.5 Local circuit features underlying period concatenation. (a) Somatosensory cortical slices generate concurrent γ and β2 rhythms in superficial (LII) and deep (LV) laminae, respectively. Reduction in excitation to cortex translates these two rhythms into a single β1 frequency rhythm seen in all laminae. Generation of a β1 rhythm from two preceding rhythms changes the phase relationship between laminae. With γ and β2 rhythms, interlaminar phase difference is nonstable showing rapid, rhythmic phase precession repeating every ~60 ms. Emergence of β1 oscillations stabilizes interlaminar phase relationship, with lag and lead times corresponding to the original γ and β2 period lengths (b) The β1 rhythm is accompanied by a pattern of spike generation in neurons suggesting concatenation: LV intrinsically bursting (IB) cells activate fast spiking (FS) interneurons with short latency. Superficial pyramids (RS) spike on the rebound of an inhibitory postsynaptic potential (IPSP) from these FS cells and proceed to activate low threshold spiking (LTS) interneurons with short latency. Following a slow inhibitory synaptic event from LTS cells, IB cells rebound to spike again. (c) The sequence of concatenated spiking of FS and LTS cells can be seen in the inhibitory synaptic potential generation in individual LTS cells. In addition, the concatenation sequence of IB and RS cell spiking is also visible in this cell type.

the raw local field potential in both layers (Figure 8.5b). Clues as to how this occurs come from the nature of this stable phase relationship: The periods of the original two rhythms (gamma and beta2 frequency) are observed in the cross correlation, suggesting that one period of the observed field potential beta1 rhythm is composed of sequential periods of the original gamma and beta2 oscillations. In other words, the two rhythms now have concatenated periods, with the simple sum of the two original periods corresponding to the new beta1 rhythm generated by reduced excitatory drive.

Two features of the local neocortical network appear vital in facilitating this concatenation process. First, as the degree of excitation of cortical neurons needs to be reduced to see this phenomenon, there is insufficient tonic drive to principal cells to sustain spontaneous action potential generation. Instead, spikes are generated through rebound excitation mediated by activation of an intrinsic conductance, Ih (Kramer et al. 2008). Second, interneurons must still be active to provide the hyperpolarization required to bring membrane potential of principal cells down to levels where Ih can participate in rebound depolarizations. Two subtypes of interneuron are required for the coordination of superficial and deep layer local circuits to generate the concatenation sequence. First, burst firing in layer 5 principal cells activates superficial fast-spiking interneurons. These interneurons provide a brief inhibitory postsynaptic potential to superficial principal cells (as they would do during gamma rhythms) inducing a rebound action potential thereafter. This superficial principal cell output is sufficient to activate the second interneuron subtype involved: low threshold spiking cells. Output from these interneurons target dendrites of principal cells, producing a slower inhibitory postsynaptic potential which causes rebound spiking approximately one beta2 period later (Figure 8.5c). Thus, as seen for coordination of gamma rhythms alone, both interneurons and intrinsic neuronal properties also combine to provide a mechanism for coordination of rhythms at different frequencies.

Why Have Dynamic Coordination of Rhythms?

The role of dynamic coordination within a frequency band has been highlighted particularly well for gamma and beta rhythms with respect to cognition and motor control, respectively. In each case, the key feature of the rhythm is that it temporally coordinates firing patterns of neurons to provide "windows of opportunity" for coactivation. The phase relationship between a rhythm at two spatially separate sites governs the relative timing of local activity at one site and distal synaptic inputs from the other. Such a relationship powerfully modulates the ability of one site to interact with the other and controls the degree of synaptic plasticity at each site, thus also influencing future interactions. Therefore, the degree of coherence of rhythms (whether field potentials or spike trains) has been shown to signal the degree of "communication" between structures: a large-scale network property with documented behavioral correlates (e.g., Baker 2007; Fries 2005). In addition, even small changes in mechanisms underlying such interactions may correlate with underlying pathology in psychiatric illness (e.g., Phillips and Silverstein 2003; Roopun, Cunningham et al. 2008). Whether we can apply that which we have learned from such studies to coordination across different coexistent frequency bands remains to be elucidated.

The observation of multiple discrete frequencies in neocortex suggests a possible general scheme of rhythm generation where different "frequency channels" are used to process different forms of information. However, thus far this has been observed mainly in association cortex, and the *in vitro* experimental models used provided a highly reduced and stable rhythm-generating environment. Transient, as opposed to persistent, cortical drive generates a broad range of frequencies, and thalamic—and external sensory—involvement in cortical function also generates variable frequencies. For example, thalamo-cortical spindle oscillations may vary in frequency from within the delta band up to beta frequencies, variable whisking frequency (around the theta/alpha bands) can be matched by oscillations in superficial barrel cortex, and cortical theta rhythms may be variably paced by the sniff cycle in rodents. In addition, frequencies of network rhythm can be observed in cortical regions other than association areas that do not fit the concatenation sequence. Gamma rhythms in primary, auditory, and visual cortices have different frequencies which do not directly correspond to that seen in parietal cortex, and entorhinal cortex can generate two gamma rhythms through different local circuits, only one of which is frequency-matched to the gamma rhythms seen in association cortex.

Questions remain: Are different discrete frequencies, and their coordination patterns, of any importance to cortical function? Do they just represent a very specific activity state of cortex whose dynamic signature we can record and recognize? Are they purely epiphenomenal? Do they form the substrate for all higher-order dynamics in brain? Given their overt expression in isolated, persistently oscillating association cortex, concatenated sequences of frequency bands may represent a spectral baseline upon which polymodal sensory and neuromodulatory influences may act. The functional significance of the ratio between multiple frequencies illustrated here for gamma and beta rhythms is not yet understood for neuroscience, though it has been much discussed in the context of other natural systems. In the current context, this ratio fosters minimal interference between coexistent frequencies, something that could not be relied upon in a system composed of multiple, continually variable rhythm generators. It has been proposed that sensory information may be handled more efficiently if it is processed along multiple "channels" with differing temporal scales (Wiskott and Sejnowski 2002). For example, evidence for the segregation of sensory information into different "frequency channels" is apparent from studies examining how different levels of detail in the visual field are processed (Smith et al. 2006). In this study, perception based on processing coarse level features occurred at theta frequencies, whereas perception of objects requiring more detailed features was associated with a faster, beta frequency rhythm—an observation which fits with the decrease in precision of tuning curves for visual cortical neurons with decreased spike frequencies. A ratio of ca. 1.6 between coexpressed frequencies may permit many "channels" for information processing to coexist with minimal temporal interference (multiplexing). In addition, a framework of discrete network frequencies, as opposed to a continuum of oscillation states, better matches the discrete anatomical organization of networks in neocortex into hierarchies and clusters (Sporns et al. 2004).

An additional advantage of coordination between discrete frequency bands is that the cortex can perform "simple math" to combine information held in different "frequency channels" (i.e., the concatenation process highlighted above). Successive concatenation steps are feasible with the ratio of discrete frequencies observed in association cortex. Thus information processed at gamma and beta2 frequencies may combine into a beta1 frequency code. This code may coordinate with other regions oscillating at beta2 frequencies to generate an alpha frequency and so on. As the concatenation process required for this implies the generation of specific phase relationships between component local circuits, multiple higher frequencies may then be "packed," albeit crudely, into a single lower frequency rhythm and unpacked for further processing at a later stage. For example, the relationship between delta rhythms and higher frequency components during sensory processing (e.g., Lakatos et al. 2005) suggests that during slow-wave sleep, subsequent delta rhythm generation may provide a substrate to replay coordinated multiple frequencies associated with previous sensory events.

Summary

The association between cortical rhythm generation and cortical function is strong for many aspects of sensory processing (particularly with gamma rhythms) and motor control (beta rhythms). The mechanisms of rhythm generation, per se, not only provide a substrate for temporal control of neuronal spike firing but are also labile to the pattern and degree of spiking of principal cells in local and spatially separate circuits. Thus, a combination of intrinsic, chemical, and electrical synaptic properties of neurons and their resulting local networks provides a powerful but highly labile means by which to translate cortical inputs and outputs into a temporal code based on principal cell spike patterns. What is becoming increasingly clear is that even very small local circuits in cortex have sufficiently diverse features to generate many network rhythms concurrently. The major challenges arising from this are to understand the temporal patterns produced by multiple frequencies, the underlying mechanisms by which they interact, and the consequences for generations of principal cell spike codes. Each challenge must be considered if we are to further understand the cortical temporal landscape as it relates to the behavior of the whole organism.

Overleaf (left to right, top to bottom):
Kevan Martin, Jörg Lücke, Mayank Mehta
Andreas Kreiter, group discussion
Terry Sejnowski, György Buzsáki
Bita Moghaddam, Anders Lansner, Danko Nikolić
May-Britt Moser, Group discussion

9

Coordination in Circuits

*Mayank Mehta, György Buzsáki, Andreas Kreiter,
Anders Lansner, Jörg Lücke, Kevan Martin,
Bita Moghaddam, May-Britt Moser,
Danko Nikolić, and Terrence J. Sejnowski*

Introduction

What are the mechanisms underlying the emergence of mind from the activity of groups of neurons? This is a difficult question that has to be addressed at many levels of neural organization, all of which need to be integrated. The following discussion of the set of neural mechanisms, neural activity patterns, and animal behaviors sketches a few simple, but general and robust, neural mechanisms at all the different levels ranging from synapses to neurons, to networks and behavior, and is illustrated using experimental observations primarily from cortical and hippocampal activity patterns during behavior. The focus is on a few key mechanisms such as energy-efficient sparse codes of mental representations, need for synchrony among sparse codes for information transmission, and the contribution of recurrent connections between excitatory and inhibitory neurons in generating synchronous activity, oscillations, and competition among networks to facilitate fast and flexible behavior.

At the level of the mind, animals can perceive different components of a rapidly changing natural scene such as luminance, contrast, local features (e.g., lines), and global features (e.g., shapes). Similarly, when an animal navigates in the world, neurons can flexibly represent the position of the animal in a given environment, the composition of the environment, the head direction, the running speed, etc. These mental representations of the world are flexible and dynamic, determined and modulated by a range of environmental, behavioral, and neural parameters.

What are the mechanisms by which the brain generates these flexible mental representations? How do these mental representations across different brain regions interact to generate perception and decision making?

What Is the Right Anatomical Level of Investigation?

The flexible behaviors outlined above depend on common features of the neuroanatomical substrates underlying these mental representations. Historically, models of cortical circuits have been designed to capture a specific feature of the experimental data. An alternative strategy that leads to a "predictive connectivity" starts with the assumption that there is a basic ("canonical") circuit common across the entire neocortex (see Figure 9.1). This assumption justifies the use of the rich cache of structure, function, and neurochemistry to build more biologically realistic models. The models can then be challenged to provide an explanation of cortical activity patterns. To the extent that the simulations are successful, the model can quantitatively predict the connectivity pattern of the circuits in that area. The ability to test a prediction about structure is a radical departure from the traditional descriptive and anatomical methods of circuit analysis. In combination with new tools for tracing pathways and combining structure with function, this predictive structural modeling will not only greatly accelerate circuit analysis in neocortex, but will provide a far more

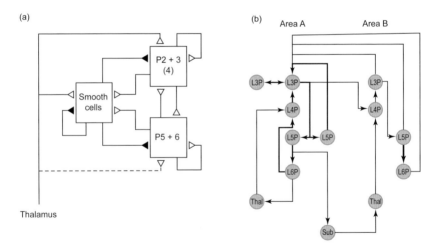

Figure 9.1 (a) Canonical circuit of neocortex. Three populations of neurons interact: the inhibitory, GABAergic population indicated by smooth cells; the excitatory population by a superficial layer population (P2+3 (4)); and a deep layer population (P5 + 6). The connections between them are indicated by edges and arrows. The functional weights of the connections are indicated by the thickness of the edges. (b) Graph of the dominant interactions between significant excitatory cell types in neocortex and their subcortical relations. The nodes of the graph are organized spatially; vertical dimension corresponds to the layers of cortex and horizontal to its lateral extent. Edges and arrows indicate the relations between excitatory neurons (P: pyramidal) in a local patch of neocortex, which are essentially those described originally by Gilbert and Wiesel (1983) and Gilbert (1983) for visual cortex. Thin edges indicate excitatory connections to and from subcortical structures and inter-areal connections. Thal: thalamus; Sub: other subcortical structures, such as the basal ganglia.

comprehensive and synthetic explanation of the computational strategies used in different cortical areas.

There are several common features of cortical anatomy shared across many brain regions. The neocortex is organized in a number of layers and columns. Major elements of the canonical cortical microcircuit of a column have been well described. For example, the excitatory, pyramidal neurons within layer 2/3 in a neocortical column are recurrently connected, as are the neurons within layer 5/6. The thalamic inputs arrive primarily in layer 4 whereas layers 5 and 6 send the output of a cortical column to other brain areas. Further, synaptic inputs are exquisitely organized on the extensive pyramidal neuronal dendrites, which have nonlinear properties. Also, there are neuromodulatory inputs specific to different layers, which could play a key role in information processing, as will be discussed later.

In addition to many other features of the canonical cortical circuit, there is one common feature in all of these layers, namely the presence of a wide variety of GABAergic inhibitory interneurons. While these inhibitory neurons comprise only about 20% of the neural population, they strongly control cortical activity because of their recurrent connections to excitatory neurons. Inhibitory synapses are often found near the soma, which can influence all excitatory inputs flowing from the dendrites to soma. Further, not only are the excitatory neurons recurrently connected to each other within and across layers, so are the inhibitory neurons. Such recurrently connected excitatory-inhibitory networks (denoted E-I networks) are ubiquitous: They are found not only in most parts of neocortex and hippocampus, but in many other structures as well.

In addition, cortical circuits receive powerful neuromodulatory inputs. Monoamine neuromodulators dopamine, norepinephrine, and serotonin are released by cells in discrete nuclei in the brainstem and midbrain that project heavily to basal ganglia and cortical regions. Most psychotherapeutic and psychoactive drugs, which have profound effects on cognition, act on receptors of monoamines. These include antidepressant and antipsychotic drugs as well as hallucinogens and stimulants such as amphetamine, cocaine, and methylphenidate. This suggests that monoamines are a critical component of neuronal machinery underlying perception and complex behaviors. The topography of their projections to cortical regions, as well as their targeted receptors, is quite diverse. For example, dopamine projections tend to be heavier to deeper cortical layers whereas norepinephrine projections are heavy in superficial layers. The receptor type and the signal transduction mechanisms used by these neuromodulators are diverse, with the exception of one of the subtypes of serotonin receptors—G-protein coupled receptors. The localization of these receptors is also specialized. For example, some subtypes of serotonin receptors are primarily localized on GABA interneurons. In addition, monoamine receptors, especially the dopamine receptors, are mostly localized extrasynaptically, suggesting that they produce slow and somewhat sustained effects on the state of cortical microenvironments. In the case of dopamine, the density of dopamine transporters in cortical areas is sparse and thus the release of dopamine has the capacity to diffuse away from presynaptic sites and act more diffusely.

For generality, the following discussion will be focused on how generic E-I networks process information and how neuromodulators influence the process, keeping in mind that this is a simplification. The precise details, such as various intrinsic properties of the different cell types, their dendritic geometry, and the exact connectivity patterns within and across cortical columns, would need to be investigated in the future to understand the neuroanatomical basis of perception. The goal is to probe the system at progressively more detailed biological levels, while deciphering the emergent properties of the system at each level, which can be robustly tested both experimentally and theoretically.

How Do E-I Neural Networks Oscillate?

Converging evidence suggests that the E-I network is balanced; that is, the total amount of excitation and inhibition are comparable most of the time, even though the total amount of activity can vary over a wide range. In the absence of stimuli, under most conditions, cortical neurons are active at a low rate with an ensemble average of about 0.1 Hz, which varies systematically across layers. When stimuli arrive, a small fraction of the excitatory stimulus-responsive neurons increase their instantaneous firing rates to 10 or even 100 Hz. This increased activation of pyramidal neurons drives the feedback inhibitory neurons, which in turn briefly shut down the pyramidal neurons. This reduces the excitatory drive onto inhibitory interneurons which generates release from inhibition synchronously across a number of pyramidal neurons. Consequently, pyramidal neurons increase their spiking activity in synchrony. A key parameter governing the frequency of such E-I network synchronized oscillations is the time constant of the inhibitory $GABA_A$ receptors of 10–30 ms, resulting in about 30–100 Hz gamma frequency oscillations. Thus, oscillations in the gamma range can be a signature of cortical activation. Notably, this simple description for generating gamma oscillations applies only to excitatory neurons connected recurrently to inhibitory ones. The additional recurrent connections within the populations of neurons of the same type would profoundly influence the strength and frequency of oscillations. Synchronization of oscillations across different E-I networks is another, even more complex process.

Thus, in a simple scenario, stimulus-driven elevation in the firing of excitatory neurons can have two concurrent effects: (a) elevated firing of the excitatory and inhibitory neurons, and (b) synchronized oscillations. Notably, both the oscillation frequencies and the degree of synchronicity between oscillations influence neural information processing.

It is important to discuss the following four points: the range of oscillation frequencies, alternative mechanisms for generating oscillations, mechanisms that modulate oscillation frequency, and synchrony without oscillations.

Oscillations with frequencies ranging from 0.1–200 Hz have been commonly observed in several neocortical areas as well as in the hippocampus and the olfactory system. Here, the focus is on oscillations that occur on the timescales relevant for processing natural stimuli (i.e., less than about half a second), so that they can modulate neural processing. This means that the focus will be on frequencies greater than about 2 Hz. Synchronized oscillations of a variety of frequencies appear in numerous brain regions during perception, attention, working memory, motor planning, sleep, epilepsy and Parkinson's disease. For example, 4–12 Hz theta oscillations are prominent in the rodent and primate hippocampus during spatial exploration. They have been reported in visual, parietal, and prefrontal cortices during maintenance of information in working memory. Somewhat higher frequencies, 10–30 Hz or beta frequency oscillations, have been reported in the visual and motor cortices. The 40–120 Hz gamma oscillations are induced by visual stimuli in numerous visual cortical areas and prefrontal cortex, and they also occur in the hippocampus. In addition, bursts of 140–250 Hz ripple oscillations occur in the hippocampus during quiet wakefulness. The focus here is on the theta and gamma oscillations that appear in the neocortex and hippocampus during cognitive tasks.

The E-I network is not the only mechanism that can generate synchronized oscillations. Neurons are endowed with a variety of conductances and intrinsic mechanisms which can also make them respond rhythmically when a fixed amount of current or neuromodulators are applied, even when isolated from a network.

The key issue is: How do groups of oscillating neurons get synchronized? Invariably, this is achieved through their coupling with the rest of the network. For example, neurons in the reticular nucleus of thalamus oscillate in isolation, whereas these oscillations are synchronized through coupling between these neurons directly or through the thalamocortical loop. This mechanism is thought to generate sleep spindles. Similarly, septal neurons oscillate in isolation and are likely synchronized by their recurrent connection to the hippocampus, resulting in synchronous theta oscillations. Finally, even when the E-I network in a cortical column oscillates at gamma frequency, an important question is: How do oscillations of different cortical columns synchronize? In all these cases, further questions arise: How do these oscillators respond to a stimulus? Does an excitatory spike from another oscillator speed up the subsequent spike from a given oscillator or delay it? In other words, how do the oscillations change as a function of the phase at which inputs arrive from other oscillators? Thus, it is important to study how oscillations change as a function of the phase at which inputs arrive. Such dependence is called a phase resetting curve and has been investigated for a variety of physical and neural systems.

The frequency of neural oscillations can be modulated by several means. For example, neuromodulators can generate a threefold change in the effective time constant of $GABA_A$ receptors, resulting in a concomitant change in the frequency of gamma oscillations. Further, cholinergic levels alter spike frequency adaptation of pyramidal neurons, which would influence the E-I balance and spike timing in an E-I network. The levels of neuromodulators change with behavioral state and attention, resulting in state-dependent modulation of amplitude, power, and synchrony of neural oscillations. For example, neuromodulators can raise the membrane potential of neurons. This can make it easier for the neurons to respond to a small amount of stimulation, and may make the E-I network more likely to oscillate.

Finally, two important features of oscillations need to be distinguished: synchrony and rhythmicity. Synchrony can occur without rhythmicity and vice versa. In particular, synchronous activation of groups of neurons occurs almost invariably, often without oscillations, when a strong stimulus is abruptly activated.

Neural synchrony is important for efficient and rapid transmission of information between brain regions. For example, individual neurons in a cortical column receive information from only about 100 thalamic neurons. To integrate this input and generate a spike, a cortical neuron typically needs to be depolarized by about 20 mV. Given the small amplitude (≤ 1 mV) and short duration (~10 ms) of cortical excitatory, AMPAR-mediated postsynaptic potentials, this small amount of thalamic neurons can only activate an entire cortical column if the inputs are synchronized within a 10 ms time window.

There are several advantages of transmitting information using synchronous activity. First, only a small number of active neurons, or a sparse code, is sufficient to transmit information from one area to another, as opposed to asynchronous transmission which would require more activity. Given that spike generation consumes energy, sparse synchronous codes are energy efficient. Second, compared to the asynchronous systems, synchronous sparse codes can be brief, allowing the system to respond rapidly to changing stimuli. Finally, the synchrony-based codes allow the system to be flexible, requiring only small changes in the relative timings of groups of neurons to make one group drive the downstream neurons more effectively than through asynchronous codes.

Synchronous activity can be generated by two different mechanisms. Synchrony can be evoked by a transient stimulus or through dynamic interactions between internal temporally organized activity patterns, such as oscillations. The latter can generate synchronous activity across multiple cycles of oscillation. Subsequent sections will discuss computational advantages of this process.

Why Aren't Gamma Oscillations Always Observed during Behavior?

The E-I network is ubiquitous, and synchronized oscillations are a likely mode of the E-I networks, yet there are instances in which oscillations are not apparent. There are several reasons for this. It is often difficult to detect gamma oscillations in the spike train of a single neuron because, even when modulated by gamma oscillations, neurons often do not spike at sufficiently high rates to be active on every gamma cycle. Also, neurons often join the population rhythm for only brief epochs. This probabilistic firing and rapidly changing neural assembly can appear to be nonrhythmic when analyzing the activity of single units in isolation. An alternative method for detecting synchronized oscillations of ensembles of neurons is through the measurement of the local field potential (LFP) and the analysis of its power spectrum. Notably, the power in the LFP spectrum decays inversely with the increase in frequency, which makes it more difficult to detect activity in the gamma band than in the lower frequencies. Here, analysis methods that compensate for this systematic tendency of power spectra can improve the ability to detect gamma oscillations. Further, the nature of the electrode used to measure the LFP influences the power of the measured oscillations: sharp, high-impedance electrodes integrate activity over a small pool of neurons, which may not be sufficient to detect synchronous oscillations above the electrical noise. Signals can be improved by using blunter electrodes, which additionally allow the detection of synchronous gamma activity in multiunit activity.

Presentation of visual stimuli within a receptive field, such as moving bars or gratings, are likely to generate synchronized gamma oscillations for at least several seconds in anesthetized animals. Similar oscillations may be more difficult to detect in behaving animals because the fixations last shorter as eyes move on average three times a second, moving the stimuli rapidly in and out of the receptive fields. This may augment the fluctuations of neural activity, resulting in rapid fluctuation of gamma power and frequency, and making detection by standard methods difficult. Time-frequency domain analyses may counter this problem by estimating the strength of gamma oscillations in smaller, relatively unperturbed windows of time.

Additionally, one should measure the gamma activity in a region that is likely to be critically involved in processing of the presented stimulus such that neurons are likely to be driven at high rates. A more strongly driven E-I circuit is more likely to be accompanied by strong gamma oscillations.

Finally, as discussed above, an E-I circuit will not always generate synchronous gamma oscillations. Nevertheless, information processing based on precise synchronization in sparse cortical circuits may take place. One possible reason is that synchronized activity patterns do not always follow limit-cycle attractors, characteristic of regular oscillations, but instead more irregular, maybe even chaotic attractors. As a consequence of the more broadband nature of these processes, auto-correlograms often show the familiar, a few milliseconds wide center peak flanked by troughs but lack satellite peaks. Another possibility is that synchronous events could occur through synfire chain mechanisms, which do not require regularly repeating activation of neurons. Both, chaotic attractors and synfire chains represent internal mechanisms of synchronization (induced synchronization) as they do not require precise locking to stimulus events. Finally, synchrony can also be evoked by an external input. For example, a flashed visual stimulus or an auditory "click" can trigger synchrony. In these cases the synchronous activity is locked in time to the stimulus and can be detected in a peri-stimulus time histogram (PSTH).

How Do Gamma Oscillations Interact with Lower Frequency Oscillations?

Generating gamma oscillations in an E-I network requires a fair amount of activity, which costs energy and would be difficult to sustain continuously. One way to bring that network to an oscillatory state is by driving these neurons externally, as discussed above. Here, it is important to synchronize gamma oscillations across multiple interacting modules. This can be a difficult task when the modules are far apart, where the transmission delays, combined with complex phase resetting curves, may make it difficult to generate synchrony. This raises the question: Are there other, transmission-delay independent ways to increase the long-range synchrony of gamma oscillations?

One possibility to make gamma oscillations more prominent is to suppress the lower frequency oscillations, thereby increasing the signal-to-noise ratio. The power spectra of neural activity show $\sim 1/f$ dependence on frequency f, with a large amount of power in low frequency signals. A small suppression of low frequency signals can significantly improve the relative contribution of the gamma power, making the gamma oscillations more effective in modulating spiking activity. This enhancement of gamma efficacy induced by low frequency suppression has been reported in several sensory cortical areas during attention and voluntary movements. Further, cortical activity is modulated by synchronous activity in the delta band (0.5–3 Hz) during quiet wakefulness and sleep, and this low frequency activity disappears during active engagement in a task and in conjunction with an increase in the gamma power.

Mechanisms also exist under which lower frequency oscillations can facilitate synchronization of gamma power fluctuations across large distances. Neurons in the hippocampus oscillate synchronously at theta frequency and project to a majority of neocortical areas. During the phase of theta oscillation, in which the hippocampus is more active, it can activate the target neocortical neurons, thereby enabling gamma frequency oscillations in those neocortical areas. The reverse happens at the phases of theta oscillations where hippocampal neurons are less active. Thus, the power of neocortical gamma oscillations would be modulated by the phase of hippocampal theta oscillations. This has been observed in behaving animals, including humans. Similarly, the phase of lower frequency oscillations modulates the power of neocortical gamma oscillations during slow-wave sleep, with higher gamma power appearing during the more depolarized phase of slow oscillations.

Thus, lower frequency oscillations can modulate gamma oscillations in two different ways. First, removal of the incoherent lower frequency oscillations can enhance gamma power. Second, the coherent lower frequency oscillations can facilitate synchronous modulation of gamma power across neocortical areas. Neuromodulators can also influence this process, in some cases removing low frequency oscillations and in others, facilitating them. This suggests that some form of optimization may occur, adjusting the balance between low and high frequency oscillations to maximize the efficacy of information processing. As a first step toward understanding how oscillations influence information processing, this interaction is discussed at the level of single neurons, of ensembles of neurons, and ensembles of neuronal networks.

How Do Oscillations Influence the Neural Representation at a Single Neuron Level?

Neurons respond in a graded fashion to sensory stimuli by altering their average activity levels: optimal stimuli evoke greater amount of activity than suboptimal stimuli. This is the classic rate code. For example, hippocampal place cells change their firing rates as a function of the spatial location of an animal, and neurons in the primary visual cortex change their mean firing rates as a function of the orientation of a stimulus. Further, hippocampal place cell activity is modulated by synchronized theta rhythm, and visual cortical activity is modulated by gamma rhythm. Thus the neural responses are modulated by two very different forces: by stimuli anchored in physical space and by internally generated oscillations. The former contain information about the external world, the latter about internal processing and timing. How do the stimulus-driven responses interact with synchronized oscillations? Would such interaction serve any purpose?

In the simplest scenario, the neuron will simply sum up the two inputs and generate a spike when this input exceeds a threshold. Thus, when the stimulus-based input is low, the neuron would spike at only that phase of oscillation when the oscillatory input is high so that the total input is sufficient to reach spike threshold. On the other hand, when the stimulus-evoked input is high, the neuron can spike even at the phase of oscillation when the oscillatory input is minimal. Thus, an interaction between the input and oscillation generates a phase code: When the inputs are strong, neurons will respond at every phase of oscillation; when the inputs are weak, neurons can respond only at the peak of oscillation when the oscillatory drive is maximal. At intermediate values of input, the outcome is a combination of phases. Thus, interaction between rate-coded inputs and synchronized oscillations would generate a phase-coded output.

Such phase-code and rate-phase, or rate-latency transformation has been observed in the hippocampus and is called phase precession (i.e., the phase of theta oscillation at which place cells spike varies systematically as a function of the animal's position). For example, on linear tracks, a place cell fires spikes near the trough of theta oscillation as the animal enters the place field, and the theta phase of spike precesses to lower values as the animal traverses farther in the place field. Phase precession has been observed in several parts of the entorhinal–hippocampal circuit, along with correlated changes in firing rates. Recent studies have shown a mathematically similar phase code in the visual system where neurons spike at an earlier phase of gamma oscillation when driven maximally by the optimal stimulus, but at later phases of gamma oscillation when driven by suboptimal stimuli. Further, rate-latency transformation can be detected in the structure of spatiotemporal receptive fields of direction-selective visual cortical neurons when probed using randomly flashed bars. Similar measurements in other structures are likely to detect similar phase codes. Further, when slowly rotating oriented bars are presented to visual cortex, they may generate gamma-phase progression, and so may single bars passing over a series of direction selective receptive fields in area MT or V1.

These phase codes have several computational and functional advantages. First, they enable the postsynaptic neuron to decode the stimulus parameters by simply measuring the phase of the oscillation at which the presynaptic neuron spiked. This phase or latency is clearly defined by the period of the oscillations. This is in contrast to a rate code where one has to specify arbitrarily the interval of time over which spike count has to be averaged to obtain an estimate of a rate code. Second, stimulus-evoked activation of groups of neurons that represent stimuli in a sequence several seconds long would generate a compressed version of the stimulus sequence within an oscillation cycle due to the rate-phase transformation, possibly allowing these stimuli to be bound together and perceived as a chunk. Third, this temporally precise sequence of activation of neurons would facilitate the induction of spike timing-dependent plasticity, thereby generating a permanent record of the group of coactivated neurons in terms of the strengths of synapses connecting them. This would not only involve strengthening of synapses, but also weakening of synapses, especially the ones that correspond to nonsequential activation.

This mechanism of rate-phase transformation can thus be used to learn temporal sequences that occur over a timescale of a second, even though synaptic plasticity mechanisms operate on timescales of milliseconds. Similar learning of sequences may occur in other scenarios as well, where oscillations are imposed by other means. For example, systematic movements of the eyes across a natural scene every third of a second could induce oscillations where sequentially perceived views of the scene are brought together to form a stable, coherent percept using short- and long-term synaptic plasticity mechanisms. In addition to the relative timing of spikes between the stimulus-selective excitatory neurons, inhibitory spikes and neuromodulatory inputs are likely to determine the pattern of synaptic modifications.

While these mechanisms would work well for learning sequences of events that occur over a period of about a second, it remains to be determined how sequences of events that occur several minutes or hours apart can be learned via hitherto unknown mechanisms. Further, the above discussion of neural responses assumed that they are fixed and can be described in terms of a receptive field. Next we question this assumption and discuss the possibility of dynamic receptive fields.

Are Neural Representations Static or Dynamic?

In a typical study of neural information processing, the experimenter measures the changes in neural activity in response to a variety of stimuli. The neurons may respond strongly to one set of stimuli and less so to others. The pattern of neural responses to stimuli defines the neuron's receptive field.

The notion of a receptive field guides our thinking on how single neurons represent information but also has several limitations:

1. There are an infinite number of possible stimuli, varying across many dimensions. Hence, it is difficult to find the stimulus or set of stimuli that drive the neuron optimally within a finite amount of time.
2. Internal variables, such as arousal and neuromodulatory state, modulate neural responses.
3. Not only the stimuli within the receptive field, but even stimuli outside the classical receptive field modulate the responses.
4. The responses of many neurons in the visual system are affected by the attentional level and the reward value of the stimulus.
5. Most importantly, during natural behavior, stimuli are not static and do not appear in isolation. Instead, a large number of visual stimuli typically appear simultaneously and the stimulus configuration changes rapidly.

As a consequence of these five influences, the classical receptive field of a neuron can change dramatically between situations. For example, transient inactivation of the somatosensory cortex or of a sensory organ generates a large reorganization of the sensory map—a process that occurs within a second. In the hippocampus, past experience can result in a complete reorganization of the spatial selectivity of place cells. This reorganization is called remapping. Remapping can occur even on short timescales (~minutes), not just over days. In addition, when stimuli are presented in a sequence, visual cortical neurons not only respond to the onset of the stimulus but the responses depend on the sequential position of the stimulus as well: some neurons fire maximally to the presentation of the first stimulus in the sequence, irrespective of the identity of that stimulus. Finally, recent experiments show that hippocampal neurons fire in a sequence even when the animal is sleeping or is running without changing its position (i.e., in a fixed running wheel).

These results suggest that neural responses are dynamic and can change rapidly with changes in stimulus configurations and internal variables, such as past experiences. This should not be surprising. As discussed earlier, to be energy efficient, neuronal codes need to be sparse and synchronous. Depending on the connectivity state of the network, recent history of neural and synaptic activity, and the nature of stimuli, different groups of neurons and synapses may become synchronously active and may hence drive different downstream neurons. Short-term dynamics of neurons and synapses can play an important role in generating such dynamic receptive fields.

This raises the question of how the downstream neurons interpret the messages sent by dynamically changing upstream neurons. Clearly, the postsynaptic neurons not only respond to just one presynaptic neuron but to an entire ensemble. Thus, dynamic reorganization of neural responses should be coordinated across an ensemble of neurons. The following discussion sketches an oscillation-based mechanism of dynamic coordination of neural codes across ensembles.

Neural Attractors, Cell Assemblies, Synaptic Assemblies, Oscillations, and Dynamic Coordination

Information is thought to be represented by the activity patterns of groups of neurons. Fault tolerant, stable, content addressable, and associative representation of stimuli across an ensemble of neurons can be implemented using the Hopfield attractor dynamics. For fixed point attractors, there are large energy barriers between the different representations, represented by local energy minima, whereas the energy required to make transitions between stimuli along some other dimension may be negligible in the case of continuous attractors. The stability and convergence of attractor dynamics are achieved through iterative processing of information in a recurrent network of excitatory neurons. Recent studies show that attractor dynamics can work even in sparsely active E-I networks. Such networks may show attractor dynamics with or without oscillations. It remains to be seen if the oscillations can facilitate the attractor dynamics.

How can the attractor dynamics generate flexible and dynamic neural responses? The answer may lie in efficient networks with short-term dynamics. As discussed above, energetically it is efficient for a small group of neurons to fire a few spikes synchronously to drive the postsynaptic neuron. Estimates show that in a period of about 20 ms, a sufficient period for the postsynaptic neuron to integrate the inputs and fire a spike, only about 500 neurons may need to be coactive out of a population of 300,000 CA1 neurons. Similarly sparse representations of stimuli are also present within neocortical circuits, given the similarly low mean firing rates of principal cells in various neocortical layers. Experiments *in vitro* indicate that within this time window, pyramidal cells can linearly integrate the activity of hundreds of presynaptic inputs and then discharge. The minimum number of presynaptic neurons may be even an order of magnitude smaller if presynaptic neurons terminate on the same dendritic segment and discharge within a time window <20 ms. Further, in the hippocampus, inhibitory interneurons respond much more effectively than principal cells and hence, a single action potential of a presynaptic principal cell may be sufficient to discharge an interneuron. Release from potent inhibition would generate synchronous computation in a population of target excitatory neurons.

The efficacy of a handful of neurons in driving the downstream neuron will therefore depend on several parameters. For example, if the synapses from these neurons are located near each other on the downstream neuron's dendrite, synchronous activation of these synaptic inputs may cooperate to generate a dendritic spike. This would increase the effective strength of such groups of synapses, thereby altering the structure of the attractor and increasing the ability of a small number of input neurons to drive the downstream neuron. Similarly, the amount of synchrony, within a 20 ms window, between the activation of these synapses would strongly influence their effective strength in driving the downstream neuron. Recent studies show that although neural responses, as a function of input strength, are threshold-linear in an asynchronous condition, neural responses are sigmoidal in the synchronous condition: low synchrony results in no response, and above some threshold amount of synchrony the result is a maximal response. Thus, small changes in the input synchrony may activate different sets of neurons. This can be rapidly reorganized by recent history, which would influence the synaptic strength via short-term plasticity, resulting in convergence to different attractors. This dynamics could explain the rapid reorganization of hippocampal and somatosensory maps with past history or with small changes in stimuli. Further, synchronous inhibition in an E-I network would synchronously release excitatory neurons from inhibition during gamma oscillations, thereby allowing the neurons to change rapidly their response to inputs in a dynamic fashion. Finally, neuromodulators could alter the efficacy and timing of these synapses, which would result in dynamic reorganization of neurons responsiveness to stimuli based on internal variables.

Neuromodulators act broadly on neural circuits, and they are typically thought to act on slow timescales. The influence of neuromodulators can be focalized and accelerated by the following hypothesized extracellular mechanism: A region of the brain with higher activity could contain a larger amount of glutamate in the extracellular medium, and the clearing of the neuromodulators (e.g., through glial processes) may be altered by the level of glutamate. This may result in rapid changes in influence of neuromodulators on neural ensembles.

In this scenario of efficient and synchronous networks, activity would propagate rapidly across processing stages, and the relevant parameter would be the group of coactive neurons within a gamma cycle or the roughly 20 ms taken to activate the postsynaptic neurons. This group of temporarily synchronous neurons is referred to here as a cell assembly. Due to the mechanisms depicted above, the cell assembly can change rapidly with stimuli and internal variables. In other words, membership in a cell assembly is highly flexible. In a Hopfield network, for example, each gamma cycle may contain a cell assembly that primes the formation of another assembly in the next cycle of iteration toward convergence. In a recurrent network with short-term synaptic dynamics, this could lead to transitions between attractors and the generation of temporally sequential activity of ensembles of neurons.

Experiments show that the assembly of coactive hippocampal neurons, defined within a period of about 20 ms, can change rapidly. This is partly related to phase precession. As the rat walks through the environment, place cells fire a series of spikes at different phases of the theta cycle. The group of coactive cells within any 20 ms period depends not only on the phase of the theta rhythm and position of the animal but also on other variables (e.g., running speed, head direction). Thus, a multimodal, dynamic, and rapidly evolving representation emerges.

Two additional mechanisms by which cell assemblies can become dynamic are asynchronous background activity and top-down influence. These factors can raise the level of depolarization of the cell, thereby altering its responsiveness to short, efficient bursts of synchronous inputs. This is particularly effective when these inputs target the fast-spiking interneurons, which can then entrain a subset of pyramidal cells, and could explain the influence of top-down inputs on the rapid reorganization of neural responses.

In such scenarios, it is conceivable that the relevant parameter for describing the network dynamics is not the group of synchronously active cells, or cell assembly, but the group of synchronously active synapses or a synapse assembly through which the information flows. Neuromodulators and their receptors, located extra-synaptically, can directly modulate the activity pattern of the synapse assembly, which is not restricted to a single cell and which may or may not result in the modulation of the cell assembly. Theoretical studies are needed to determine how such a dynamic synaptic assembly can also be stable and noise tolerant. In addition, experimental studies are needed to determine the structure of synaptic assemblies.

These cellular and synaptic assemblies in the oscillating and balanced E-I networks are examples of dynamic equilibrium, in which a network is kept maximally responsive to changing patterns of inputs, while keeping energy expenditures low. Such self-organized systems are often characterized by a power-law spectrum of event amplitudes. Supporting evidence may be found in the neural systems in terms of the power-law-shaped spectra of the activity of ensembles, such as the LFP or EEG. However, these systems occasionally produce very large events, which would be catastrophic for neural systems. Perhaps the strong, fast, and reliable feedback within an E-I circuit can help prevent such runaway events while keeping the system more responsive, near a critical point, through the generation of neural oscillations.

How Does the Brain Dynamically Select Cell Assemblies to Make a Decision?

Decisions are likely to occur by coordinated activity patterns in neural ensembles across different regions. These cell assemblies across regions can coordinate or compete in a number of ways to generate a winner assembly that drives the decision. First, the long-range excitatory-excitatory connections between the E-I assemblies in different regions can synchronize their activities and increase their firing rates, thereby making a selected group more active than others. Second, the excitatory-inhibitory connections between two different E-I assemblies could raise the level of inhibition and suppress the activity of the inhibited population. Third, the long-range inhibitory-inhibitory connections could either decrease or increase the firing rates of an E-I circuit. It would seem that this inhibitory transmission between cell assemblies may result in reduced activity of the second assembly. However, it has been shown that in recurrent E-I networks, under some parameter regimes, increased inhibitory inputs result in increased overall activity due to suppression of inhibition via recurrent inhibitory synapses. Thus, inhibitory connection between two cell assemblies may serve a dual purpose of (a) synchronizing their gamma rhythmic activity and (b) increasing their overall firing rates, thereby allowing this group of cell assemblies to drive the downstream group of neurons toward decision.

In addition to these mean firing rate-based effects and mechanisms, precisely timed inhibitory inputs could synchronously release distant cell assemblies from inhibition, thereby making them coactive in brief windows of time. Here, oscillations could facilitate this rapid synchrony and competition by synchronously activating and inactivating large neural ensembles across multiple brain areas. Thus, the decision-making process may be a phase-dependent rather than a rate-dependent process.

Neuromodulators could play a key role in these processes by altering the E-I balance and rhythms, thereby generating state-dependent synchrony of cell assemblies that determine the winner ensemble. The above mechanisms address direct competition between assemblies. An advantage is energy efficiency. However, it is possible that competition between assemblies occurs at the level of their efficacy in driving a downstream structure, such as the prefrontal cortex, resulting in competition between synaptic assemblies, which in turn can bias information processing in the upstream network. Such a recurrent process of decision making across networks could be slow, but can be speeded up through the use of a phase code, where the top-down and bottom-up influences can arrive at different phases of oscillations, thereby determining a winner ensemble within an oscillation cycle.

Conclusions

Our discussion emphasizes a few key mechanisms of neural information processing. At the heart of this discussion is the ubiquitous neural circuit of recurrently connected groups of excitatory and inhibitory neurons. The E-I circuit can remain at a dynamic equilibrium, allowing it to respond rapidly to inputs. The E-I module can be easily replicated to generate larger circuits, perhaps during evolution, with each component using a similar language. Further, the dynamic equilibrium would allow a small number of inputs to alter the state of the network and make neurons respond. Thus, the network could have sparse activity, thereby making it energy efficient. In such sparsely active E-I networks, synchronous activity would be transmitted efficiently and rapidly.

Under many conditions, this E-I circuit oscillates. Interaction between these oscillations and inputs from external stimuli would generate a phase-coded representation of input that is rapid, efficient, and malleable; one that can facilitate learning via mechanisms of spike time-dependent synaptic plasticity. Such phase codes could bind multiple neural representations for brief periods in a flexible fashion and determine the computational interactions between different signals. The duration of each syllable in an E-I network phase code, called a cell assembly, would be about 20 ms, corresponding to a gamma cycle. Cell assemblies across multiple gamma cycles can either converge to a Hopfield-like attractor, in the presence of stationary stimuli, or generate history-dependent responses with dynamic stimuli; alternatively, in the absence of stimuli, it can spontaneously transition across a sequence of cell assemblies due to short-term dynamics. The oscillations of E-I circuit can be synchronized across different regions, allowing dynamic coordination of phase codes across brain regions. Other processes, such as lower frequency oscillations and neuromodulators, can influence coordination and competition between the E-I assemblies, generating a state-dependent winning ensemble or decision. While some tantalizing support is available for these mechanisms of the emergence of mind from neurons, much remains to be theoretically understood and experimentally tested.

10

Coordination

What It Is and Why We Need It

Christoph von der Malsburg

Abstract

Trying to apply our everyday concept of coordination to the brain raises a number of fundamental questions: What is the nature and meaning of local brain states that are to be brought together? On what grounds are they to be coactivated and connected? What is the nature of meaningful structural relationships, and how does the brain learn them? What is the role of focal attention? How does the brain assess its current level of coordination? How do brain states address goals? What is the nature of our environment's statistics, and how is it captured by the brain? What mechanisms endow brain dynamics with a tendency to fall into coordinated states? Some of these questions seem to be difficult to address within the current experimental paradigm.

Introduction

What is coordination? There are many domains in which coordination plays a central role: from the preparation of a meal or setting of the dinner table to the writing of a literary novel or musical composition. The aesthetic feelings conjured up by a work of art in our mind are often no more than a reflection of the level of coordination that takes place between our sensations. The establishment and running of a company or the organization of a conference requires structural elements to relate to each other in a meaningful way: they must fit together to form a whole if the result is to function well. The coordination inherent in institutions or manufactured structures reflects necessarily that which is present in their creators' brains. Thus, any insight that we can glean from coordinated artifacts is relevant for understanding coordination in the brain.

What Is Coordination in the Brain?

Coordination means putting things together that belong together. Each term in this sentence raises questions, which I will address in turn. Throughout this essay, I will focus more on posing questions than on providing answers. Because things in our brain[1] are assembled on a slow timescale of learning as well as on a fast timescale of thinking, it is important for us to heed both.

What Are the Things That Must Be Put Together?

The brain has a great variety of modalities or subsystems, corresponding to internal and external senses, to motor control, to memory, to emotions, to behavioral control, and more. Minsky (1988) spoke of the "Society of Mind." Superficial modalities are formed by wiring with the sensory, motor, or humoral periphery and are thus defined (onto-)genetically. More centrally, there is considerable plasticity, at least in the cortical system, as shown by many cases of neurological recovery (Weiller and Chollet 1994). Different modalities focus on different reflections of phenomena and contain patterns that are similar, so that they can be mapped and stacked onto each other. This pervasive phenomenon of cortical localization requires a natural measure of similarity between neural patterns, the need for which we will encounter repeatedly later.

Each subsystem of the brain is able to create a large variety of alternate activity patterns, of which at any given time only one, or a few, can be clearly and unambiguously expressed. For much of the time, neural activity in a subsystem expresses an ambiguous superposition of different states, which is reduced in stages under the influence of signal exchange. This reduction must be coordinated, so that patterns which belong together are coactivated. Such reduction of uncertainty is, for example, modeled in probabilistic formulations (Pearl 1988; Bishop 2006).

What Does It Mean to Belong Together?

Neural patterns belong together because they are generated by stimuli that are statistically linked in the environment (often, a common cause is responsible), or because they successfully interact to attain goals or to exercise and develop capabilities playfully. "Belonging together" is thus both defined in a passive, recording mode and in an active, creative mode. The creative mode is responsible for the brain's ability to handle new situations and has been grossly neglected in the neuroscientific literature.

---

[1] I am adopting the attitude of Spinoza and am viewing brain and mind as two sides of the same coin. When using electrode or microscope, we observe the brain. Psychophysics or introspection, by contrast, lets us view the mind.

How Does the System Learn What Belongs Together?

The initial motive for patterns to be related is similarity. Different modalities, by contrast, speak different languages and contain neural patterns that are not similar. How, then, can the brain distinguish which patterns from different modalities belong together? The only means of establishing *de novo* pattern association is simultaneity; that is, significant correlation in time. Unfortunately, however, it is not useful to associate all neural patterns in the brain with each other merely on the basis of simultaneity, as is implied in associative memory models (e.g., Hopfield 1982). To an overwhelming extent, pattern simultaneity in the real world is accidental and not of lasting value. If synaptic plasticity had the form of indiscriminate stickiness, the brain would soon be cluttered with myriad connections. It is thus important to single out associations that are significant.

The natural definition of pattern significance—recurrence—suggests a strategy that has actually been adopted by much of the artificial neural network literature (Haykin 1994); namely, sifting through the input for those patterns that appear repeatedly with statistical significance. Unfortunately, this strategy fails for input fields of any realistic size, because the number of patterns which must be tracked is simply too large, and literal repetition of a given pattern is too unlikely. The system, therefore, needs to possess a similarity measure and powerful prejudices concerning the nature of significant patterns. Yet, what is the nature of these prejudices?

Once different modalities have accumulated a sufficient mass of pattern associations between them, they can use general laws of composition to generate creatively novel, modality-spanning composite patterns. An important issue is the nature of these laws of composition. A typical (or perhaps *the* typical) law may be that overlapping patterns in one modality must be associated with overlapping patterns in another.

The Detection of Significant Patterns by Focal Attention

Why is the information content of attention limited? Does this represent an imperfection of the brain, or is this even functionally significant? Focal attention powerfully (though not exclusively) restricts learning to a small subset of active neurons at any one time (for further discussion, see Jiménez 2003). Key questions focus on how this restriction is expressed and what effects this has on learning. One proposal is based on gamma rhythms (Fell et al. 2003) and on the ensuing concentration of neural spikes into narrow temporal windows, as a boost to synaptic plasticity. This reduces the input and the memory domains to small sectors (as modeled in Jacobs et al. 1991). The restriction addresses a fundamental problem of present-day models of learning (i.e., the scaling-up to realistically large input and storage domains) by restricting system modification to the narrow focus of attention. Accordingly, the reason why informational content of attention is limited results not from an imperfect mechanism but from the necessity to preclude confusion.

What is the informational basis on which focus of attention is formed? Attention is very much at the core of the problem of coordination: it has to bring together sets of patterns that belong together. If it only unites patterns that have already been associated in the past—in passive mode, so to speak—it does not aid the problem of learning. For that, it has to find new associations in a creative manner.

External events, signaled by temporally isolated sensory signals (which are emphasized by the filter properties of our senses), can bring together specific patterns that light up simultaneously in different modalities. Patterns can also be brought together on the basis of abstract properties with which they are tagged. For example, if the senses can be focused on the same point in space, for which the colliculus superior seems to be well equipped (at least in the mouse; see Dräger and Hubel 1975), patterns aroused from that point can be associated. The general Gestalt laws (Koffka 1935; Crick 1994) can define significant patterns as figures set apart from a background. An object moving against a static background can, for example, be made to stand out as the focus of attention on the basis of common motion (Spelke 1998). Thus, the Gestalt laws formulate abstract properties (common motion, color, stereo depth, spatial grouping, good form, and edge continuity), which help to tag novel patterns as significant.

A statistical definition of pattern significance is not sufficient; a biological definition is also required. Important classes of patterns or events must be genetically defined to inform an individual on what is required for success in life. Such a definition must be laid down in some abstract fashion so that concrete occurrences can be recognized and selected. Ethologists have described many such schemata, such as the facial schema with which the human infant is born, the definition of mother goose for the gosling, or a red dot on the beak of the seagull to indicate to the chick the source of food (Toates 1980). Upon recognition of the releaser for an innate cue, attention is focused on the recognized stimulus, which is then separated from the ground; an appropriate action is induced (e.g., an orienting reflex or grasping); and an appropriate sector of memory is selected for modification by the stimulus. This sequence of events may be referred to as schema-based learning.

Important questions remain, however, regarding the technical implementation of these processes.

How Is Coordination Evaluated?

People have a keen sense of the level of coordination that goes on in their brains. Sometimes we feel distracted or confused; we cannot make up our mind or feel that something is awry, or not quite right. Other times, we experience the sensation of being sharply focused, highly concentrated, or fully conscious of a situation. Then there are those precious moments, when we suddenly feel that we have it; everything has fallen into place and we shout Eureka!

To a large extent, aesthetic pleasure is due to the level of coordination generated in our brain by the object of our attention. It is not conscious insight into the structure of the work of art that we experience, but rather some direct feeling of the level of coordination in our brain. This *Aha!* effect, this falling into a state of organization, is what Gestaltists refer to as reorganization of the perceptual field and insight (Köhler 1925).

Which structures in our brain are responsible for the evaluation of this measure of coordination, and what is the nature of this measure? Obviously, it cannot be attributed to some superior intellectual entity whose insight into the subject matter serves as judge. Rather, it must be some signal that can be "mechanically" generated and globally evaluated. The essence of it may be the level of nontrivial agreement of independent signals at convergence points, evaluated over the whole brain.

How Is Purpose Defined, Enforced, and Achieved?

In an active mode, the brain must coordinate patterns to achieve its purposes. Here, the central issue is that purpose (e.g., I am hungry and am looking for food) as well as the generation of neural patterns that serve to achieve that purpose (opening the fridge, or calling the pizza delivery service) are defined at very different levels of detail, and generally in different parts of the brain. A newborn possesses, presumably in the midbrain, the schematic definition of a set of fundamental goals. These form a hierarchy, the honing of which keeps us busy over much of the course of our life. Goals are activated either spontaneously or in response to some stimulus, like "danger" or "thirst." Goals tend to be mutually exclusive and come equipped with powerful mechanisms of enforcement. Complex tasks require the attainment of goals and subgoals in hierarchical fashion, and there are profound questions concerning the nature, establishment, and implementation of goals in our brain.

Behavioral patterns usually have a number of functional components, each of which has a range of possible role fillers. In the looking-for-food scenario, relevant roles include possible foodstuffs, sources of food (e.g., the refrigerator, delivery service), and potential modes of acquisition. In a specific situation with a concrete goal, the system must select the appropriate role fillers which will interact functionally to attain the goal. How is this type of coordination achieved through the interaction between a goal schema that contains a set of role descriptions, the possible role fillers that have the ability to combine appropriately, and the sensory patterns that describe the situation? In addition, we need to know the way in which behavioral schemata and goal descriptions are implemented neurally, the mechanisms by which these schemata are triggered and prioritized, the reward mechanisms by which the achievement of goals is evaluated, the mechanisms by which the activity of the brain is biased in the direction of goal fulfillment, and the mechanisms by which, over the course of our life, goal schemata are elaborated in richer and richer ways in terms of detailed sensory and motor patterns.

How Are the Environment's Statistics to Be Captured?

The brain receives signals from the environment over many millions of fibers and, in turn, influences the environment through multiple output fibers. All our brain can ever know and learn is contained in the statistics of these activity patterns. From all possible combinations of individual neural input or output signals—a space of vast volume—only a minute subvolume is ever realized in terms of actual signals. Exhaustive recording of global activity patterns is not possible, nor would it make sense as no sensorimotor activity pattern ever has a chance to recur. Only by extracting significant subpatterns, by ordering them in groups of similar patterns, and by developing schemata for their arrangement is it possible to capture the environment's statistics and cope with the ever-changing, novel situations that humans encounter daily. This requires a prejudice to define significant patterns for extraction; it demands a general similarity measure by which these patterns are to be grouped; and it requires a preestablished format for the representation of pattern arrangements.

One is caught, so to speak, between a rock and a hard place. If the prejudices are too weak, the system is overwhelmed by variance that cannot be captured in a realistic finite system. If the prejudices are too narrow, the reality of the environment may be missed (the bias-variance dilemma; Geman et al. 1992). Another indication that the system's prejudices must be tuned to the environment are the no-free-lunch theorems (Wolpert and Macready 1997), according to which any learning or optimization mechanism can be totally vitiated by an environment that does not fit its a priori assumptions. In summary, the brain needs powerful a priori assumptions, and these must fit the actual environment! What, then, are these a priori assumptions?

What Is the Nature of Our Environment's Statistics?

Of all the questions, this is probably the most crucial, since, as argued, the mode of operation of the brain must be tuned to the environment. In fact, the brain must coordinate with the environment.

Some important aspects of sensory pattern statistics result from the media through which they are transmitted. The visual medium, for example, is the optical radiation field captured by the eye, and the patterns that appear on our retinae are shaped by the laws of reflection and propagation of light, geometry, and motion. These transformations need to be inverted for the brain to decipher the structure of the patterns that are there. The laws of transformation are to a large extent independent of the environmental patterns themselves. The first sensory stages of the brain can reduce the complexity of the input patterns tremendously by inverting these transformations, thus reducing large sets of patterns to invariance classes (Wiskott 2006).

What is the regularity of the world beyond that? What are the repeating patterns? Or, rather, in what general format do they appear? According to Kant's analysis, we come equipped with "categories" (i.e., a priori structures that permit us to absorb information). Among these he counted space, time, and causality. We take for granted that repeating spatial and temporal patterns and causal sequences of events play an important role in our environment. It is a wide-open question, however, as to how whole scenes are to be decomposed to find repeating patterns, and how the general rules of composition by which our environment generates its configurations in ever-new ways are formulated. Coordination, to relate back to our theme, is the ability to create internal scenes that capture the reality of the environment. The brain's task, then, is to extract environmental patterns and their relations, together with a measure of likelihood for their relevance, so as to acquire the ability to complement partial information in a given scene with additional details familiar from the past, to generate a more complete description of the scene. Our challenge is to second-guess the general form—the architecture—on the basis of which this is possible.

What Is the Nature of Structural Relationships?

The patterns that we experience never repeat precisely. When recording a novel pattern, it is thus important to be able to define a spectrum of other patterns that are similar to it. This implies a similarity measure, or some definition of the likelihood that a sensory pattern is to be identified with a stored pattern. If properly constructed, the stored structure and the similarity measure can decide to a high statistical significance whether a perceived pattern is an accidental arrangement of elements, or whether it is the same pattern repeating itself.

What is this similarity measure? The simplest idea of pattern recognition is template matching, where a rigid pattern, the template, is moved over an image to find an identical fit. The "motion" takes care of the invariance aspect if it includes all possible transformations (e.g., translation, scaling, and rotation). Template matching has long since fallen out of favor because identical fits are never found in real images. A first step toward solving the problem is to dismantle the "template" and endow the resulting pieces with flexible relationships so that distortion can be addressed. Thereafter, pieces of the model must be replaced with statistical models of possible variants. A version of this is the leading mechanism of face recognition, as described, for example, by Wolfrum et al. (2008). Finally, the pieces themselves can then be replaced by composite models to create a hierarchical structure.

What has been described here for vision applies analogously to other modalities. Motor patterns (including speech) form patterns within patterns, each being a role filler, each permitting a range of variants, the whole put together flexibly to permit continuous time warping and, of course, further nesting. Hidden Markov models (Rabiner and Juang 1986) capture essential aspects of this. I contend that this kind of architecture applies to all modalities of the brain individually, and the brain as a whole.

Out of these considerations arises a picture according to which mental objects are hierarchical graph structures, with concrete patterns arranged in spatial and temporal relationships to each other. Graphs are, in general, embedded in or linked to more abstract graphs (which, due to their abstractness, are called schemata), whose nodes refer to classes of exchangeable subpatterns, and whose links describe permitted relations. Recognition is the process by which abstract graphs are mapped to concrete patterns homeomorphically (i.e., with equivalent parts mapping to each other under preservation of relations). Coordination is the process by which concrete brain states are generated under the guidance of abstract descriptions (including formulation of goals) and of sensory input. Usually, several abstract schemata conspire to create a detailed description.

Important questions include: How is the repertoire of neural behavior tuned to the construction of such hierarchical descriptions? How are hierarchical descriptions developed in the brain on the basis of experience? How can this architecture be described in concrete mathematical terms?

How Are Neural Patterns Put Together?

The brain is endowed with an architecture that tends to fall into globally ordered patterns, structured accordingly to the world in which we live. This ability can be likened to the process of crystallization, in which constituent atoms or molecules create global order out of local interactions, by exerting their preferences according to the shape of the local environment. For crystallization to be initiated, a seed (or minimal structure) is required such that further molecules quickly find a niche into which to fall.

In the case of the brain, the constituent elements are neurons, and knowing how their behavioral repertoire is structured, so as to favor global order, is crucial. Some aspects of this are already emerging. Outgrowing processes are guided by chemical or electrical signals so as to favor ordered connectivity patterns, as exemplified by the ontogenesis of retinotopy (Goodhill 2007), whereas intrinsic plasticity (Butko and Triesch 2007) regulates the duty cycle of cellular activity.

If the behavioral repertoire of neurons (or of a collection of neural types) were known to any degree of precision, it should be possible to simulate on the computer the growth of ordered connectivity and activity states. Thus far, this venture has had some level of success, especially in modeling the ontogenesis of ordered connectivity structures. However, modeling the generation of neural states that can be interpreted as mental objects is still out of reach. Thus, questions remain regarding the repertoire of neural behaviors, and, possibly equally important, the equivalent of seed structures with which the process of coordination is initiated.

Concluding Remarks

To coordinate "stuff" in our mind, even while performing routine tasks, we regularly have to put things together that have previously never been assembled. These associations are creative acts that are not imposed on us by our environment. In recognizing an object, for example, we apply a schematic description of this type of object to a concrete image, thereby associating abstract features with concrete instantiations. When we manipulate the object or describe it verbally, this enables us to relate directly to the specific character of the feature instances. Certainly, our brain contains the neural pathways to connect what is to be connected; however, these pathways are embedded, or are drowned, in a virtual continuum of others which, at present, are irrelevant but are all required and useful when their time comes. Although I intended simply to pose questions in this essay, I could not do so without interjecting my conviction that the task of our brain, in any given situation, is not just to activate a subset of all neurons, but to select a tiny subset from the vast numbers of physical connections. Synapses outnumber neurons, and the task of selecting connections is larger than that of selecting neurons by a factor of about ten thousand. If we ignore this task, we are, in my view, ignoring 99.99% of the information in our brain's state.

If indeed the brain, in its rapid state changes, is mainly concerned with selecting structured connectivity patterns, and if we need to study and understand synaptic dynamics in addition to neural dynamics to bridge the chasm between mind and brain, then we have a problem. Experimental technology is highly developed to study neural dynamics and, to a lesser extent, static or slowly changing connectivity. It can even record short-term modification of synaptic effects for individual connections, but the imaging of whole, rapidly changing, connectivity patterns is presently beyond our imagination. When Ludwig Boltzmann first established statistical mechanics, he was ridiculed by his colleagues Ernst Mach, Wilhelm Ostwald, and others for his atomistic ideas, and it took three decades until experiments made the reality of atoms and molecules concrete enough to convince the community. Must the neurosciences also wait for decades for the necessary revolution? Dedicating years of effort to experimental exploration is too risky if it is not supported by community convictions, but the community will not be convinced without focused efforts. Let us hope that this vicious circle can be broken with the help of concrete computer models of cognitive functions whose demonstrable success rests on outlandish physiological assumptions.

Acknowledgment

I acknowledge support received from the BMBF project Bernstein, FKZ 01GQ0840.

11

Neocortical Rhythms

An Overview

Wolf Singer

Abstract

Information processing systems need to be able to identify and encode relations. Relations can be defined in space and time. Nervous systems exploit both dimensions for the handling of relations. Their anatomical layout is characterized by selective convergence of connections on target cells, allowing the establishment of relations among signals of different origin. In addition, relations can be expressed more dynamically by adjusting the temporal rather than the spatial contiguity of signals. It is proposed that this is achieved by rhythmic modulation of neuronal activity and context as well as task-dependent modulation of oscillation frequencies and phases.

The Encoding of Relations

Analyses of neuronal responses in sensory systems reveal a characteristic sequence of processing steps. At subcortical levels, neurons encode local properties of perceptual objects and barely any relations among these elementary features. As one proceeds along the hierarchically arranged processing areas of the neocortex, responses become increasingly selective for complex constellations of elementary features. This sensitivity is the result of interactions mediated by intracortical tangential connections and iterative recombination of feedforward connections from lower- to higher-order neurons. Thus, responses of higher-order neurons signal not only the presence of certain sets of component features but also prespecified relations among these features.

This binding of features by recombination of feedforward connections (labeled-line coding) needs to be complemented by mechanisms that permit flexible encoding of relations to resolve the ambiguities inherent in the responses to natural environments. Natural scenes, for example, usually contain a large number of different objects, the contours of which may be overlapping, partially occluded, or superimposed on those of the background. A neuron may thus be stimulated by contours belonging to different objects. To avoid the encoding of false relations (conjunctions), these ambiguities need to be resolved before signals can be subject to grouping in the respective convergent pathways. This sorting of appropriately groupable responses must occur in a context-dependent way at each level of processing. At low levels, it is indispensable for scene segmentation; at high levels, it is required for the disambiguation of simultaneously configured distributed representations (assemblies) (Wang 2005).

Experimental data and theoretical considerations suggest that perceptual objects are represented not only by individual neurons that respond selectively to single objects but also by distributed assemblies of cells (Singer 1999; Tsunoda et al. 2001). First, object-specific neurons are rare and seem to exist only for highly overlearned objects or for objects of particular behavioral relevance (Quiroga et al. 2005; Logothetis et al. 1994). Second, novel objects cannot be represented by preestablished neurons because the required feedforward architectures would have to be specified a priori to support formation of the appropriate conjunctions. Third, objects that are simultaneously encoded in different sensory modalities elicit responses from several different sensory systems, and these need to be interrelated to arrive at a comprehensive polymodal description of a particular object. Here, the space of possible conjunctions is so large that it cannot be covered by applying the "labeled line" strategy alone. These considerations suggest that objects which cannot be represented by individual conjunction-sensitive neurons are encoded by assemblies of distributed neurons, each of which represents only a particular component of the object. In assembly coding, however, a relation-defining mechanism is required that tags responses which are evoked by the components of the same object as related. In essence, neurons have to convey two orthogonal messages in parallel: (a) they must signal whether the feature or conjunction of features for which they stand as a "labeled line" symbol is present, and (b) they must indicate, from instance to instance, with which subset of the myriads of simultaneously active neurons they are actually forming a coherent assembly. There is consensus that the first message is conveyed by a rate code: the higher the discharge frequency, the higher the probability that the neuron's preferred feature is present. The second message, so the proposal defended in the chapter, is conveyed by precise temporal relations among the discharges of neurons and, thus, by a relational temporal code.

Synchrony as a Tag of Relatedness

Although some of the Gestalt grouping occurs preattentively, it is commonly held that attentional mechanisms also play an important role in relation-defining grouping operations, both at low levels, where scene segmentation is accomplished, and at higher levels, where objects are thought to be represented (Treisman 1999). However, the mechanisms underlying these functions are still poorly understood. One proposal states that responses are selected for further joint processing by joint increases in saliency, i.e., in discharge rate (Cook and Maunsell 2002). Neurons recruited into the same assembly would be distinguished from all others by their higher discharge frequency. This interpretation has been challenged by the argument that response amplitude may be an ambiguous signature of relatedness for two reasons: (a) because it depends on stimulus variables, such as intensity, that are inappropriate grouping cues, and (b) because it requires long readout times, which makes it difficult to segregate assemblies from one another that are simultaneously configured within the same neuronal network (Singer 1999). Therefore, it has been proposed that neurons should be able to signal independently of their actual discharge rate to which of the other simultaneously active neurons their responses are related. This latter code should ensure that responses tagged as related are processed jointly at subsequent stages, i.e., are routed together into the appropriate channels and/or are unambiguously recognizable as originating from cells of the same assembly.

Following the discovery that neurons in the primary visual cortex can synchronize their spike discharges in a context-dependent way and with a precision in the millisecond range (Gray and Singer 1989), it was proposed that the synchronization of responses could serve as the required tag for relatedness (Gray et al. 1989) and thus, for binding them together. Precise temporal synchronization of spike discharges is as effective as joint rate increases for selectively raising the saliency of neuronal responses (Biederlack et al. 2006). The reason is that synchronized input to target neurons has a stronger impact than temporally uncoordinated input. For example, simultaneously arriving excitatory postsynaptic potentials (EPSPs) summate more effectively than temporally dispersed EPSPs; active dendritic conductances amplify fast-rising depolarizations of large amplitude (Ariav et al. 2003); the frequency adaptation of synaptic release and postsynaptic receptors attenuates the effects of rate increases (Markram and Tsodyks 1996); and the dependence of firing threshold on the rising slope of depolarizations favors responses to synchronized inputs (Azouz and Gray 2003). Because synchronization capitalizes on spatial rather than temporal summation, it modulates the efficiency of individual synaptic events and therefore operates with a temporal resolution in the millisecond range. Thus, relations can be defined within narrow time windows (<10 ms), and different relations can be encoded in rapid succession.

The notion that precise temporal synchrony serves as a tag of relatedness agrees well with the temporal sensitivity of correlation-based learning mechanisms. One mechanism, known as spike timing-dependent plasticity (STDP), classifies discharges as related (unrelated) and causes synapses to strengthen (or weaken) as a function of the precise temporal relations among pre- and postsynaptic activity patterns. This mechanism operates with a temporal resolution in the range of milliseconds; it also operates with millisecond precision (Markram et al. 1997; Zhang et al. 1998; Wespatat et al. 2004). Thus, there is a perfect match between the signatures of relatedness used in signal processing and Hebbian learning. This cannot be otherwise because both processes must rely on the same relation-defining code to avoid learning false conjunctions.

Oscillations as a Timing Mechanism

Precise synchronization of discharges is often associated with an oscillatory patterning of the neuronal responses (Gray and Singer 1989). Because individual cells tend to skip cycles, these oscillations are rarely detectable in the spike trains of single cells, but they are readily seen in data representing the responses of large populations of neurons, as in multiunit recordings or recordings of local field potentials (LFPs). The periodic patterning of these responses is the result of oscillations generated within the various pools of inhibitory interneurons. These interneurons are coupled through chemical and electrical synapses and are capable of sustaining oscillatory activity patterns (Kopell et al. 2000; Whittington et al. 2001; Cardin et al. 2009; Sohal et al. 2009). Oscillatory inhibitory inputs to pyramidal cells veto the latter's discharges during the inhibitory troughs and favor discharges at the depolarizing peaks, thus causing synchrony in firing. Surprisingly, these locally synchronized oscillatory responses can become phase-locked, with zero delay over large distances, despite considerable conduction delays in the reciprocal excitatory corticocortical connections that mediate long-range synchrony (Engel, König, Kreiter et al. 1991).

Several mechanisms have been proposed that are capable of establishing zero-phase lag synchronization despite conduction delays. One mechanism relies on the fact that interneurons discharge with spike doublets when the networks engage in beta and gamma oscillations (Kopell et al. 2000). Another mechanism exploits the special topology of coupling connections and the nonlinear properties of networks of coupled oscillators (Vicente et al. 2008). The precision with which spike timing can be adjusted increases with oscillation frequency (Volgushev et al. 1998) and, often, one observes a relation between oscillation frequency and the distance over which synchronization is maintained. Synchronization among remote groups of neurons, or among large assemblies of neurons, tends to occur at oscillations in the theta or beta frequency range, whereas the highly precise synchronization of local clusters of cells is carried out by gamma oscillations. Often, oscillations in different frequency bands coexist and exhibit complex phase relations (Roopun, Kramer et al. 2008). This concatenation of rhythms offers the attractive option of establishing graded correlations between neuronal assemblies of different size, thereby encoding nested relations. Such encoding is required for the representation of both composite perceptual objects and composite movement trajectories. This hypothesis, however, awaits further experimental testing.

The Duration of Synchronized Events

Early studies were based mostly on conventional cross-correlation analysis of cell discharges and/or LFPs. This method reliably detects synchronous firing if it is sustained over prolonged periods, but it fails if synchronous events are concentrated within narrow temporal windows. Novel measures have thus been developed to allow brief events of coincident firing to be assessed. One of these methods—the unitary event analysis—uses statistical methods to identify single, nonaccidental incidences of coincident firing (Pipa et al. 2007, 2008); the other evaluates consistent phase relations between the discharges of individual neurons and LFP oscillations (spike-field coherence; Fries et al. 2002). Applying these methods to data obtained from the visual cortex of monkeys exploring natural scenes has revealed that episodes of excess synchronized firing occur shortly after the onset of visual fixation and are restricted to epochs as short as a few tens of milliseconds (Maldonado et al. 2008). This finding agrees with the evidence that scene segmentation and object identification can be accomplished within less than 200 ms, leaving only 10–20 ms per processing stage to accomplish the required grouping operations (Thorpe et al. 1996; VanRullen and Thorpe 2001). Because information encoded in variations of discharge rates is limited (since cells can generate only a few spikes within such short time windows), a substantial amount of information is likely encoded in the precise timing relations between individual discharges of distributed neurons.

Functions Attributed to Synchronization

Binding

Evidence from studies of the visual system suggests that response synchronization may be used throughout all processing stages, from the retina to the highest cortical areas. Its purpose is to establish relations among distributed responses (i.e., to bias grouping of responses for subsequent joint processing) and to tag responses of assembly members as related (Kreiter and Singer 1996; Neuenschwander and Singer 1996; Castelo-Branco et al. 1998; Castelo-Branco et al. 2000). In all cases, synchronization probability reflects some of the Gestalt criteria that are used for scene segmentation and perceptual grouping. In the retina, ganglion cell responses synchronize with millisecond precision if evoked by continuous contours (Neuenschwander and Singer 1996). This synchronization is associated with high frequency oscillations (up to 90 Hz) and is based on horizontal interactions within the network of coupled amacrine cells. In the visual cortex, synchrony is often associated, especially when it is observed over larger distances, with an oscillatory pattern of spike discharges in the beta and gamma frequency range (30–60 Hz).

Synchronization correlates well with elementary Gestalt rules such as continuity, colinearity, and common fate (Engel, König, and Singer 1991; Engel et al. 2001; Castelo-Branco et al. 2000; Samonds et al. 2006). The substrate for this context-dependent synchronization is made up of tangential intracortical connections that preferentially link columns encoding features, which, according to common Gestalt criteria, tend to be grouped (Löwel and Singer 1992). In the inferior temporal cortex of the primate brain (the likely site for the generation of representations of visual objects), synchronization probability appears to reflect the formation of object-representing assemblies (Tsunoda et al. 2001). Neurons responding to the components of faces (e.g., eyes, nose, mouth) synchronized their responses when the arrangement of these components was such that the animals signaled having recognized a face; however, they did not synchronize when the components were scrambled or presented in a way the animal deemed incompatible with the appearance of a normal face. Interestingly, the distinction between face and nonface could not be derived from changes in the discharge rate. These findings are compatible with the interpretation that discharge rate signals the presence of particular features, whereas the correlations among the discharges indicate the relatedness of these features.

Attention and Stimulus Selection

Grouping operations based on elementary Gestalt rules and the binding of the stereotyped feature constellations of highly familiar objects can occur preattentively and are observable in anesthetized preparations. This automatic, attention-independent grouping is likely based on binding by convergence in fixed feedforward architectures and on the synchronizing effects of the intraareal tangential fiber systems (discussed above).

In addition to this evidence for attention-independent grouping by synchrony, more recent results clearly indicate that synchronization is highly susceptible to top-down, attention-dependent influences. They also show that it plays an important role in attention and expectancy-dependent response selection (Fries, Reynolds et al. 2001; Fries et al. 2002). An in-depth discussion of the mechanisms involved is provided by Whittington et al. (this volume) and Börgers and Kopell (2008).

Various measures have been used to assess the influence of selective attention on neuronal synchrony: correlations among spike discharges, spike-field coherence, correlations in phase locking between oscillatory field potentials, and, finally, the amplitude and phase locking of oscillatory responses as seen in MEG and EEG recordings. Because the amplitude of these latter signals depends not only on the number of active neurons but also on the degree of synchronicity of the captured activity, both phase locking and the power of oscillations can be taken as measures of synchrony. At all levels of analysis, evidence indicates that focusing attention on a particular stimulus or modality increases the synchrony of responses in the gamma and beta frequency range in the neuronal networks that are devoted to the processing of the attended stimulus (Roelfsema et al. 1997; Schoffelen et al. 2005). Interestingly, in these and many other cases, synchronization of oscillatory activity is not necessarily associated with major changes in the discharge activity of neurons. This observation supports the notion that synchronization and rate of discharges can be adjusted independently, and that precise synchronization can be used to raise the saliency of responses independently of discharge rate (Fries, Neuenschwander et al. 2001; Fries et al. 2007). Recent results from multisite recordings in cats and monkeys take this proposal one step further and suggest that the (often anticipatory) induction of coherent oscillations across distributed cortical areas and executive structures facilitates selective routing of activity and rapid handshaking among the involved processing stages (Womelsdorf et al. 2007). However, at present, it is not known which mechanisms coordinate these preparatory phase adjustments of oscillatory activity.

Association with Consciousness

Probability increases that signals become part of consciousness when they are attended to, either because they are salient and attract attention, or because they are selected by focused attention. Moreover, contents appearing in consciousness are usually interrelated (unity of consciousness). Because synchronization in the gamma range enhances saliency (Biederlack et al. 2006), supports selection by attention (Fries, Reynolds et al. 2001), and establishes relations (Gray et al. 1989), it is a prime candidate for being a neural correlate of consciousness. Evidence from studies on binocular rivalry in cats (Fries et al. 1997; Fries et al. 2002) and from masking experiments in human subjects (Melloni et al. 2007) indicates that there is indeed a close correlation between gamma synchronization and conscious processing. In cat primary visual cortex, responses to the respective perceived stimulus differed from those to the suppressed stimulus because the former were more synchronized, not because they were more vigorous. In human subjects, conscious processing of stimuli has been associated with precise phase locking of gamma oscillations across widely distributed cortical areas, whereas unconsciously processed stimuli evoked only local gamma oscillations (Melloni et al. 2007).

Abnormal Synchrony and Mental Disorders

Several clinical conditions (e.g., schizophrenia, autism, and Alzheimer's disease) are associated with cognitive impairments that suggest disturbed coordination of distributed brain processes. This concept has received support from recent EEG and MEG studies (Uhlhaas and Singer 2006; Vierling-Claassen et al. 2008), as well as from *in vitro* pharmacology (Roopun, Cunningham et al. 2008). When challenged with cognitive tasks that require feature binding (Uhlhaas, Linden et al. 2006; Uhlhaas and Singer 2010) or storage of visual information in working memory (Haenschel et al. 2009), schizophrenic patients exhibited several deficits: reduced synchronization of early evoked responses, reduced power of evoked and induced gamma oscillations, and a dramatic loss of the ability to synchronize gamma oscillations across distant cortical areas. These abnormalities were also seen (albeit in a less pronounced manner) in unmedicated, first admission patients. Impaired synchronization of oscillatory activity in the beta and gamma frequency range was also found in patients with autism (Wilson et al. 2007) and Alzheimer's disease (Koenig et al. 2005; Stam et al. 2007).

These findings suggest that some of the dissociative symptoms characteristic of these disorders could result from abnormalities in the precise temporal coordination and binding of distributed cortical processes. This insight offers researchers the option to use synchronization of oscillatory activity as an endophenotype for further investigations into the pathophysiology of these severe and hitherto incurable brain diseases.

The Stance of a Sceptic

No contemporary neurobiologist would deny the existence and necessity of "dynamic coordination" of distributed processes in the brain. Such coordination requires fast, temporary, and mostly reversible modification of neuronal interactions within the fixed or only slowly changing anatomical architecture. This implies that effective connectivity must be modifiable on the fly in a context-, attention-, or task-dependent way. This can be achieved by a host of well-established mechanisms, raising the question as to why oscillations, synchrony, and temporal codes should matter at all. The gain of connections is effectively modified by changes of firing rate, since this increases or decreases the saliency of responses. This suffices to select signals for further processing, especially if saliency maps enhance contrast between selected and non-selected responses. Experiments on attention-dependent rate enhancements suggest such a mechanism. To define relations and support selective grouping, it is sufficient to increase jointly the rate of the responses that are to be associated with each other. Since coding is sparse, activity is low, and since there is topological (spatial) organization, the risk of confusion and grouping of unrelated but simultaneously enhanced responses is small. At least it is not greater than misinterpreting spurious correlations among simultaneously active neurons as meaningful. The flow and routing of activity can further be controlled very effectively by short-lasting and reversible changes in synaptic gain, by active dendritic conductances that introduce nonlinearities in the summation of synaptic inputs, by shunting inhibition which can switch off entire dendritic segments, by modulatory inputs that alter time and length constants, and by top-down influences which, in principle, can gate all these processes. The problem of feeding learning mechanisms with activity that is sufficiently structured in time can probably be solved by statistics, assuming simply that the conditions required for gain increases occur more often, if coupled neurons are more active. There have been negative findings. Some labs have failed to find oscillatory activity and response synchronization in brain areas where others had observed them, or found no relation between the occurrence of synchrony and perceptual functions that should involve dynamic grouping of responses. Some of the conflicts between controversial findings could be resolved. Thus, it is well established that gamma oscillations and the associated synchronization of spikes are extremely state and task dependent. In light anesthesia they occur only during states of activated EEG (Herculano-Houzel et al. 1999), whereas in the awake brain they are strongly dependent on attentional mechanisms (Fries, Reynolds et al. 2001). Moreover, cell discharges, even if synchronized to an oscillatory process, may fail to exhibit an oscillatory firing pattern because of irregular cycle skipping. Finally, some of the negative findings may have to do with the fact that different cortical areas accomplish different functions. Just as one would obtain negative evidence for a rate code if one searched for face-specific responses in area MT rather than IT, one would obtain negative evidence for synchrony as tag of relatedness if one searched for higher-order binding functions in V1. Still, there are negative findings that require a continuous and critical evaluation of results supporting the temporal coding hypothesis.

One could argue that we have come a long way by assessing the rate variations of individual neurons without coming across explananda that would require us to search for something additional. Why then should we look for fine-grained temporal relations between the discharges of distributed neurons, since this necessitates technically much more challenging multisite recordings? The fact that oscillations, synchrony, and fine-grained temporal relations are observed and interesting functions can be associated with such phenomena is not sufficient to assign a function to them. What is the argument behind the view that conventional mechanisms do not suffice to account for what we observe and need to understand? What evidence exists to show that all these temporal dynamics are not just an epiphenomenon of the conventional, purely rate-based processes?

Obviously, to address these questions in an ultimate fashion, causal rather than correlative evidence is required. Such evidence is equally difficult to obtain for conventional rate codes and oscillation-based temporal codes, because both are constitutive attributes of neuronal processing. Interfering with rates most often entrains changes in oscillations and synchrony. Moreover, manipulating rates interferes in an often trival way with the cell's sole signaling mechanism. Conversely, blocking oscillations and synchrony often leads to changes in discharge rate. However, studies are available that can claim having obtained causal evidence for either rate or temporal codes. Assuming that high frequency microstimulation acts by increasing rates rather than by rendering discharges more synchronous, the many studies that show relations between stimulation frequency and the vigor of perception or motor responses can be taken as support of rate codes (Salzman et al. 1992). Support for a coding role of synchrony comes from studies in which synchronization has been reduced or abolished without interfering with discharge rate. In frogs, an escape response was abolished when the synchronization of retinal ganglian cells was prevented (Ishikane et al. 2005), and in locusts the discrimination of odor mixtures was impaired when synchrony was abolished among projection cells in the olfactory lobes (Stopfer et al. 1997). Finally, optogenetic methods have recently been applied to enhance synchronization in the gamma frequency range, and this led to predicted and functionally relevant changes in the performance of neuronal networks (Cardin et al. 2009; Sohal et al. 2009).

In conclusion, it appears that there is ample and equally convincing correlative evidence that the brain uses rates as well as temporal relations between spike trains in parallel to encode complementary information. However, if the brain does indeed exploit temporal relations among the firing sequences of large assemblies of neurons, new challenges must be overcome before we can decipher these temporal codes. Multisite recordings will be imperative. Moreover, analysis of relations must go beyond pairwise correlations, comprise all frequency bands, and consider all phase relations in case of oscillatory activity or relative delays between spike times of non-oscillatory activity. This leads to a dramatic expansion of search space, just as exploitation of temporal relations for information processing—if the brain applied that strategy—would dramatically increase coding space.

Conclusion

Temporal relations among distributed neuronal responses can be assessed only with multisite recordings. Because this approach has a relatively short history, we are just beginning to understand coding strategies based on the dynamic interactions among large numbers of neurons. It may turn out that precise synchronization is only one, albeit very important signature of the many potentially significant dynamical states. Precisely timed phase offsets between oscillating cell assemblies, concatenations of different rhythms, and sequences of patterns defined by specific temporal relations are likely to play an equally important role (Fries et al. 2007). To analyze these more complex patterns, and to examine whether they contain information that can be related to behavior, remains one of the great challenges for future systems neuroscience.

12

Stimulus-driven Coordination of Cortical Cell Assemblies and Propagation of Gestalt Belief in V1

Yves Frégnac, Pedro V. Carelli,
Marc Pananceau, and Cyril Monier

Abstract

This chapter reviews the concept of dynamic coordination in the mammalian primary sensory cortex during low-level (non-attention-related) perception. Among critical issues, it questions the necessity to keep the relational information (the "whole") separable from the information carried initially by each stimulus component (the "parts"). It also underlines the need for documenting in higher mammals the possible existence of subcortical or cortical supervisors, whose firing activation or suppression would condition the merging and segmentation of functional cortical assemblies. Emphasis is given to cases where coordination is generated by the sensory drive itself and amplified by built-in anisotropies in the network connectivity. The joint comparison of synaptic functional imaging (at the intracellular level) and real-time voltage-sensitive dye network imaging (at the functional map level) is used to demonstrate the role of intracortical depolarizing waves, broadcasting an elementary form of collective belief. The functional features of these slow waves support the hypothesis of a dynamic association field, propagating synaptic modulation in space and time through lateral (and possibly feedback) connectivity, which accounts for the emergence of illusory motion percepts predicted by the Gestalt theory.

For a Taxonomy of Dynamic Coordination

Coordination of distributed elementary dynamic processes into a coherent "whole" is the organizational hallmark of brain cognitive activity. Dynamic coordination is required so that, by using time as an additional coding dimension, the "whole" can be distinguished from the "static" sum of the parts. Failure to do so results in superposition catastrophe (von der Malsburg 1986).

Flexibility in coordination ability is necessary so that the functional outcome of the ensemble remains adapted to the ever-changing goals which govern relationships between the "self" and the "outer" world.

Historically, dynamic reconfiguration of neural activities has been considered by many explorers of the brain—philosophers, psychologists, and physiologists—as a likely substrate for thoughts, dreams, and other mental processes, with particular relevance to perception and memory recall. Without much knowledge of the circuits involved, and inspired by the brilliant concept of cell assemblies introduced by Yves Delage (1919) and Donald Hebb (1949; reviewed in Frégnac 2002), theoreticians of the brain later imagined the waxing and waning of synchronized oscillations or synfire chains as the dynamic signature of coordination (Milner 1974; von der Malsburg 1981/1994; Abeles 1982). As beautifully worded by Delage (1919), the concept of "parasynchronization" implies that "every modification engraved in the neuron's vibratory mode as a result of its co-action with others leaves a trace that is more or less permanent in the vibratory mode resulting from its hereditary structure and from the effects of its previous co-actions. Thus its current vibratory mode reflects the entire history of its previous participations in diverse representations." It is therefore not surprising that numerous contemporaneous brain scientists, under the flagship of Wolf Singer and Christoph von der Malsburg, recognize the fugitive emergence of associative percepts in the gamma wave signature of V1 local field potentials (Fries 2009; Tallon-Baudry, this volume).

In its simplest form, dynamic coordination is defined by the multiple interrelations in space and time that can be drawn between elements of any given assembly. Its phenomenological expression is signaled by the reconfiguration of elementary dynamics and their potential phase relations as a function of an externally defined context or an internally generated goal. Phillips et al. (this volume) constrain the issue further by adding: "in general, coordinating interactions are those that produce coherent and relevant overall patterns of activity, while preserving the essential individual entities and functions of the activities coordinated." This additional constraint implies that the dynamic binding process itself should not interfere with local properties of the interlinked elements (which these authors refer to as "meaning" tokens). This hypothesis has its own virtue since numerous cognitive operations seem to preserve at the same time the representation of the "parts" (segmentation) and the "whole" (binding). In sensory perception, this applies to vision, where fusion and perception coexist together, and to a certain extent to audition, but probably less to olfaction. In this latter case, the perception of the "whole" overrides that of each component, especially when the subject is given heteromodality priors (e.g., the "color" of smell in oenology; Morrot et al. 2001). At a more abstract level, such condition is needed in the compositionality framework set by Fodor and Pylyshyn (1988), where elementary features have a fixed symbolic value (a letter in an alphabet). In the conceptual view defended by Bill Phillips, compositionality operates as a relational grammar which does not interfere with the semantic value of the elements. However, there is no strong reason for reducing dynamic coordination in neural systems to linguistic compositionality, although abstract models of logogenesis have been proposed where the brain composes language from the fast reversible binding of synfire chains (Doursat 1991; Bienenstock 1996).

The obvious, critical issue that should be discussed is that coordination or "binding" in the brain operates not between fixed entities called "neurons" but rather between local dynamic integrative processes, each being specific to the considered "neuron." In classical *in vitro* electrophysiological studies, neurons have an identification profile that is empirically defined at three levels: (a) morphological/structural identity, (b) excitability spiking pattern in response to intracellularly injected current (a kind of crude input/output curve) which characterizes the intrinsic conductance repertoire, and (c) genomic expression pattern revealed by multiplex PCR of the cytoplasmic content (Toledo-Rodriguez et al. 2004; Markram 2006; reviewed in Frégnac et al. 2006). In sensory cortical networks, the elementary neuronal integrative function is also characterized *in vivo* by a static receptive field organization (equivalent to a stimulus/response curve). The discharge field organization revealed by impulse-like sensory stimulus is largely dominated by the convergence patterns of labeled lines extrinsic to the studied structure (for V1, coming from the dorsal lateral geniculate nucleus, LGNd, in the thalamus). The strict application of Bill Phillips' criteria, as present in many of the early pioneering experiments by Wolf Singer's team (e.g., Gray et al. 1989; Kreiter and Singer 1996), would suggest that the processing realized by each neuron on elementary parts of the composite stimulus reflects fixed discharge field properties and remains unchanged during associative center-surround stimulation protocols. The underlying assumption is that the relational information should remain separable from the information carried initially by each stimulus component taken in isolation.

To debate this issue further, let us turn from the literature of mammalian cortex to that of invertebrate sensorimotor ganglion: the particular case of the stomatogastric ganglion of the lobster constitutes a striking example of assembly dynamic reconfiguration correlated with changes in behavior where, during the coordination, the electrical input/output properties of individual elements are not preserved. This paucineuronal net is an assembly of giant cells with invariant morphology, and their number is limited enough such that the total blockade of the afferent connectivity to any given cell can be obtained by photo-inactivating all the putative synaptic partners. Early experiments in the invertebrate revealed that isolated neurons are, in all cases, conditional oscillators, displaying a large variety of intrinsic membrane potential patterns such as bursting, plateau, postinhibitory rebound, and spike-frequency adaptation. Further work showed that the repertoire of intrinsic conductances dramatically changes in the presence of neuromodulatory signals secreted by specific broadcasting units in the afferent sensory network (Dickinson and

Nagy 1983). These cells play the role of "orchestra leaders" whose activity triggers the widespread diffusion of neuromodulators. This neuromodulation impacts on the intrinsic reactivity of the other cells by changing reversibly the expressed repertoire of membrane conductances. Consequently, the individual excitability patterns of any given cell will change depending on the context (before, during, or after the orchestra leader cell has fired). This flexibility in intrinsic properties explains why switching on and off neuromodulation reorganizes the dynamics of the full network in distinct functional central motor program generator assemblies, each associated with different motor behaviors, such as swallowing, crunching in the gastric bag, expulsion of processed food for the considered example (Hooper and Moulins 1989; Dickinson et al. 1990; Meyrand et al. 1991) (see Figure 12.1). In this paucineuronal biological network, where all partners are known, the coordinator is identified and the causal link between the temporal assembly motifs and the behavioral actions as well as their functional significance are clearly defined.

These data, often ignored from the vertebrate neuroscience community, support the hypothesis of the existence of "orchestra leader"-like neurons which, through the presence or absence of firing activity, condition and format merging and segmentation across functional assemblies. The existence of giant identified coordinators with invariant morphology and widespread projections has been confirmed in other invertebrate species and, in the bee, associated with reward; however, evidence in larger central networks, such as the mammalian brain, is still lacking. Nevertheless, there is already ample evidence that aminergic and cholinergic subcortical brainstem nuclei release neuromodulator *en passant* along their long axonal projections that travel through all cortical areas, from the occipital to the frontal lobe, and change the repertoire of expressed conductances. Like central pattern generators, thalamic circuits are subject to neuromodulatory influences (Steriade 1996). In this case, neuromodulators, such as acetylcholine, norepinephrine, or serotonin, affect intrinsic currents and switch the circuit from an oscillatory mode to a "relay mode" in which oscillations are abolished (McCormick 1992). These neuromodulators are present in activated states, promoting the relay of sensory information by the thalamus, while their diminished levels during slow-wave sleep unmask participation by the thalamus in the genesis of large-scale synchronized oscillations involving the entire thalamocortical system. We conclude from this brief review that, to a certain extent, both in invertebrate ganglia and the vertebrate brain, the dogma of separability between intrinsic and extrinsic factors in the control of cellular excitability is doomed to fail. Thus, the "whole" cannot be the sum of the "parts," and segmentation does not always coexist with perceptual binding.

Other examples can be considered where the coordinating agent is not part of the biological system but rather the product of high-order statistical features present in the sensory input stream. Changing the statistical regularities of the environment produces a drastic reorganization of ensemble activity patterns

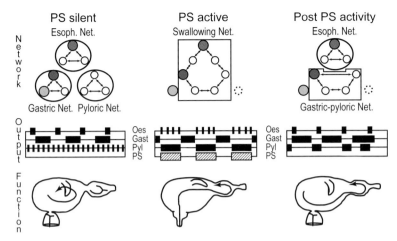

Figure 12.1 Example of dynamic coordination: Reconfiguration of network dynamics by internal "orchestra leaders" in the stomatogastric ganglion of the lobster. The same sensorimotor network is functionally reorganized in independent assemblies depending of the activity state of neuromodulating cells (PS cell) (top row). These assemblies are characterized by specific excitability patterns expressed by each of the composing cells and specific phase relationships between them (middle row). The functional role of each assembly during the swallowing stomatogastric cycle is depicted (bottom row). When PS is silent, the esophageal, gastric, and pyloric networks (top) generate independent rhythmic output patterns (middle) involved in regionally specific and separate behavioral tasks (bottom). When PS is rhythmically active, it drives the opening of the esophageal valve (bottom), and by disrupting the preexisting network(s) and recruiting only part of the neurons, it constructs a novel assembly (top) that generates a coordinated motor pattern (middle) appropriate for swallowing behavior. When PS is again silent post-activity, the esophageal valve closes (bottom) and motor units immediately resume their original network activity while units (i.e., gastric and pyloric) controlling regions more caudal to the sphincter continue to generate a single pattern before resuming their separate activities (adapted from Meyrand et al. 1991). Note that the same cell (color coded) can switch from assembly to the next; this dynamic reconfigurability accounts for behavioral changes.

and their stimulus-locked reliability in the early visual system of mammals. For instance, the presentation of drifting gratings in a V1 receptive field (Figure 12.2) evokes dense but highly unreliable responses across individual trials, both at the spiking and subthreshold levels. In contrast, virtual eye-movement animation of natural scenes temporally evokes in the same cell precise sparse spike responses and stimulus-locked membrane potential dynamics, which are highly reproducible from one trial to the next (Frégnac et al. 2005; Marre et al. 2005; Baudot et al., submitted). Importantly, the fast components of the membrane potential trajectory show a high trial-to-trial reproducibility, even when the cell is not firing, for silent periods extending for several hundred of milliseconds prior and after the reliable spiking event (Figure 12.2).

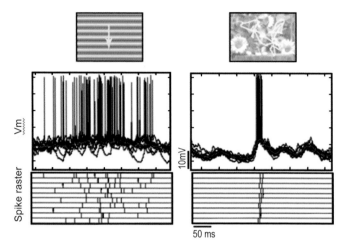

Figure 12.2 Example of dynamic coordination: Coordination by the statistics of the sensory drive. Reconfiguration of spike timing precision as a function of input statistics (adapted from Frégnac et al. 2005). The left and right panels represent evoked subthreshold (top) and spiking activity patterns (bottom), evoked in the same V1 cell, respectively, for a drifting grating and a natural scene animation. Trial-by-trial membrane potential trajectories are overlaid in black; red indicates mean Vm response. The raster dot displays of the spiking activity are built on ten trials. All records are synchronized with the stimulus onset, and the visualization window is adjusted to illustrate a time period in the movie where evoked spikes are observed. The response of the same cell is dense and noisy for the grating, whereas it is sparse and highly reliable from trial-to-trial for the natural scene.

In this second example, coordination is unrelated to the behavioral outcome driven by perception or neuromodulation, since it is observed in the anesthetized and paralyzed preparation (Frégnac et al. 2005) as well as in the attentive-behaving monkey (Vinje and Gallant 2000). Stimulus-driven recruitment of dynamic nonlinearities distributed across the network results in the sparsening and reordering of spiking events. This self-organized process adapts the temporal precision of the sensory code to the statistics of the input (for further definitions of self-organized coordination, see Engel et al., this volume). The more complex or the closer to a natural environment the input is, the higher the temporal precision of stimulus-locked events are and the more deterministic-like the network dynamics behave. However, and in contrast with the first example, this adaptive form of temporal coordination is done in the absence of an internal executive or supervision units. Multiscale cooperation across the network is still needed, and long distance center-surround interactions across the visual field—which implies long-distance interactions across the retinotopic cortical map—suppress or facilitate interneuronal binding. Note also that, as demonstrated in the first example, the full field stimulation (i.e., "whole" condition) will affect *in fine* the functional characteristics of the recorded unit (i.e., the individual receptive fields of the V1 cells). The classical discharge field defined with low dimensionality stimuli (Fourier input, sinusoidal luminance grating) does not predict, at least in our hands, the subthreshold dynamical responses of the same cell stimulated with richer input statistics (temporal modulation reproducing eye-movement effects, spatial $1/f^\alpha$ spectrum for natural scenes).

A third class of examples focuses on relational coding in cortex. Long-duration single-frequency tones do not evoke tonic changes in the firing rate of single neurons in auditory cortex. At most, transient burst responses are detected in A1 cortex, at the onset or offset of the tone. The presence of the stimulus, however, is signaled in the same area by a dramatic and tonic elevation in the correlation between cortical units coding for the sound frequency, without any apparent change of firing rate (deCharms and Merzenich 1996). Figure 12.3 illustrates the case where the information that can be stored or recalled on the basis of coordinated activity is separable from the rate responses of single neurons. Furthermore, the cortical domains of shared activation can be expanded or contracted through the coordinating action of a neuromodulator (acetylcholine) or by stimulating the ascending afferent cholinergic corticofugal projection from the nucleus basalis (Kilgard and Merzenich 1988).

A fourth example of coordination, partially of the same class, can be revealed by stimulus-locked analysis of the covariation of the firing rates of several simultaneously recorded units in premotor cortex of awake animals

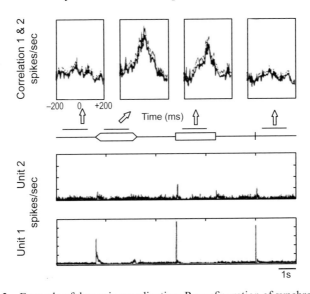

Figure 12.3 Example of dynamic coordination: Reconfiguration of synchrony during sensory processing (adapted from deCharms and Merzenich 1996). A1 cortical units have been simultaneously recorded during a sequence alternating control (silent) and stimulation periods with different tones and temporal profiles. Bottom: PSTHs show fast transient responses in the evoked mode for the onset and offset of the tones, but no tonic activity. Top: Correlograms show decorrelation between units in the ongoing mode and tonic coordination maintained as long as the tone is present.

submitted to a GO/NO-GO paradigm (Figure 12.4). The computation of the time course of the joint activity between units gives access to the dynamics of effective connectivity between simultaneously recorded units (Aertsen et al. 1989). This calculus requires the averaging of joint peristimulus time histograms (PSTHs) over trials corresponding to the same behavioral task (GO vs. NO-GO). The remarkable result is that the stimulus-locked time course of the effective connectivity is flexible and differs significantly between the two task phases (Vaadia et al. 1995). This implies that some task-related coordination switches the functional allegiance of the two recorded neurons between two different relational assemblies. Again, in this example, the rate mean of each unit is unchanged whereas the rate covariation signals information related to the behavioral significance of the cognitive task (GO vs. NO-GO).

These examples of dynamic coordination show a diversity of processes where the coordinating agent can take several forms: in Figure 12.1 and 12.4 it is an internal supervisor embedded in the network; in Figure 12.2 and 12.3 it is the sensory drive or, in Figure 12.4, an external prior. Other examples could have been given where the coordinator is generated by the correlations

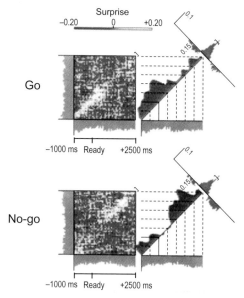

Figure 12.4 Example of dynamic coordination: Reconfiguration of effective connectivity during operant conditioning GO/NO-GO paradigm (adapted from Vaadia et al. 1995). Effective connectivity between simultaneously recorded cortical units is measured by joint PSTHs and statistical surprise measure matrix. The time profile of the synchrony level (green) between the two simultaneously recorded units is represented as a function of delay from trial onset, respectively, in the GO and NO-GO condition. Note that the time course of the effective connectivity between the two same units differs between the two conditions, whereas no modulation is observed in the mean PSTHs (pink) flat histograms represented along the two unit-related axes.

imposed by the behavioral task; for instance, by manipulating differentially the shared attention between different sensory modalities and the occurrence of a reward. During this Ernst Strüngmann Forum, some experts expressed the strong expectation that coordination requires an internalized supervisor or executive agent associated with a well-defined computation or function, whereas others considered that a sufficient condition might be the genesis of a symmetry-breaking of activity within the network, whatever its substrate. In this latter condition, the assembly cooperativity defines a bias or "prior" which can take forms as diverse as a propagating depolarizing wave (Chavane et al. 2000; Jancke et al. 2004; Roland 2002; Frégnac et al. 2010), local "up" states (Frégnac et al. 2006), synchrony enhancement, or frequency changes in the local field potential (LFP) spectrum (Meyrand et al. 1991; Fries 2009). In the low-level Gestalt association paradigms reviewed below, such asymmetry-breaking in cortical activity propagation takes the form of self-generated intra-V1 wave which may be related to the perceptual outcome. This introductory section underlines the importance of defining further a taxonomy of coordination where the underlying mechanisms of each phenomenological form can be clearly separated (see also Moser et al. and Engel et al., this volume).

Measuring Network Coordination at a Single Recording Site

One often posits that simultaneous multiple recordings in different sites are required to track coordination (Wolf Singer, pers. comm.), and this condition becomes a necessity when coordination is engaged between processes operating simultaneously in several distant cortical areas. However, single-site recordings provide meaningful relational information, as long as the electrophysiological signal itself includes a measure of local network activity, integrated across a sufficiently wide range of space and time constants. LFP studies show that one recording site is enough to detect oscillatory synchronized activity. For instance, gamma chattering (30–80 Hz) in V1 is present in most LFP recordings during visual stimulation, whereas the probability of detecting it at the single-unit extracellular recording level remains very low, at least in the anesthetized animal (Frégnac 1991).

Another experimental strategy, reviewed in greater depth here, uses the intracellular membrane fluctuations of a single neuron as a readout probe of the correlation structure of the network afferent to the recorded cell (Figure 12.5). Intracellular recording of neocortical neurons provides a unique opportunity to characterize some statistical signature of the synaptic bombardment to which it is submitted. Indeed, the membrane potential (Vm) displays intense noise-like fluctuations, which reflect the cumulative impact arising from the coordinated activity of thousands of input neurons. It has an advantage over LFP in that the selection of the recorded assembly is not based on the distance relative to the recording site (the "visibility radius" of the LFP electrode is estimated to

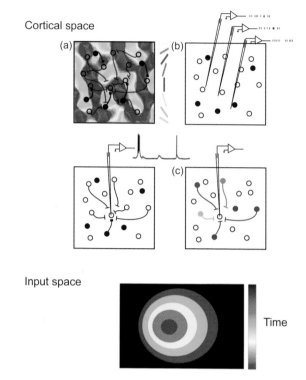

Figure 12.5 Network dynamics imaging versus synaptic functional imaging. Three methods for visualizing network dynamics are compared: (a) Optical functional imaging allows charting of cortical domains of iso-functional preference (color coded) on the basis of metabolic or hemodynamic signals. (b) Multiple simultaneous extracellular recordings are used to evaluate correlated activity patterns through the blind selection of potentially interconnected neurons. (c) The reverse analysis of subthreshold activity during long duration intracellular recordings of the same single cell can be used to retrieve the effective network afferent to the recorded cell (see text).

be 250 μm; Nauhaus et al. 2009) but is realized by the target cell itself, which anatomically "selects" its input lines. By analyzing these fluctuations, preferably in voltage clamp mode, it should be possible to "decode" in a reverse way the global afferent activity of the network in which the cell is embedded. Despite the inherent difficulty of space clamping *in vivo* (Spruston et al. 1993), it is likely that the input lines that will be seen are the ones which have an effective impact at the soma (where the recording is done most often) and hence influence the spiking process.

For the past 15 years, we have been developing a reverse engineering approach in current clamp (Bringuier et al. 1997; Bringuier et al. 1999; Frégnac and Bringuier 1996; Frégnac et al. 2010) and voltage clamp modes (Monier et al. 2003) which allows, in principle, the retrieval of the effective connection graph in which the cell is embedded at any point in time (Figure 12.5).

The analysis is based on the synaptic rumor recorded in a single cell. Its principle is similar to that of echography in the etymological sense (transcription of echoes) and is referred to as "functional synaptic imaging" (Frégnac and Bringuier 1996). During sensory activation, the cortex is considered as a chamber of echoes produced by the thalamocortical input. The readout of the sources is based on the extraction of correlations in space and time of synaptic events with specific features of the stimulus (e.g., orientation, direction, ocular dominance). This method is equivalent to the principle of time reversal mirrors in acoustic physics and medicine (Fink 1996). The success of this demultiplexing computation relies on the underlying assumption that the input sources are separable in space and their synaptic influence travels in time with similar speed. This condition is rarely met in the general case, but seems to be valid for sparse stimulation regimes or during ongoing activity. Functional synaptic imaging gives a prediction of the macroscopic activation of the network in space and time, which can be confronted with the direct observation of the spatiotemporal cortical dynamics evoked in the superficial layers of cortex, using voltage-sensitive dyes (VSDs) (Frégnac et al. 2010).

A variety of spectral analysis methods can be applied to measure the impact of stimulus-driven coordination on the synchrony state of the synaptic bombardment afferent to the recorded cell, which can change over different time-scales, ranging from phasic (few milliseconds) to steady state (maintained for several seconds). Here we wish to underline two methods that have been used successfully to demonstrate a dependency of the cortical intracellular "hum" with the input statistics.

The first method, time-frequency wavelet analysis, has been applied with great success to reveal the stimulus-locked and stimulus-induced (which may vary in phase between trials) synchrony and oscillatory structure of assembly spiking patterns (Tallon-Baudry and Bertrand 1999; Tallon-Baudry, this volume). It provides an efficient means of achieving a trial-by-trial time-frequency analysis of the signal, at any temporal delay following stimulus onset, through an array of temporal (Gabor or Fourier) wavelets ranging from 1 to several hundred Hz (Varela et al. 2001). For intracellularly recorded cells, this multiscale filtering method can also be applied to both its supra- and subthreshold activity (Vm) by considering an array (see raster in Figure 12.6a) of repeated responses of the same unit for different trials, precisely realigned with the stimulus onset. By using signal-noise (SNR) measures (Croner et al. 1993), highly reliable stimulus-locked events are revealed by the time-frequency SNR matrix of the subthreshold activity. These events are seen as hot peaks (Figure 12.6a) which straddle from low (wide band) to high (thin band) frequencies. In the chosen illustration, they indicate the dense presence, since repeated at many discrete times, of synchronous volleys of synaptic inputs despite a sparse postsynaptic discharge. This technique shows that the precision of time coding in V1 is dependent on the dimensionality of the stimulus. Reliability is poor for low-dimensional stimuli whereas the afferent network shows a highly

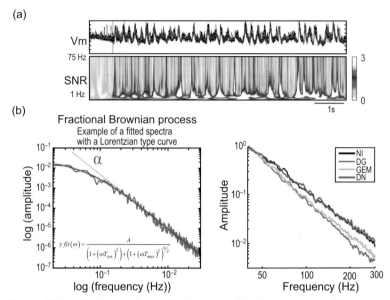

Figure 12.6 Read-out of synaptic input coordination in a single V1 neuron using time-frequency and spectral fractal analysis. (a) Synaptic input coordination measured by wavelet analysis of subthreshold intracellular dynamics. Top panel: trial-by-trial membrane potential trajectories are overlaid in black; in red, mean Vm response. Lower panel: time-frequency wavelet analysis of the Vm dynamics and signal-to-noise (SNR) power matrix in a V1 Simple cell (Frégnac et al. 2005; Baudot et al., submitted). The time axis represents 400 ms of ongoing activity followed by several seconds of continuous visual activation with a pseudorandom animation (kept identical across trials) of natural scenes. The colored SNR peaks straddling between 1–75 Hz signal the reliability of the evoked synaptic bombardment and the presence of highly temporally structured input when processing natural scenes. (b) Extracting the contextual network correlation state from the spectral properties of the membrane potential of a single cell. Left: scaling invariance of the power law for a Lorentzian process and asymptotic linear fall-off (slope α) in a log-log coordinate plot. Same analysis applied to the subthreshold dynamics of a single V1 cell for various input statistics (DG: drifting grating; GEM: grating with eye movements; NI: natural image; DN: dense noise). The fractal slope component shows a strong stimulus dependency in the coordination effectiveness of the network activity "seen" by the recorded cell (adapted from El Boustani et al. 2009).

structured input for natural scene statistics (Frégnac et al. 2005; Marre et al. 2005; Baudot et al., submitted).

The power spectrum of the membrane potential of any given cell provides another valuable tool to extract information on the second order statistics of the synaptic input bombardment. A recent study from our lab has demonstrated that the power spectral density of the subthreshold membrane potential (Vm) of visual cortical neurons can be fitted by a power function $1/f^{\alpha}$, at least in the upper temporal frequency range between 80–200 Hz (El Boustani et al. 2009). This observation holds both during ongoing activity and evoked states, although the power law slope changes as a function of input statistics: the fractional exponent α is given by the fall-off slope of the Vm spectrum in a log-log representation and provides a measure proportional to the range of the temporal correlations (the larger the longer). The linear asymptotic behavior of the log-log plot, called power law scaling, is shared by many complex dynamic or self-organizing processes, both in physics and biology. The origin of their shape invariance with frequency rescaling (f → kf) is a matter of debate for multiple recordings of LFPs and spiking assemblies (Beggs and Plenz 2003; Plenz and Thiagarajan 2007). Until recently, it was thought that the slope component was dominated by the intrinsic conductance repertoire of the cells (Bedard and Destexhe 2008). *In vivo* recordings from our lab show that the fractional slope component value varies with the complexity of the sensory input statistics (El Boustani et al. 2009; see Figure 12.6b) and that the short-range correlations present during natural scene viewing are of the same range as those found in ongoing activity. These results have been emulated by computational models, which demonstrate that the fractional exponent is determined by the mean level of correlation imposed in the recurrent network activity. Similar relationships have also been reproduced in cortical neurons recorded *in vitro* with artificial synaptic inputs by controlling *in computo* the level of correlation in real time (using dynamic clamp techniques). From this confrontation between theory and electrophysiological recordings, we conclude that the frequency-scaling exponent of subthreshold Vm dynamics provides a reliable measure to monitor changes in the coordination state of neural networks, when they are maintained for several seconds or longer. This multiscale analysis could be potentially generalized with other types of signals which achieve integration of neural activity at more meso- or macroscopic scales.

Reconstruction of Lateral Propagation Waves from Synaptic Echoes

The combination of intracellular electrophysiology and VSD network imaging in the study of visual cortical processing has given unprecedented access to binding and coordinating processes that operate at a subthreshold level, and which cannot be detected by solely studying spiking activity. These techniques give evidence for propagation along long intracortical distances of depolarizing waves, which may contribute to facilitate synaptic integration anisotropically in the cortical network. On one hand, VSD imaging reveals the spatio-temporal signature of information propagation across the target network, and reflects anatomical constraints in the "divergence" of the connectivity. On the other, intracellular recordings characterize the "convergence" of the connectivity to any given cortical locus. They provide evidence that the receptive field of visual cortical neurons is not limited to a tubular view but extends over a large region of the visual field. The mean discharge field (MDF) size of V1 cells defines a spiking receptive area that does not extend beyond 1–2 degrees of visual angle for vision around the area centralis in the cat (the equivalent of our fovea). In contrast, the synaptic integration field, reconstructed from intracellular recordings, corresponds to the region of space from which a local input still evokes a significant excitatory or inhibitory response. It captures synaptic echoes that originate from the far "silent" surround of the classical discharge field (Figure 12.7). Schematically, the subthreshold receptive field is characterized by a hill of spatial sensitivity (falling off 5–10° away from the MDF center) and a basin of latency: synaptic responses elicited by stimuli placed far from the center of the discharge field show increasing delays (up to several tens of milliseconds) with the relative eccentricity (Bringuier et al. 1999).

Functional synaptic imaging offers a link between these two modes of visualization of apparently distinct connectivities ("divergence" vs. "convergence"): although our method is based only on inferences established from intracellular records, it allows the reconstruction—first in space, then in cortex—of the source location distribution corresponding to the recorded synaptic echoes produced by the sensory drive. The hypothesis of a traveling wave is made on the assumption of symmetry in exuberant intrinsic connectivity: since

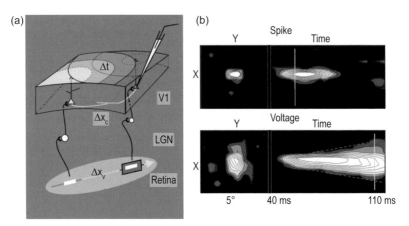

Figure 12.7 Functional synaptic imaging. (a) Schematic representation of the hypothesis of reciprocal horizontal connections between two cortical cells (left, "sender"; right "receiver"). This schema allows reconstruction of a propagating wave (circles) from the intracellular measure of evoked latencies of synaptic responses in the "receiver" cell (right, electrode). Δx_v is the eccentricity of the distal stimulus (white rectangle, no outline) from the central stimulus shown in the discharge field center (grey rectangle, black outline). Δt is the latency between the synaptic response onsets evoked through the two pathways. Δx_c is the intracortical distance between the cortical feedforward impacts produced by the two stimuli, inferred from the know retinocortical magnification factor (see text for details). (b) Spatiotemporal (left: X–Y; right: X–t) maps of suprathreshold (spike, upper panels) and subthreshold (voltage, lower panels) activations in the same V1 cell. The X–Y maps are presented for two specific delays corresponding, respectively, to the maximal extents of the discharge field (upper) and subthreshold integration field (lower). The pink dotted lines show the average speed of the propagation of the reconstructed wave (0.2–0.4 m s^{-1}) (adapted from Frégnac et al. 2010).

V1 is a highly recurrent network, we assume that each cell is connected reciprocally to any other cell, with identical propagation delays from and to the cell (Figure 12.7a). In simpler terms, it should take exactly the same amount of time for any given cell to receive a signal from a distant cortical source as to send it back to the same cortical locus. This theoretical shortcut allows the inference of propagation patterns (the cell being seen as a "wave emitter") solely on the basis of the spatiotemporal maps of stimulus-locked synaptic responses recorded in a single cell (the cell being seen as an "echo receiver"). As illustrated in Figure 12.7b, the slopes (dotted lines) seen in the spatiotemporal pattern (X-t) associated with the receptive field (X-Y) corresponds to the subthreshold latency basin of the recorded cell (Bringuier et al. 1999). This suggests that the information received from the receptive field center in the cortex through the feedforward afferents is then propagated radially by the horizontal connectivity to neighboring regions of the visual cortex over a distance that may correspond to up to 10 degrees of visual angle. These data led successfully to the functional identification and reconstruction of a propagating wave of visual activity relayed by the horizontal connectivity.

The principle of computation of the propagation speed from the intracellular recording is straightforward (Figure 12.7a): we compare the synaptic effects of two elementary stimuli (white bar), one in the core of the minimal discharge field, the other in the "silent" surround. The distance between the primary points of the feedforward impact produced in cortex by the two stimuli can be predicted on the basis of their relative retinal eccentricity Δx_v and the value of the retino-cortical magnification factor (RCMF). This factor can be measured electrophysiologically in cat (Albus 1975), by 2-deoxyglucose metabolic labeling in monkey (Tootell et al. 1982), by intrinsic imaging in mouse (Kalatsky and Stryker 2003), and even by fMRI in humans (Warnking et al. 2002). Thus, beyond a certain scale of spatial integration (larger than the columnar grain), any distance in visual space, Δx_v, can be converted to a distance in visual cortex, Δx_c. The spatial range of the subthreshold field extent agrees with the anatomical description of 4–7 mm horizontal axons running across superficial layers in cat V1 (Mitchison and Crick 1982). Although the RCMF factor is dependent on the eccentricity from the fovea in primates and humans, this is not the case in cats and ferrets, where 1° of visual angle corresponds roughly to 1 mm in cortex within the 10° of the area centralis. Furthermore, the electrophysiological recordings give access to the delay Δt_c between the two synaptic echoes obtained through the feedforward and the horizontally mediated pathways. By dividing the inferred cortical distance Δx_c in cortex by the recorded delay Δt_c, an apparent horizontal propagation speed can be computed within the cortical map, hence in the plane of the layers of V1. The propagation speeds we inferred range from 0.02–2 m s^{-1}, with a peak between 0.1–0.3 m s^{-1}.

These velocity values have since been confirmed for other sensory cortical structures, such as somatosensory cortex (Moore and Nelson 1998). They are thus ten times slower than X-type thalamic axonal propagation and feedback from higher cortical areas (2 m s^{-1}; Nowak and Bullier 1997) and one hundred times slower than the fast Y-pathway (8–40 m s^{-1}; Hoffman and Stone 1971). They are, in fact, within the order of magnitude of conduction speeds measured *in vitro* and *in vivo* along nonmyelinated horizontal cortical axon fibers. In view of the difference in size that exists between the subthreshold receptive field and the discharge field, propagation most likely involves long monosynaptic horizontal connections, although the contribution of rolling waves of postsynaptic activity cannot be entirely excluded. Recent reports based on cortical LFPs triggered on LGN spike activity rule out the possibility that divergence of LGN axons may also contribute to the buildup of the observed latency shifts as discussed by Nauhaus et al. (2009).

Thus, our intracellular study of synaptic echoes detects the propagation signature of an intracortical wave of visual activation traveling along long-distance horizontal connections. One obvious consequence is that the V1 network should not be considered as an ordered mosaic of independent "tubular" analyzers, but rather as a constellation of wide field integrators, which simultaneously integrate input sources that arise from much larger regions of visual space than previously thought. The collective behavior of these integrators is coordinated during sensory processing by the anisotropic propagation of stimulus-induced facilitatory waves traveling at slow speed within the superficial cortical layers. Primary visual cortical neurons thus would have the capacity to combine information issuing from different points of the visual field, in a spatiotemporal reference frame centered on the discharge field itself. This ability imposes precise constraints in time and in space on the efficacy of the summation process of elementary synaptic responses, and specific functional predictions linked to the intracortical genesis of coordinating waves will be reviewed below.

The macroscopic reconstruction of intra-V1 waves on the basis of microscopic echoes remains, however, an extrapolation made between two scales of spatial organization differing by two orders of magnitude (neuron vs. map). Brain imaging methods, and more specifically VSD techniques, give an unprecedented view of the state of the cortical network, best detected as a depolarizing field in the terminal tuft of the dendrites of layer 2/3 pyramidal cells (Roland 2002), with a time sensitivity close to that of intracellular recordings. Since the pioneering study of cortical spread function by the group of Amiram Grinvald (1994), numerous groups have confirmed the propagation of spontaneous and evoked waves across the cortical laminar planes in visual primary and secondary cortical areas of rodents and higher mammals: rat, from 0.05–0.07 m s^{-1} (Xu et al. 2007); cat, 0.09 m s^{-1} (Jancke et al. 2004); cat, 0.3 m s^{-1} (Benucci et al. 2007); monkey: 0.20 m s^{-1} (Grinvald et al. 1994).

The group of Matteo Carandini measured, in both cat and monkey V1, the spatial distribution and the temporal phase of the second harmonic VSD response to the contrast reversal of a one-dimensional bar (Benucci et al. 2007).

The spatial spread attenuation constant of the cortical response was found to correspond quantitatively, in each species studied, to the mean extent of horizontal axons (2–3 mm in monkey, 5–8 mm in cat). The response phase (i.e., the temporal delay of the evoked oscillation in each pixel with respect to the inducer stimulus) was shown to increase linearly as a function of the lateral distance from the feedforward impact zone of the bar. These observations, which have since been replicated by LFP studies applied to multielectrode grid recordings (Nauhaus et al. 2009), confirm that the focal stimulus induced a traveling wave across cortex, with an apparent speed of propagation estimated at around 0.30 m s^{-1}. As shown in Figure 12.8, these VSD studies fully corroborate the predictions we extracted from our intracellular recordings more than ten years ago (Bringuier et al. 1999), and most remarkably both methods

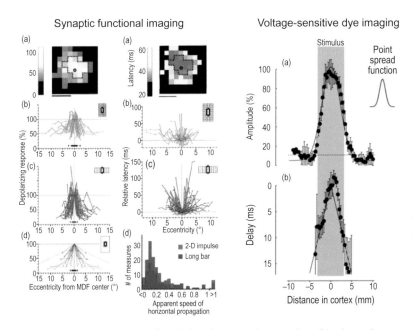

Figure 12.8 Comparison of spatial and temporal properties of horizontal propagation, using synaptic functional imaging (left) and network voltage-sensitive dye (VSD) imaging (right). See text for details. Synaptic functional imaging is used to infer propagation waves from the readout of synaptic echoes recorded in the same single cell in response to sparse localized inputs. VSD imaging monitors the cortical lateral spread of a depolarizing wave triggered by a ultrathin phase reversed contrast grating. Left panel: horizontal propagation inferred with intracellular synaptic functional imaging (adapted from Bringuier et al. 1999). Right panel: horizontal propagation wave monitored in the superficial layers of cat V1 with the with the f2 component of he VSD imaging signal (adapted from Benucci et al. 2007). Note the similarities between the spatial hill sensitivity profiles (left column) obtained with intracellular recordings and with VSD imaging (top right panel). The same propagation speed (0.1–0.3 m s^{-1}) is measured by the two imaging techniques (second column from the left and bottom right panels).

give the same mean estimate of propagation speed (0.30 m s^{-1}) although they are based on different measurement methods and analysis. As nicely worded by Nauhaus (2009:72), one may conclude with some confidence that "the synaptic input to neurons during spontaneous activity can be thought of as the superposition of a myriad of traveling waves originating from individual spikes distributed over an extended region of cortex."

Visualizing Correlates of Gestalt Illusions in V1

The multiscale comparison of these various different imaging techniques (synaptic functional imaging, VSD network imaging) opens up a new field of study, where it becomes possible to compare real-time imaging in cortical networks with membrane dynamic recording in single cells, on the one hand, and psychophysical performance measures, on the other. Almost a century ago, psychologists and philosophers proposed a theory of perceptual grouping, Gestalt (Koffka 1935; Köhler 1947), which predicts the emergence of coherent percepts of global shape and motion from the temporal superposition of static presentations of elementary spatial features. This theory assumes the existence of psychic processes that favor associations in space (according to spatial proximity and similarity in contrast polarity) as well as in time (continuity, common fate).

Those predictions inspired a series of psychophysical studies, whose results strongly support the following working hypothesis: the temporal characteristics of the recruitment of the "horizontal" intracortical connectivity could affect the perception of motion. Among various demonstrators, the "Phi" apparent motion protocol, originally called the "beta phenomenon" by Wertheimer (1912), induces a powerful illusion when the same target is repeatedly flashed at different moments in time in different positions in the visual field ordered along an imaginary trajectory (Figure 12.9, left). Although at each moment in time the observer sees only a static image, he reports the perception of continuous motion of the same object along the trajectory defined by the "association" path linking the various positions explored in succession. The strength of the percept depends on the complexity of the test stimulus (shape and texture), the duration of the static presentations, the interstimulus interval, and the spatial offset between positions (Anstis et al. 1998). The "line motion" illusion is also based on the same induction process of asynchronous static presentations. In this latter case, the cue feature is a uniform luminance square, followed by a bar of the same luminance, one polar end of which encroaches on the previously flashed square. For adequate interstimulus intervals and presentation durations, the human subject reports a continuous movement of one border, perceived as a smooth morphing of the square into the elongated bar (Hikosàka et al. 1993).

If the spatial contextual effect can be easily interpreted in the framework of the perceptual "association field" of Field et al. (1993), the temporal

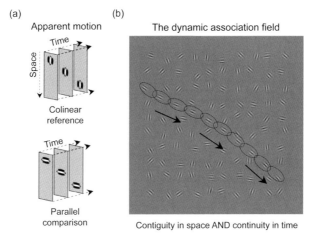

Figure 12.9 Apparent motion and the hypothesis of the "dynamic association field." (a) Two forced-choice apparent motion protocol, where the human observer has to report which sequence of oriented elements is seen "faster." Two configurations are compared (collinear and parallel), in which the orientation of each element is respectively collinear or orthogonal to the downward motion axis. (b) The "dynamic association field" hypothesis. Local oriented inputs (Gabor patches) induce a facilitation wave of activity traveling along horizontal connections intra-V1. This wave binds in space and time proximal receptive fields with co-linear preferred orientations, thus creating a contiguous path of temporal integration. The associative strength of the perceptual effect is maximal when the asynchronous feedforward sequence produced by joint strokes of apparent motion (arrow) travels in phase in the cortical network with the visually evoked horizontal propagation.

determinants and the preferred co-linearity configuration of these effects suggest a strong dependency on the intra-V1 horizontal connectivity and its propagation speed constraint. VSD imaging correlates of the "line-motion" illusion have been found in area 18 of the anesthetized cat (Jancke et al. 2004). Grinvald's team compared the spatiotemporal patterns of the cortical responses to a flashed small square and a long bar alone, and the configuration of the line-motion illusion: a square briefly preceding by a few tens of milliseconds the presentation of a light bar (co-linear with the square). In the (last) associative condition, the VSD pattern demonstrated the spread of a low-amplitude wave in the cortical layer plane, extending far beyond the retinotopic representation of the initial "cue." This spread was most visible along the main orientation axis of the bar, with a horizontal propagation speed around 0.10 m s^{-1}. This pattern was indistinguishable from the spatiotemporal pattern produced by the continuous motion of the same square at a few tens of degrees per second. Thus, most remarkably, the anisotropic spread of the cortical activity pattern relayed by the horizontal connectivity observed in primary visual areas was isomorphic to the percept of the continuous square-bar morphing reported by the human observer.

Visualizing Propagation of Orientation Belief

An important issue in our understanding of coordination in sensory cortical areas during low-level perception is to determine how much of the perceptual biases results from structural built-in constraints and how much derives from contextual activity or coordination effects defined by the stimulus configuration itself. For instance, psychophysical studies show that a two-stroke flashed sequence of oriented elements displaced along their orientation axis appears to move "faster" to human observers than the same stimulus sequence at an angle to the motion axis. This effect peaks at 64°/s in humans and decreases for higher and lower speeds (Georges et al. 2002). This speedup is highly sensitive to orientation anisotropy, strongly depends on the relative angle between the orientation of the moving elements and the motion axis, and is still observed for curvilinear trajectories. This suggests that it involves units highly sensitive to orientation, a property mainly expressed by neurons in areas V1 and V2. Two features of the perceptual effect are closely related to V1 physiology and anatomy: (a) the sensitivity of the speedup effect to orientation resembles that of the recently uncovered "association field" (Field et al. 1993) presumably involved in contour integration; (b) the speed at which the speedup is maximum is comparable to the speed at which neural activity propagates along long-range horizontal connections. This structure–perception match may be adapted to a specific oculomotor exploration strategy since the conversion of the horizontal propagation speed in degrees of visual angle per second in the visual field (which depends on the species-specific cortical magnification factor) corresponds in most species to the saccadic range of eye movements (50–500°/s).

As stated above, it is generally assumed that orientation binding results from anisotropies in the intracortical connectivity, where long-distance horizontal axons in visual cortex have been reported to link columns sharing similar orientation preference. However, the anatomical evidence in favor of such bias is rather scarce in the cat cortex (the strongest evidence for a structure–function correlation has been obtained in the tupaïa glis and the ferret V1). Combinations of optical imaging and intracellular labeling show indeed a diversity of potential links established between orientation columns which do not obey, at least at the statistical significance level, the rule "like couples to like" (Monier et al. 2003). Quantitative reanalysis of published data correlating axonal bouton distribution with target orientation preference (relative to that of the parent cell) revealed by intrinsic imaging reveals that the tendency for horizontal axons to connect iso-orientation loci is not exclusive and interconnection probability is only about 1.5 times greater than chance level. This bias has been mostly observed for supralaminar pyramidal neurons. However, inhibitory interneurons and neurons in layer IV or close to pinwheel centers have also been reported to connect lateral orientation columns in a cross-oriented or unselective way (Karube and Kisvarday, pers. comm.). As a consequence, at a more integrated mesoscopic level, the net functional effect cannot be predicted.

Other processes, such as activity-driven coordination, may amplify a small structural bias in a strong perceptual effect, when enough temporal synergy and spatial summation are recruited from the start by the sensory drive. A recent collaborative work between our lab and the group of Amiram Grinvald attempted to achieve a multiscale analysis of the visually driven horizontal network activation, using population and single-cell measures of postsynaptic integration. VSD imaging showed that a local-oriented stimulus evoked an orientation-selective activity component that remained confined to the feedforward cortical imprint of the stimulus. Orientation selectivity decreased exponentially along the horizontal spread (space constant ~1 mm). To dissect the local connectivity rules, we also made intracellular recordings during dense-oriented Gabor noise stimulation to identify the orientation selectivity and preference of converging horizontal inputs onto the same target cell. The combination of the imaging and electrophysiological results suggests, somewhat surprisingly, that the horizontal connectivity does not obey iso-orientation rules beyond the hypercolumn scale. In contrast, when increasing spatial and temporal summation, both optical imaging and intracellular measurements showed the emergence of an iso-orientation selective spread. We conclude that stimulus-induced cooperativity is a necessary constraint for the emergence of iso-functional Gestalt-like binding (Chavane et al., submitted).

This last study, combining network VSD imaging and synaptic functional imaging, shows two contrasted dynamic behaviors of the same network for two distinct levels of coordination driven by the stimulus configurations: a single local stimulus does not propagate orientation preference through the long-range horizontal cortical connections whereas stimulation imposing spatial summation and temporal coherence facilitates the buildup of orientation preference propagation. These observations do not forcibly contradict each other. On one hand, for the local-oriented stimulus, the divergent connectivity pattern may facilitate detection of high-order topological properties (e.g., orientation discontinuities, corners, geons). On the other hand, for stimulation protocols involving a larger extent of stimulation, summation of multiple-oriented sources in the far "silent" surround can optimize the emergence of iso-orientation preference links. Configurations such as oriented annular stimuli may, for instance, recruit iso-oriented sources collinearly organized with the orientation preference axis of the target column/cell; similar synergy may be obtained when sources, independent of their exact location, share the same motion direction sensitivity as the target grating. Both of these configurations, which are confounded in annular aperture protocols, correspond to the neural implementation of the Gestalt's continuity and common fate principles, but other more dynamical principles could also emerge from such network configuration.

These different experimental observations have led us to formulate the concept of the "dynamic association field" (Frégnac et al. 2010), which adds a temporal coordination dimension to the static "association field" introduced originally by Hess and Field (Field et al. 1993). In its dynamic version, the revised concept assumes that local-oriented inputs (Gabor patches) induce a facilitation wave of activity traveling along horizontal connections intra-V1 (Figure 12.9a). This coordination wave tends to bind proximal receptive fields with co-linear preferred orientations, thus creating a contiguous path of temporal integration. The associative strength of the perceptual effect is maximal when the asynchronous feedforward sequence produced by joint strokes of apparent motion (arrow in Figure 12.9b) travels in phase with the visually evoked horizontal intracortical propagation. Several arguments strongly support the relevance of introducing a time coordination. Intracellular recordings show that centripetal apparent motion produced by iso-oriented Gabor stimuli (co-aligned with the motion axis and presented from periphery to center) at saccadic speed (250°/sec in the cat) is more efficient than the simultaneous static presentation of the same stimuli (or the reverse centrifugal sequence) at evoking subthreshold synaptic responses from the "silent" periphery. Recent unpublished intracellular work from our lab shows that sparse apparent motion two-stroke noise appears as a powerful stimulus condition to trigger the coordination of synaptic activity along motion streaks attuned to the orientation preference of the target cells (Carelli, Pananceau, Monier, and Frégnac, in preparation).

Conclusion

The coordination processes that we have reviewed are generated mostly by recurrent and lateral intrinsic connections in the same cortical area, with a possible contribution of feedback control from higher cortical areas (although it may be minored by deep anesthesia). In terms of cognition/perception relevance, these processes are low level and not linked to attention since they are observed in humans during forced choice tasks as well as in the anesthetized mammal. These various interdisciplinary studies, based on intracellular electrophysiology, network imaging, and psychophysics, all point to the emergence of cooperative Gestalt-like interactions, when the stimulus carries a sufficient level of spatial and temporal coherence. Above a given activation threshold (yet to be quantitatively defined), a cooperative depolarizing or facilitatory wave becomes detectable in primary and secondary visual areas. The trigger zone can be considered as the initial point in cortical space where the symmetry of the compound effect of neural activity creates symmetry-breaking (see Engel et al., this volume). This process, which can occur spontaneously or be evoked during sparse sensory stimulation regime, initiates a wave that travels at low speed in the plane of the superficial cortical layers (0.10–0.30 m s^{-1}) and most likely becomes anisotropic for oriented inducer stimuli. The physiological features of the spatiotemporal propagation–coordination pattern recorded in V1 are highly correlated with the percept reported by the conscious human observer (e.g., Georges et al. 2002) and agree with predictions derived from the Gestalt theory. In the two cases of motion illusion reviewed here (apparent motion and line motion), a wave of perceptual binding modulates the integration of feedforward inputs yet to come: this wave can be seen as the propagation of the network belief of the possible presence of a global percept (the "whole": here, continuous motion of a space-invariant shape) before the illusory percept becomes validated by the sequential presentation of the "parts" (signaled by direct focal feedforward waves). This neuronal dynamics obeys closely the Gestalt prediction that the emergence of the "whole" should precede in time the detection of the "parts."

These cortical processes result, at the perceptual level, in the propagation of functional biases binding of contour and motion, which goes beyond the scale of the columnar orientation and ocular dominance network. It remains to be determined whether the correlations we report between perception and horizontal propagation are the sole result of neural processes intrinsic to V1, or whether they reflect the reverberation in V1 of a collective feedback originating from multiple secondary cortical areas, each encoding for a distinct functional representation of the visual field. It may be indeed envisioned that the primary visual cortex plays the role of a generalized echo chamber fed by other cortical areas (visual or not) that participates in the coding of shape and motion in space: accordingly, the waves traveling across V1 would signal the emergence of perceptual coherence when a synergy is reached between the different cortical analyzers. Synaptic functional imaging provides a new way to explore the genesis and propagation of such slow coordination, which may be instrumental to low-level cortical-mediated cognition.

Acknowledgments

This work has been supported by the CNRS, and grants from the French ANR and the European Community (FACETS (FET- Bio-I3: 015879) and BRAIN-I-NETS (FET-Open: 243914).

Overleaf (left to right, top to bottom):
Maurizio Corbetta, Bob Desimone, Yves Frégnac
Pascal Fries, Edvard Moser, Christoph von der Malsburg
Informal discussion, Ann Graybiel
Hannah Monyer, Laurent Itti, Wolf Singer
Lucia Melloni, John-Dylan Haynes, Matt Wilson

13

Coordination in Brain Systems

Edvard I. Moser, Maurizio Corbetta,
Robert Desimone, Yves Frégnac, Pascal Fries,
Ann M. Graybiel, John-Dylan Haynes, Laurent Itti,
Lucia Melloni, Hannah Monyer, Wolf Singer,
Christoph von der Malsburg, and Matthew A. Wilson

Abstract

This chapter reviews the concept of dynamic coordination, its mechanistic implementation in brain circuits, and the extent to which dynamic coordination, and specific manifestations of it, have the power to account for functions performed by interacting brain systems. In our discussions, we addressed how on-the-fly changes in coupling between neural subpopulations might enable the brain to handle the fast-changing recombination of processing elements thought to underlie cognition. Such changes in coupling should be apparent, first and foremost, in the statistical relationship between activity in interconnected brain systems, rather than in the individual firing patterns of each subsystem. Dynamic coordination may manifest itself through a variety of mechanisms, of which oscillation-based synchronization is likely to play an important but not exclusive role. Also discussed is how modulation of phase relationships of oscillations in different brain systems, in neocortex and hippocampus of the mammalian brain, may change functional coupling, and how such changes may play a role in routing of signals at cross sections between cortical areas and hippocampal subdivisions. Possible mechanisms for oscillation-based synchronization, particularly in the gamma frequency range, are explored. It is acknowledged that the brain is likely capable of producing zero-phase lag between spatially dispersed cell populations by way of rather simple coupling mechanisms, primarily when neuronal groups are coupled symmetrically. Synchronization with remote areas may be most efficient with phase differences that match the conduction delays. Fast-conducting, long-range projecting interneurons are identified as a potential substrate for synchronizing one neural circuit with another. A number of research strategies are identified to enhance our understanding of dynamic coordination of brain systems and how it might contribute to the implementation of the functions of those systems.

Introduction

In its simplest form, coordination is defined by the multiple interrelations that can be drawn between elements of any given assembly, and its phenomenological expression is signaled by the reconfiguration of elementary dynamics. The potential relations are viewed as functions of an externally defined context or an internally self-generated goal. In their introductory chapter, Phillips et al. (this volume) further constrain the issue by adding: "In general, coordinating interactions are those that produce coherent and relevant overall patterns of activity, while preserving the essential individual identities and functions of the activities coordinated."

We tried to define the process of dynamic coordination by contrasting it with non-vacillating alterations, ever aware of the fact that virtually all neuronal processes are coordinated in one way or another. Much of the coordination of activity required for the formation of specific response properties, such as the receptive fields of neurons in sensory systems or the generation of sequences necessary for the execution of movements, can be achieved by appropriate neuronal architectures that allow for the recombination and sequencing of signals in processing cascades based on divergence, selective convergence, and feedback. Such architectures allow for highly complex coordination and association of signals, even if these originate in separate processing streams, provided that there are adequate connections between the various stages of these processing cascades. To define relations and support selective grouping, it would be sufficient to increase jointly the rate of the responses that are to be associated with each other. Sparse coding and topographic coding would further enhance the salience of rate changes and reduce the risk of grouping unrelated but simultaneously enhanced responses. The fact that a number of phenomena can be predicted by firing rate-based models raises the question whether the fine-grained temporal structure of neural activity in different brain regions has additional explanatory power. Thus, we discussed whether more dynamic mechanisms are required to allow the flexibility, robustness, and speed at which cognition operates in the performing brain.

Dynamic coordination is required when the results of computations achieved in different processing cascades need to be recombined and associated in a flexible, nonstereotyped way. A paradigmatic case is working memory, where ever-changing items have to be temporarily associated with each other. It is unlikely that fixed neuronal architectures would be sufficient to anticipate and cope with the virtually infinite number of possible constellations of associable contents. For each possible constellation, one would need a devoted set of neurons receiving the appropriately selected converging inputs. Because of this limitation, the mechanisms required would need to be capable of establishing transitorily, and in a highly flexible way, relations among signals originating in different processing cascades. These mechanisms would need to be able to select in a dynamic, task- or goal-directed way signals from different, spatially segregated processing streams and assure selective interactions between, and further joint processing of, these signals. In our deliberations, we considered the possibility that if several such signals are to be coordinated for the first time, that is, before anything is known about their relatedness, a possible mechanism for such relatedness (presumably due to common cause) might be in the form of temporal correlation. This transient coordination may be necessary to establish new linking paths, although its subsequent recruitment might not always be required.

Furthermore, we recognized that most behaviors, and relations between different processing streams, are not generated *ex novo* but are highly predictive based on the history of activation and learning of the system. We rarely reach for a visual object by allowing an arm to cross the midline; we rarely raise a leg and an arm simultaneously unless we are dancing. These relationships cannot be coded just in the anatomy, but must be implemented in the statistical properties of the intrinsic (not task-driven) functional connectivity, which may also require mechanisms of temporal correlation. Dynamic coordination thus does not arise from noise, but from a patterned baseline landscape that may constrain to some extent its flexibility.

The question then arises: How can the presence of dynamic coordination be diagnosed? We agreed that dynamic coordination should be apparent in the patterns of interactions between defined neuronal populations. For all coordinated processes it should be the case that more information can be retrieved by considering the joint activity of neurons belonging to different processing cascades than evaluating the activities of the respective neurons in isolation. In nondynamically coordinated processes, the relations among the respective firing sequences will be stereotyped across trials, if stimulus conditions remain constant. The additional information contained in the relations between the respective firing patterns can therefore be retrieved in sequential recordings from these neurons and with averaging across trials. Such analysis has been applied, for example, to the motor cortex and has led to the discovery of population codes for movement trajectories (see Georgopoulos et al. 1986; Frégnac et al., this volume). However, this approach may be insensitive to more fine-grained temporal relations between the discharges of dynamically grouped neurons. These relations can only be determined by simultaneously recording from the neurons whose activity is suspected to be coordinated. What one might observe in these cases is that the individual responses change only little, if at all, in terms of average rate, while measures of relations between the responses change in a systematic way.

A further constraint is that, in dynamic coordination, the information containing relations must change in a context-, task-, and goal-dependent way. For example, temporal structures may depend on whether a trial was successful or an error trial, and they may change as a function of stimulus context or shifts in attention or goal definitions. To understand the functions of dynamic coordination on the basis of extracellular recordings, it is imperative to perform multisite recordings and to search for real-time correspondences between temporal structure and task demands. Such approaches may require analytical methods that extract activity patterns in many cells at the same time and take their trial-by-trial covariation into account. Note, however, that it is possible to infer coordination-related processes from single site recordings by appropriate multiscale analysis of intracellular membrane recordings, which reflect the impact of long-distance feedback and lateral connectivity (see Frégnac et al. 2009; also Frégnac et al., this volume).

Experimental evidence for such conditions of dynamic coordination is available. In neuronal recordings during binocular rivalry, for example, whether the considered neuron is part of the processing cascade leading to conscious experience cannot be predicted from the rates of individual neurons in V1. In contrast, a measure of synchrony, in this case precise synchronization of periodic activity in the gamma frequency range, predicts correctly whether or not a given pair of neurons participates in the processing cascade which, in this very moment, conveys the activity that reaches consciousness and is perceived (Fries et al. 1997). In inferotemporal cortex, this information can be retrieved from the rates of individual neurons, suggesting that synchronization (dynamic coordination) at early processing stages was used to select responses for further joint processing by enhancing their saliency, which then facilitated transmission of this synchronized activity to higher levels where it then evoked increased rate responses.

Examples for highly specific inter-regional coordination can also be found in human neuroimaging. Functional neuroimaging is particularly suitable to reveal large-scale dynamic coordination processes that span the entire brain. Several methods are available that allow researchers to investigate how cognitive processes change the interactions between remote brain regions. These include psychophysiological interactions (Friston et al. 1997), dynamic causal modeling (Friston et al. 2003), and Granger causality mapping (Roebroeck et al. 2005). Haynes et al. (2005), for example, have investigated the effects of attention on the connectivity between representations of attended locations. They let subjects attend to two out of four spiral stimuli and report whether they had the same or different handedness. Thereafter they measured the functional connectivity between the individual representations of these stimuli in retinotopic visual cortex. They found that functional connectivity was increased between the retinotopic representations of jointly attended stimuli, both within regions (i.e., V1–V1, V2–V2) and between regions (V1–V2, V2–V1).

A topic of our discussion was whether the definition of dynamic coordination as structured temporal relationships between neuronal populations excludes simultaneous changes in the individual populations to be synchronized. A strict application of the criteria implied for dynamic coordination by Phillips et al. (this volume) would suggest that local processes, such as the discharge fields of V1 neurons, would remain unchanged under associative stimulation protocols. The underlying assumption is that the relational information should remain separable from the information carried initially by each stimulus component taken in isolation. It was agreed that a strict application of the definition may not be valid as there are numbers of examples which suggest that changes in dynamic coordination give rise to, or are associated with, changes in individual subcircuits. Three examples illustrate the diversity of interactions between coordinated processes and activity in single cells and single populations.

The first example was taken from the stomatogastric ganglion of the lobster. This particular sensorimotor network constitutes a striking example of assembly dynamic reconfiguration correlated with changes in behavior where, during the coordination, the electrical input–output properties of individual elements are not preserved. Coordination is controlled by "orchestra leaders" (PS cell), whose activity triggers the widespread broadcast of neuromodulators. This neuromodulation impacts on the intrinsic reactivity of the other cells by changing reversibly the expressed repertoire of membrane conductances. Consequently, the individual excitability patterns of any given cell will change depending on the context (before, during, or after the orchestra leader cell has fired). Note that in this paucineuronal biological network where all partners are known, the coordinator is identified and the causal link between temporal assembly motifs and the behavioral actions, as well as their functional significance (food swallowing, crunching, and expulsion), are clearly defined (Meyrand et al. 1991).

In a second example, the coordinating agent was part of the high-order statistical features present in the sensory input stream. Changing the statistical regularities of the environment may produce drastic reorganization of ensemble activity patterns and their stimulus-locked reliability. For instance, it is well known that repeated presentation of drifting luminance gratings in V1 receptive fields evokes strong but highly unreliable responses, both at the spiking and subthreshold levels. In contrast, in the same cells, virtual eye-movement animation of natural scenes evokes temporally precise sparse spike responses and stimulus-locked membrane potential dynamics which are highly reproducible from one trial to the next (Frégnac et al. 2005). In this second example, coordination is unrelated to the behavioral outcome or neuromodulation since it is observed in the anesthetized and paralyzed preparation as well as in the attentive-behaving monkey (Vinje and Gallant 2000). This self-organized process adapts the temporal precision of the sensory code to the statistics of the input. However, in contrast to the first example, this adaptive form of temporal coordination is done in the absence of internal executive or supervision units. As demonstrated in the first example, the full field "whole" condition will affect the functional identity of the recorded unit (i.e., the individual receptive fields of the V1 cells).

These two examples illustrate conditions where properties of the individual units of a circuit clearly change in parallel with coordinating processes; however, the literature also contains illustrations where the information that can be stored or recalled on the basis of coordinated activity is separable from the rate responses of single neurons. This can be seen in recordings of responses to long-duration single frequency tones from the auditory cortex, where highly transient burst responses are detected at the onset or offset of the tone whereas the mean activity is unchanged during the tonic phase of the stimulation. The presence of the stimulus is here signaled by a dramatic and tonic elevation in the correlation between cortical units coding for the sound frequency, without any apparent change of firing rate (deCharms and Merzenich 1996).

It remains a challenge to define a taxonomy of coordination where the underlying mechanisms of each phenomenological form can be clearly separated. Nonetheless, we agreed that dynamic coordination is apparent in a number of studies that show changes in the temporal structure of the joint activity of two or more neuronal populations that differ in character to those taking place at the single-population level.

A Possible Need for Fast-changing Neuronal Architectures

All coordination requires the definition of relations. In the nervous system, relations are established by the anatomical connections among neurons, the anatomical architecture of the networks, and the patterns of inter-regional spontaneous activity, the baseline or intrinsic functional architecture of the networks. Work in nonhuman and human primates indicates that the anatomical connectivity matrix has small world properties allowing for the coexistence of local processing and long-range integration (Kotter 2004; Hagmann et al. 2008). This small world architecture also gives rise to space-time structures of coupling and time delays, which in the presence of noise defines a dynamic framework for the emergence of spontaneous and task-driven cortical dynamics at different temporal scales (minutes, seconds, hundreds of milliseconds) and could support both long- and short-term changes in functional connectivity. To allow for dynamic coordination in behavior (task-dependent selection of responses for joint processing, selective association of subsystems to be engaged, etc.), the functional architecture must be modifiable at the same rapid pace as cognitive and executive processes can change. This requires fast changes in effective coupling among neurons; that is, the gain or the efficiency of a connection must be modifiable. The brain is likely to have a number of mechanisms for achieving such changes in coupling, operating at different timescales.

Coordination by Gain Modulation

Dynamic coupling can to some extent be accomplished by well-characterized gain-modulation mechanisms. Synaptic gain changes can be induced within tens of milliseconds, they may be (but do not have to be) associative and can last from a few tens of milliseconds (e.g., during frequency-dependent changes in transmitter release) to many decades (e.g., when activity is stored in long-term memory). Effective coupling can also be changed by purely activity-dependent gating, such as when dendritic segments are switched off by shunting inhibition, or when the sequence of activated synapses along a dendrite is changed so that excitatory postsynaptic potentials (EPSPs) either summate effectively or shunt each other, or when the nonlinear amplifying effect of NMDA receptors is enabled or vetoed by local or global adjustments of membrane potential. The question raised in our discussions was whether such gain-modulating mechanisms would be sufficient to account for the speed and flexibility of cognitive operations.

Coordination by Synchronization of Oscillation Patterns

A candidate mechanism for effective change of the coupling among neurons involves rhythmic modulation of discharge activity (neuronal oscillations). Oscillating networks facilitate the establishment of synchrony because they can capitalize on the effects of entrainment and resonance. Oscillators that are tuned to similar frequencies have the tendency to engage in synchronous oscillations if reciprocally coupled. This is the case even if coupling is very weak and even if their frequency tuning is broad and the preferred frequencies are not identical.

An oscillatory modulation of membrane potential, such as occurs in oscillating cell assemblies, confines spiking to the rising slope of the depolarizing phase. Thus, spikes emitted by networks engaged in synchronous oscillations become synchronized. The temporal precision of this synchronization increases with oscillation frequency. In the case of gamma oscillations, output spikes can be synchronized with a precision in the range of a few milliseconds. Because of the coincidence sensitivity of neurons, this synchronization greatly increases the impact that the output of synchronized cell assemblies has on subsequent target neurons.

Another virtue of oscillations is that they allow the exploitation of phase (relative timing) for coding (see discussions on phase precession in the hippocampus in Mehta et al., this volume). In oscillating, synchronized cell populations, responses to strong excitatory inputs will occur earlier on the rising phase of the oscillation than responses to weak inputs. Thus, intensity can be encoded in the time of spiking relative to the oscillation phase. This is a convenient way of coding since the latency of first spikes already contains all information about the amplitude of the driving input. Early studies on retinal coding by Kuffler (1953) showed that relative intensities of visual stimuli can readily be assessed from the relative latencies of the first spikes of ganglion cells. Later studies showed that image reconstruction from first spike latencies is as good as counting rates over several hundred milliseconds (VanRullen and

Thorpe 2001). Thus, readout time for this temporal code is much faster than for the rate code. In the case of the retina, these intensity-dependent differences of spike latencies are of course caused by receptor kinetics. In central processing, the same conversion of an amplitude code into a temporal code can be achieved, in principle, by oscillatory modulation of cell assemblies.

These considerations provide answers to the question: Under which circumstances are oscillations needed? They are needed or at least highly advantageous if (a) spikes have to be synchronized with high precision to support their propagation in sparsely connected networks (see synfire chains of Abeles 1991); (b) spike timing has to be adjusted with high precision for the definition of relations in learning processes such as spike timing-dependent plasticity (STDP); or (c) phase is used as coding space, i.e., if timing relations between spikes or between spikes and the phase of a population oscillation convey information about input amplitude or the relatedness of distributed processes.

There was consensus in our group that several pieces of experimental data are consistent with a role for oscillation-based synchronization in cognitive processes. For example, researchers have shown that attention during visual search correlates with increases in coherence between local field potentials (LFPs) from the frontal cortex and the parietal cortex (Buschman and Miller 2007, 2009). Around the time when monkeys find and shift attention to a visual target, there is an increase in coherence in two different frequency bands: an upper frequency band (35–55 Hz) for bottom-up attention (pop-out), and a lower frequency band (22–34 Hz) for top-down attention (conjunction search). During search for conjunctions, the monkeys shift the location of their attention every 40 ms. The attention-related shifts in frontal eye field spiking activity were correlated with increased power in the lower frequency band, suggesting that the oscillations act as a "clocking" signal that controls when attention is shifted (Fries 2009). The study suggests that serial covert shifts of attention are directed by the frontal eye field and that synchronization between cortical systems may regulate the timing of cognitive processing. Task-induced changes in synchronization or coherence have been reported at the level of individual regions during sensory integration (Roelfsema et al. 1997), selective attention (Fries, Reynolds et al. 2001), working memory (Pesaran et al. 2002; Howard et al. 2003), and motor control (Crone et al. 1998). Between distant cortical regions they have been reported during object recognition (Varela et al. 2001), working memory (Jones and Wilson 2005), long-term memory encoding (Fell et al. 2001), visual attention (Gregoriou et al. 2009), and sensorimotor integration (Roelfsema et al. 1997).

Oscillations and Dynamic Routing

Oscillations may influence routes of communication within structurally constrained brain networks. Consider two groups (A and B) of neurons that provide converging synaptic input to a common target group (*C*) and compete for influence on this target group. If there is rhythmic synchronization among the neurons in group *A* and among the neurons in group *B*, but not between *A* and *B*, then *C* will most likely synchronize to either *A* or *B*, but not to both at the same time (Börgers and Kopell 2008). The locking of *C* to either *A* or *B* implements a winner-takes-all between the competing inputs of *A* and *B* and establishes an exclusive communication link between the target *C* and the more strongly synchronized input (Fries 2005; Fries et al. 2008).

The described constellation of two neuronal groups converging onto one target group is a fundamental motif in cortex. While this motif renders the postsynaptic neurons selective to diagnostic features of the learned input pattern, it also renders them nonselective or invariant to nondiagnostic accidental features. This invariance is an advantage, because it might provide the basis for object recognition in the face of changes to irrelevant stimulus aspects; however it is also a curse, because a given stimulus will never cover the complete input space of a given neuron, leaving room for competing stimuli. It would be beneficial if the effective input of a given neuron at a given moment in time were limited to functional subsets corresponding to one actual object. This selective efficacy of subsets of a neuron's input might be implemented through the above mentioned exclusive communication link, possibly by synchronization in the gamma frequency band. For this solution to work, two conditions have to be met simultaneously: First, inputs driven by a given stimulus need to be rhythmically synchronized to each other, but not to inputs driven by other stimuli. This corresponds to the binding-by-synchronization hypothesis (Singer and Gray 1995; von der Malsburg 1981/1994). Second, one of the input segments has to be given a competitive advantage over the other through an enhancement. This corresponds to the hypothesis of biased competition through enhanced synchronization (Fries 2005). Thus, the way to use structural convergence in order to harvest both selectivity and invariance seems to lie in the interplay between structural neuronal connectivity and dynamic neuronal synchronization.

In proposing a role for oscillatory activity in dynamic coordination of neuronal populations, our group agreed that one should not forget that oscillatory activity, which may certainly be considered a signature or a manifestation of dynamic coordination, does not necessarily explain the causes or mechanisms by which such coordination arises. Consider a simple dynamic routing example like the one described above, where information in a low-level sensory processing area could be routed toward one of two possible targets in a higher area, with the choice of direction being endogenous, i.e., not dictated by the stimulus. When information is routed one way (choice *A*), some neurons may oscillate in one manner; when routed the other way (choice *B*), the same or other neurons may oscillate in a different manner. The mere existence of these oscillations does not explain how the selection was implemented in spatially specific synchronization patterns. The result, signature, or manifestation of choosing between *A* and *B*, and communicating it to lower areas, may be what is expressed in patterns of oscillations in the functionally connected areas, but how the decision is made and what mechanisms and pathways are employed to communicate it remain pressing questions.

Oscillations and Phase Relationships

Coupling between cell populations is heavily dependent on the phase relationship of the cells in the groups to be linked. By adjusting phase angles, coupling can be modified over the whole range from ineffective to maximal. Controlled changes in firing phases can also be used for dynamic routing if sender and receiver are oscillating at a similar frequency and phases are adjusted. Because oscillations can occur over a wide frequency range, many routes can be specified at the same time without interference. Finally, because coupling in oscillatory networks depends on phase and because we observe the coexistence of oscillations in different frequency bands (theta, beta, gamma), many different and graded adjustments of couplings can be structured, providing opportunities for establishing dynamically graded and nested relations, which could be advantageous for the encoding of compositionality.

Consistent with a role for oscillations in routing of information, experimental data suggest that the phase of the ongoing oscillation can establish preferential windows for information processing. Inputs that arrive in the "good phase" of the ongoing oscillation will be processed preferentially, whereas those arriving at the "bad phase" will be suppressed. For a long time it has been known that the ability to perceive weak signals fluctuates slowly over several seconds (streaking effect). A recent study showed that infra-slow (0.01–0.1 Hz) fluctuations of ongoing brain activity correlate with this behavioral dynamics. In this study (Monto et al. 2008), the probability of detecting a tactile target at threshold was 55% more likely in the rising phase of the fluctuation, and strongly correlated with the power amplitude of higher frequency (1–40 Hz) EEG fluctuations. Support for the same hypothesis comes from two more recent studies showing a relationship between visual detection and phase in the theta–alpha range (Busch et al. 2009; Mathewson et al. 2009).

Finally, we discussed the potential impact of precise phase relationships on learning mechanisms. This seems important because processing architectures have to be adjusted to the requirements of mechanisms establishing durable relations (e.g., in associative learning); that is, they have to transform the (semantic) relations defined during processing into permanent changes in coupling that represent these relations. If any of the mechanisms of associative synaptic plasticity known to date (LTP, LTD, STDP) have anything to do with learning, it would seem that processing architectures need to be capable of defining relations in the temporal domain and that they will have to do so by adjusting the timing of individual spikes with a precision of a few milliseconds.

In STDP, for example, it matters whether an EPSP arrives just a few milliseconds before or after a spike to increase or decrease the efficacy of a connection. It would seem, therefore, that in signal processing and in dynamical coordination, relations should be specified with a similar temporal resolution and precision. Synchronization and phase adjustments in the gamma frequency range could provide the time frame for the precise adjustment of spike timing required for STDP. One should note, however, that the evidence for STDP *in vivo* is relatively scarce and that it is easier to demonstrate the negative (depression) than the positive (potentiation) parts of the STDP curve in the adult cortex (Jacob et al. 2007).

A Wider Repertoire of Coordination Mechanisms

Oscillations cannot be the sole mechanism of dynamic coordination. The neuroscience literature contains a number of examples of dynamic coordination where brain circuits communicate using precise temporal codes not expressed as lasting synchronization in oscillatory patterns in the LFP. The diversity of mechanisms can be illustrated by patterns of hippocampal–neocortical interactions in slow-wave sleep. Slow-wave associated transitions in excitability from low firing rate (putative down-state) to high firing rate (putative up-state) exhibit a systematic timing relationship in which the neocortex leads the hippocampus. During the elevated firing rate period that follows that transition, the hippocampus expresses a series of sharp wave-ripple burst events that replay sequential spatial memory information in which the timing relationship is reversed, with the hippocampus thought to be leading the neocortex. The dialog that may be reflected in shifting timing relationships may reflect the dynamic coordination of oscillatory modes during memory processing.

The possibility of a wider repertoire of mechanisms was further illustrated by discussion of the mechanisms for rapid object recognition in the visual cortex. A given scene may be analyzed in terms of complex arrays of relations by focal attention visiting here and there, and even if the relations thus identified are indeed represented in terms of correlated oscillations, a more permanent trace of these relations must be left behind to be available at later visits by focal attention. If, for instance, there is a number of objects and a number of persons present in a scene, then sequential focal attention may discover which object belongs to which person, one by one, as a result of some inference; when coming back to one of the persons or objects, this result should be available immediately without the necessity of going through the process of inference from scratch. In addition, there is the necessity of maintaining ongoing relations or links between different neuronal ensembles over longer timescales. Many patterns of behavior are predictable, although not necessarily across individuals, repeated over and over with little variation, while at the same time novel behaviors can be stabilized with learning.

What could the mechanism of such short- and longer-term storage of relations be? It was proposed that in addition to the elementary symbols represented by neurons (or groups of neurons), there might be a large network of dynamic links. These links correspond to permanent neural connections, which can, however, be modified (e.g., made ineffective) temporarily. In this view the brain's representations would not have the form of vectors of activity, or neural signals, as in classical conceptualizations, but they would have the form of dynamical graphs. There are various mechanisms by which the efficiency of connections can be rapidly modified. There is synaptic plasticity on a continuous range of timescales, starting from a few milliseconds, and the effective connectivity (Aertsen et al. 1989) of a network can be changed by a variety of presynaptic and postsynaptic influences.

Recent work in human brain imaging shows that spontaneous activity, as measured by fluctuations of the blood oxygenation level dependent (BOLD) signal, is not random but organized in specific spatiotemporal patterns (Deco and Corbetta 2010; Fox and Raichle 2007) that resemble functional networks recruited during active behavior. These correlations occur at a very slow temporal scale (<0.1 Hz), which correspond to fluctuations of slow cortical potential (0.1–4 Hz) and band limited power fluctuations of the gamma band (He et al. 2007). These patterns of spatiotemporal correlation at rest reflect not only the underlying anatomy, but are gated by their recruitment during tasks. The leading hypothesis, supported by studies showing changes with learning and lesions, is that these patterns of spontaneous activity code for relations in the cortex that are related to the history of network activation and learning. They may represent attractor states that constrain and potentially bias the recruitment of brain networks during active behavior.

If these views are correct, then neuroscience is currently ignoring a large part of the representational machinery of the brain—very large indeed, as there are many more connections than there are neurons in the brain.

If coordination is expressed largely by dynamic connections, then what is the importance of signal correlations? We agreed that signal correlations are likely to be indispensable when a set of neurons are to be coordinated for the first time; that is, when the downstream circuits have not yet encoded this relatedness in their link structure. In this way, neural oscillations could play a vanguard role, appearing only early in some learning task, disappearing as soon as the coordination pattern is encoded in some connectivity structure.

Computer modeling work will be particularly useful in shaping our thoughts about neural operations if the model can be related in a convincing way to neural operations, instead of just using the brute force of high-speed computers, and if the performance of the model can be proven superior in public benchmark tests. Such tests are available for face recognition (e.g., FRVT 2002; Messer et al. 2004; Phillips et al. 2005). The consistently winning systems were all correspondence-based; that is, they are based on representations of faces in terms of two-dimensional arrays of local features (mostly of Gabor type, i.e., modeled after receptive field types found in primary visual cortex) and on finding correspondences between local feature sets in model and image. As a large number of such correspondences are to be found for a given match, temporal coding is bound to be rather time-consuming. A model of correspondence-finding by temporal coding (Wiskott and von der Malsburg 1996) would, if implemented with realistic neurons, have needed more than 10 s to recognize a face, two orders of magnitude slower than human performance. If, however, connectivity patterns were installed that allowed for the fast dynamic installation of topographic fiber maps with the help of control units (introduced by Olshausen et al. 1993), high-performance face recognition by correspondence-finding within physiological times of 100 ms was feasible (Wolfrum et al. 2008). In a related study (Bergmann and von der Malsburg, submitted), it is shown that the necessary control unit circuitry can be developed on the basis of synaptic plasticity controlled by synchrony coding.

Correspondence-based object recognition models have explicit representation of shape. It was argued, however, that pure feedforward models (such as Serre et al. 2007; Poggio and Edelman 1990), which do not make use of dynamic coordination, are also able to represent shape. The Chorus of Prototypes algorithm (Duvdevani-Bar and Edelman 1999) and the Chorus of Fragments model (Edelman and Intrator 2003) may represent shape if endowed with a mechanism for relating together the responses of the ensemble of neurons that represent, in a distributed yet low-dimensional manner, the current input. Temporal binding by synchrony may be just such a mechanism (as was proposed in Hummel and Biederman 1992).

The Mechanisms for Synchronization between Neural Populations

Although there was consensus that the brain has a wide repertoire of mechanisms for achieving dynamic coordination, we chose to discuss in more detail coordination by synchronization of oscillatory activity across neuronal populations. This form of coordination has support in the experimental literature, as suggested above, and there is now a considerable literature exploring mechanisms of synchronization at the level of cellular assemblies.

We began the discussion by reviewing models for synchronization between cell populations. Several models have been proposed as mechanisms for achieving zero-phase lag between same-frequency oscillatory activity in different populations, which by definition might be seen as the ultimate expression of synchronization. Evidence indicates that zero-phase lag synchronization is ubiquitous and can occur over surprisingly large distances, such as between the hemispheres (Engel, König, Kreiter et al. 1991), despite the rather considerable conduction delays of pathways connecting the synchronized assemblies. At first it may seem that such synchronization between widely dispersed populations could be achieved only by common input from a central oscillator that slaves the respective synchronized assemblies. However, since callosotomy abolishes interhemispheric synchrony, it may instead rely on interactions between the synchronized assemblies. Several models for such interactions were considered.

One class of models relies on spike doublets of inhibitory interneurons (Traub et al. 1996; Ermentrout and Kopell 1998). When neurons synchronize locally in the gamma band, there is a characteristic interaction between excitatory and inhibitory neurons: excitatory neurons spike first and trigger inhibitory neuron spiking with short delays. The ensuing inhibition shuts down the local network until inhibition fades and the cycle starts again with the firing of excitatory neurons. Long-range synchronization between two such gamma oscillatory groups can occur when excitatory neurons of group A excite interneurons of group B, even if this entails a conduction delay of a few milliseconds between A and B. Essentially, the excitatory input from A to B triggers a second inhibitory spike in B and thereby prolongs the inhibition inside B by the conduction delay. The two interneuron spikes in rapid succession gave this model the name "spike doublet mechanism." One prediction of this model is that local gamma band synchronizations decrease in frequency when coupled across long distances and the frequency decrease is proportional to the conduction delay.

Synchronization across long distances might also be supported by other configurations of reciprocal interaction between the subcircuits. Evidence is now available which shows that zero-phase synchrony can be established despite conduction delays in the coupling connections both from experiments with coupled lasers (Fischer et al. 2006) and modeling of networks with spiking neurons (Vicente et al. 2008). As long as at least three reciprocally coupled systems are allowed to interact (triangular configurations), zero-phase synchrony is easily established and very robust against scatter in conduction times of coupling connections.

A useful mathematical perspective on the phenomenon of zero-phase synchronization comes from the study of coupled map lattices and globally coupled maps. These are systems of coupled nonlinear dynamical systems, whose long-term (ergodic) behavior can show some universal properties under some simplifying assumptions (Tsuda 2001). One of these assumptions is that the system is globally and symmetrically coupled with a single coupling strength. Under these constraints, it can be shown that the states of every coupled dynamical system come eventually to occupy a synchronization manifold. Crucially, because of the symmetry constraints on the dynamic equations, the set of all solutions must obey the same symmetry. Zero-phase synchronization represents a symmetrical solution. Due to spontaneous symmetry breaking, however, individual solutions might violate symmetry (i.e., exhibit nonzero-phase synchronization). A simple example is a ball sitting on top of a hill in a completely symmetric state. However, as this state is unstable, the slightest perturbation will cause the ball to roll downhill. This movement will not occur symmetrically in all directions but in each case in some particular direction. Each single solution (i.e., rolling in a specific direction) breaks the symmetry of the initial problem. Only the set of all solutions and the probability for the ball to roll in a specific direction is symmetric. This has been used in a model investigating oscillatory interactions in primary visual cortex (Schillen and König 1991). Here a specific type of excitatory tangential connection avoids the trivial solution of global synchronization. With the additional assumption of ergodicity (i.e., the system that evolves over a long timescale and visits all regions of state space), individual solutions have to obey symmetry constraints. For systems with more than two coupled oscillators, this reduced the solution for the entire system to global synchrony with zero-lag quasi periodic or chaotic oscillations. At low but nonnegligible coupling strength, the synchronization manifold is "riddled" with unstable points that "eject" the trajectory away from the synchronization manifold to produce intermittent bursts of localized activity (Breakspear et al. 2009).

The main message from these theoretical treatments is that there is nothing mysterious about zero-phase lag synchronization among three or more populations. Indeed, under the constraints of the model, it is impossible to get any synchronization other than zero lag. To get consistent (nonzero) phase coupling, one has to break the symmetry, in terms of the intrinsic parameters of the system or its coupling parameters. This basic phenomenon has been illustrated using neuronally plausible simulations by Chawla et al. (2001), where it proved difficult to break the symmetry provided by three or more neuronal systems that are interconnected in a roughly symmetrical fashion.

Although zero-phase synchronization could serve as a useful guide for understanding the mechanisms underlying long-range synchronization between neural circuits, it was argued that the current models for producing zero-phase lag have relied on unrealistic architectures and that the physiological properties of the model neurons do not match those of the performing brain (e.g., neurons do not regularly fire in doublets during gamma oscillations). It was proposed that zero-phase lags may not even be desirable for synchronization when information is communicated over long distances. It may often be advantageous to introduce systematic phase shifts to coordinate convergence of distributed information from different sources or to enforce timing relationships that would establish specific patterns of dynamic routing. The actual phase lags between oscillating populations in two regions may vary across task conditions and network states. One example for modulating the efficiency of interareal coupling by systematic phase shifts between oscillatory activity is cortico-tectal communication (Brecht et al. 1998). It was also recognized that regulation of spike timing through systematic phase locking can be used to encode temporal relationships (such as spatial behavioral sequence encoding through theta phase precession).

An example of time-shifted synchronization across brain areas was recently reported in a study of frontal eye field (FEF)–V4 interactions in an attention task (Gregoriou et al. 2009). There is considerable evidence that FEF plays an important role in the top-down control of attention in visual cortex, including V4. In the Gregoriou et al. study, spikes and LFPs were recorded simultaneously from FEF and V4, in monkeys trained in a covert attention task. One stimulus always appeared inside the shared receptive field and two others appeared outside; the monkey was cued to attend to a different stimulus on each trial. Spike-field coherence in the gamma band increased with attention in V4 and FEF. The effect was particularly strong when cells in the two areas had overlapping receptive fields. However, there was almost a 180° phase lag in synchrony in the gamma frequency band between FEF and V4, corresponding to a time delay of about 10 ms. The same 10 ms time shift was found in other frequency bands of the V4–FEF synchronous activity, suggesting that there is a constant 10 ms time shift between the time while cells spike in one area and cells are maximally depolarized in the other. It was suggested that this time shift may be accounted for by conduction and synaptic delays between the two areas. If so, then spikes from one area would actually arrive in the connected area at a time when the receiving cells were most prepared (depolarized) to receive them, which is consistent with the strong effects of FEF activity on the top-down attentional modulation of V4 responses. The study illustrates the potential role of time-shifted synchrony between areas as a common mechanism for functional interactions between cortical areas and raises the possibility that zero-lag synchronization may be implemented primarily in local circuits.

To add to the complexity, a neuronal population may have different phase lags to different subsets of a population with which it interacts. The recent description of traveling theta frequency waves in the hippocampus (Lubenov and Siapas 2009) suggests that neurons in regions that communicate with the hippocampus may be synchronized with a subset of the hippocampal population across a wide range of the oscillatory cycle, but the identity of the neurons with which synchronization occurs may change with phase. These phase lags may influence the wider patterns of coherence between the hippocampus and other structures, such as the striatum, for which phase angle changes with task and with learning (Tort et al. 2008; DeCoteau et al. 2007).

There was general consensus that the mechanisms enabling synchronous firing across widespread brain regions are poorly understood, especially for the higher frequency (e.g., gamma), and that alternative solutions should be considered. One possibility considered involves long-range axonal collaterals. Synchronization between two oscillating populations might be achieved if the collaterals of gamma-modulated pyramidal cells in one location phase reset the basket neurons in the other location. A fundamental problem with this model is the limited axonal conduction velocities of pyramidal cells. An alternative fast-conducting conduit between distant sites may instead be provided by axon collaterals of so-called "long-range" interneurons. The anatomical "short cuts" provided by the long-range interneurons may offer the interarea fast transmission that is required to phase-synchronize gamma oscillations between distant cortical regions. Such far-projecting fast-conducting interneurons have been described in the hippocampus (Sik et al. 1994). The axons of this interneuron family innervate multiple regions of the hippocampus and can project to multiple external regions, including medial septum, subiculum, presubiculum, entorhinal cortex, induseum griseum, and possibly other cortical regions. Similarly, GABAergic interneurons in the medial septum project to the hippocampus, preferentially to GABAergic interneurons (Freund and Antal 1988), hippocampal GABAergic neurons provide long-range projection back to the septum (Gulyás et al. 2003), the basal forebrain has GABAergic neurons that project widely across the cortex (Sarter and Bruno 2002), and long-range GABAergic interneurons are known to connect remote areas in the ipsilateral and contralateral cortex (Buhl and Singer 1989; Gonchar et al. 1995; Kimura and Baughman 1997; Tomioka et al. 2005). A common property of many of the long-range interneurons is that their axon caliber is nearly twice as large as that of parallel conduits from pyramidal cells connecting the same regions and the diameter of the surrounding myelin is three times thicker (Jinno et al. 2007). The estimated volume of the total axon arbor of a long-range interneuron is several times larger than the volume occupied by the axon tree of pyramidal cells, suggesting that only few such neurons may be needed to establish coherence between regions. There was consensus in the group about the need for further investigation of the potential role of long-range fast-transmitting inhibitory interneurons in fast inter-area cortical synchronization.

We also discussed the potential role of ascending neuromodulatory systems in synchronization of activity across brain regions. The broad terminal fields of axonal projections from monoaminergic and cholinergic cell groups generally speak against a role in controlling dynamic changes in specific subsets of interacting cell clusters, as does the slowness of many receptors for such transmitters (e.g., dopamine) and the long time that it takes to clear the transmitter from the synaptic cleft. These facts do not, however, exclude a key permissive function for ascending neuromodulatory systems in providing necessary conditions for inducing oscillatory activity. The discharge patterns of cholinergic as well as monoaminergic cell groups change radically during transitions between brain states (e.g., when subjects switch between awake states and sleep), and such changes are temporally correlated with massive changes in the oscillatory properties of cortical networks. Observations suggest that, although terminal fields are broad, subtypes of intermingled interneurons are innervated by different neuromodulatory systems (e.g., 5-HT axons terminate on CCK-expressing interneurons but not parvalbumin-expressing cells, whereas cholinergic projections primarily terminate on basket cells). The specificity of the neuromodulatory innervation, as well as the specific combinations of receptor subtypes expressed by different classes of interneurons, and the ability of neuromodulators to change the time constants of GABA receptor potentials are likely to have significant impact on the generation of oscillations and synchrony across brain regions, although the exact mechanisms remain to be determined.

In our discussions, we briefly straddled the issue as to whether areas in the neocortex only exchange information once they have finished their respective computations and then transmit the result (discontinuous communication) or whether they permanently interact (continuous processing) until they converge to a collective result. We felt that the latter scenario is more realistic, although some ERP studies seem to suggest that information is transmitted in discrete packages.

Steps into the Future

Computational Models

How might neuroscientists improve their understanding of the brain's mechanisms for dynamic coordination? Our discussion of models and experiments will be presented sequentially, although the consensus is that advances require an integrated approach.

Models will play a critical role in interpreting the many disparate empirical findings regarding coordination in neural systems. Cortical models for dynamic coordination across brain systems can be roughly categorized according to whether they are focused on the role of large population influences on single neuron properties versus models centered on the nature of interactions across two or more specific cortical structures or layers.

There are numerous examples of models that examine the effects of attentional feedback or task demands on single neuron properties. The feedback in these models comes from unspecified sources, and in most cases the models only consider the effects of feedback on average firing rates. In the field of attention, for example, biased competition (Desimone and Duncan 1995; now described as normalization models, Reynolds and Heeger 2009), feature-similarity gain (Maunsell and Treue 2006), and response gain models all attempt to explain how attentional feedback cause the enhancement of responses to attended targets and the suppression of responses to unattended distracters. Normalization models explain and predict the large majority of attentional effects that have been reported on single neuron properties.

In contrast to these attentional models, based on average firing rates, some models also address the role of spike timing and synchrony in neural populations. It is claimed that only spiking neuron models that incorporate gamma synchrony can explain the effects of attention on competing stimuli within the same receptive field (Börgers et al. 2008), although direct tests of competing models on these data are missing. In the future, it will be critical to make differential predictions from models based on static firing rates versus synchrony and population dynamics, which can then be tested empirically in neurophysiological studies.

Fewer quantitative models take on the daunting task of modeling the interactions among two or more cortical areas. Efforts are ongoing to collect data on a large number of individuals (upward of 2,000 healthy subjects) to characterize the anatomical, functional, and electrophysiological neuromatrix of the human brain (The Human Connectome Project). The goal of this project is to provide the neuroscience community with a public data set, which will hopefully describe for the first time the entire array of cortical areas, as well as their anatomical and functional links. This will allow quantitative mathematical modeling of their properties and exploration of the range of dynamics and interactions that are possible within these networks both in healthy and damaged brains. Presently, more limited systems models are being considered. In attention, Hamker (2005) proposed a model that incorporates interactions among a large number of visual areas and the "attentional control" system that provides feedback. Quantitative models of object recognition typically incorporate the receptive field properties of neurons located along the ventral stream. Examples of these types of models are ones developed by Poggio, Edelman and colleagues (Serre et al. 2007; Edelman and Duvdevani-Bar 1997; Duvdevani-Bar and Edelman 1999). These models are strictly feedforward, based on findings that inferotemporal neurons show object-selective responses at times so short that they seem to preclude multiple recursive cycling up and down the visual pathways. When trained on a large database of images, these models are able to achieve recognition performance of human observers who classify images based on very brief stimulus presentation times. For more complex, cluttered scenes that require more recognition time, the latest version of the Poggio model incorporates attentional feedback (Chikkerur et al. 2009). By contrast, the face recognition model of von der Malsburg incorporates feedback to visual cortex from neurons holding stored representations of faces (see above). This feedback model achieves a high level of performance on published databases of faces. However, it was argued that in all of these system models, only average firing rates are considered and the timescale of the feedback is still relatively slow. A critical goal for the future is to find out whether the proven success of object classification and face recognition models are only first steps and that models based on binding mechanisms can be expanded into a broad range of functional models for dynamically coordinated perception.

It was agreed that an essential element for evaluating models is their performance on large, publically available image databases. Although some databases exist, there is a need for more realistic conditions in the databases, including the recognition of objects at different scales and embedded in complex scenes. Furthermore, beyond simple recognition, there is a need for models that can answer at least basic questions about the objects, such as shape, size, and location. The development of such models will help in understanding how and why synchronous interactions may be important for perception and memory.

Experiments

We considered a number of experimental approaches to the testing of the role of dynamic coordination in cognitive performance. Because oscillations and synchronization are currently the best-explored mechanistic paradigms, our discussion focused on possible ways to test whether such phenomena are necessary and sufficient for the cognitive functions performed by those brain regions where synchrony is observed. There was consensus that such experiments must monitor activity from two or more cell populations at the same time; as discussed in the introductory section, changes in the joint activity of two or more cell ensembles can be seen as a defining criterion for dynamic coordination.

We agreed that much of the current evidence linking synchronization of oscillatory patterns to coordination functions is correlation-based—a concern that is shared with most other fields of study in systems neuroscience—and that results thus, in principle, might be explained by other models, including those based solely on rate changes. However, the literature does contain some interventional studies which at least partly address the question of whether synchronization between cell populations is necessary for behavioral functions relying on the synchronized assemblies. In a study with multisite recordings from the frog retina, for example, activity was recorded from cells that respond to changes in shadows on the retina. Interventions that disrupted the synchrony of firing across the recording electrodes disrupted escape behavior elicited by shadow stimuli under conditions that did not change the average rates of the cells (Ishikane et al. 2005). Other experiments, performed in the hippocampus of the rat, have shown that using cannabinoids or other approaches to disrupt temporal order in hippocampal place cells, in a manner that does not change the average firing rates of the neurons, is sufficient to disrupt navigational performance in a spatial memory task (e.g., Robbe et al. 2006). In awake-behaving monkeys and healthy human subjects, some experiments have modified activity in visual cortex during stimulus detection by stimulating putative attention control regions in frontal cortex (Ruff et al. 2006; Ekstrom et al. 2008). Interference with frontal or parietal regions by TMS has been shown to alter, in behaviorally significant ways, anticipatory alpha rhythms in occipital visual cortex (Capotosto et al. 2009). The invention of optogenetic tools for selective stimulation or silencing of genetically defined cell populations is likely to result in a number of experiments along these lines within the next few years. It is clear that synchrony can be interrupted experimentally, and those data that exist so far suggest that such interventions may disrupt the functions performed by the affected cell populations.

Although interventional approaches represent the gold standard for studies of causal relations between coordination and brain function, we agreed that the caveats of such studies should not be forgotten. Interventions such as stimulation or inhibition of target cell populations may have additional effects on top of the intended ones; for example, disrupting synchronization between brain areas may also affect the proximal activity of each subpopulation, such as firing rates or precise local phase relationships. We also spent some time discussing strategies for gaining insight about coordination mechanisms under circumstances where physiological variables cannot be manipulated directly. One possible approach exploits the fact that human subjects often confuse the color and shape of different objects. Such "illusory conjunctions" can be used as a diagnostic tool to investigate which neural mechanism breaks down during binding errors. To investigate the role of response synchronization in feature binding, one could ask patients with intracranial electrodes to report on the color and shape of multiple objects in the visual field under conditions that lead to occasional misbindings (e.g., when stimuli are presented very briefly). One would need to record from cells that encode two distinct properties of two different objects in the visual field (e.g., one could choose color and motion as features and then record from color-sensitive cells in V4 and from motion-sensitive cells in MT). If synchronization is indeed the neural mechanism for feature binding, one would expect that the action potentials of cells belonging to the same object are synchronized when perception is successful, and that synchronization reflects illusory conjunctions when they occur. The same recordings could be used as well to test a different model, where the positional information encoded in V4 and MT signals maps corresponding features together. In this case, the positional information might be disrupted or shifted in either of these populations, thus providing a potential alternative account for the misassignment of features and spatial positions. If intrinsic dynamic connectivity turns out to be an important mechanism to code relations, especially for behaviors that are predictable or well-learned, then new investigations should be directed toward manipulating the ongoing intrinsic connectivity, either through behavioral paradigms or interventions like stimulation or disruption, and then correlate these changes to behavioral performance or task-driven activity.

We concluded that a variety of experimental approaches and systems are available to explore the function of oscillation-based synchronization and other possible mechanisms of dynamic coordination between neuronal populations. A common factor of all experiments that aim to test these functions should be the recording of activity from two or more brain regions at the same time; this is the only way to study changes in inter-regional temporal structure that may or may not be accompanied by activity changes in each of the areas locally. A number of brain systems, each with their unique advantages, should be used to extract the mechanisms of coordination. The study of temporal structure in large dispersed neuronal populations is likely to require an arsenal of new analytical and statistical techniques. Finally, there is a strong need for interaction between computational models and experimental testing; models should make clear predictions about activity changes in realistic neuronal architectures, and experimental strategies should be developed to test specific predictions from the models.

14

"Hot Spots" and Dynamic Coordination in Gestalt Perception

Ilona Kovács

Abstract

How is light that is transduced by retinal receptors and interpreted by neurons converted into visual information? To answer this question, vision science typically employs two types of scientists: those interested in local receptors and neurons that analyze very small pieces of the retinal image, and those interested in global visual information (surfaces, objects, scenes, and events that are meaningful to people). While both types may function well within their own areas of research, "translating" the results between the areas is a problem. Two explicit issues are discussed where strictly local processing stops short: (a) the problem of accumulating local errors and (b) the trade-off between spatial and temporal resolution in pictorial representations. To illustrate the first issue, which is architectural, an old and wonderful architectural mystery, the enigma of the Florence Dome, is used. An example from the history of photography illustrates the second issue, which is representational. Both problems have an important aspect in common: the solutions are both based on global geometry. Both classic examples will be accompanied by visual phenomena demonstrating the relevance of symmetry-based representations in the dynamic coordination of visual perception.

In What Way Is a Gestalt More Than Its Elementary Features?

> If a line forms a closed, or almost closed, figure, we see no longer merely a line on a homogeneous background, but a surface figure bounded by the line. This fact is so familiar that unfortunately it has, to my knowledge, never been made a subject of special investigation. And yet, it is a very startling fact, once we strip it of its familiarity. — Koffka (1935:150)

Gestalt, shape, *prägnanz* constitute the core mysteries of perception. Gestalt theorists considered the formation of perceptual pattern a dynamic process, best demonstrated by the various ambiguous figures with which they have entertained the world. Edgar Rubin's popular face-vase reversal image demonstrates that even boundary ownership is flexible, and the percept can quickly reorganize although the physical stimulus is constant. The compelling multistability of ambiguous images is induced by carefully balanced stimuli, where the two interpretations are equally "salient." One of the simplest multistable stimuli is shown in Figure 14.1. The two superimposed gratings are equiluminant and "orthogonal," both in terms of orientation and color. By simply staring at this stimulus, one observes monocular rivalry, and the two gratings start to alternate spontaneously.

The exciting instability of the perceptual system elicited by such images is not only among the most popular topics within the modern quest for neural correlates of conscious percepts (Crick and Koch 1995; Kovács et al. 1996; Leopold and Logothetis 1996), it also clearly demonstrates that our usual sense of perceptual stability is an illusion, and that the brain has many different ways to assemble new "realities" from competing pieces of concurrent external and internal events. Each competing interpretation entails a certain segmentation of the image into figure and ground.

The Unsolved Problem of Segmentation

Among the many unsolved issues of vision, the issue of segmentation may be one of the toughest. Although it does not sound very difficult to parse an

Figure 14.1 By staring at the image, spontaneous alternation between the two gratings is observed and monocular rivalry is stimulated.

image into different regions that correspond to objects and ground, machine or computer vision systems still cannot match the capabilities of the human visual system, not to mention categorization and image comprehension, which are strongly linked to segmentation. Human segmentation of visual images might depend on acquired knowledge and proceed interactively with object recognition (e.g., Ullman 2007). However, the explicit manner in which these higher-level knowledge systems communicate with low-level feature extractors has not yet been clarified.

There is obviously some higher-order structure, globally organized by the brain. However, there is continuous input from the environment, analyzed by low-level, local feature extractors of some sort. Just how do these computationally very different levels of processing (i.e., global organization vs. local analysis) meet and interact to provide us with our visual world?

According to the "standard" view of visual processing, visual information is first transmitted from the retina through several parallel pathways to the brain in a compressed version, emphasizing edge information at a number of spatial scales. A crucial second step is carried out by cortical area 17 (V1, or primary visual cortex), which is assumed to extract a set of local features based on the retinal input. Although the standard view then proceeds to progressively more complex representations, let us focus on the second step and the cortical "mosaic" generated by the primary visual cortex. It has been known for over forty years that the receptive fields of the primary visual cortex are composed of elongated antagonistic zones (Hubel and Wiesel 1959). The shape and layout of these receptive fields furnish the cells with selectivity for oriented line segments, and receptive field size determines the spatial scale of orientation information.

The primary visual cortex thus provides a neural description of oriented edge primitives and their locations at a number of spatial scales. This can be viewed as an enormous puzzle containing millions of pieces to be put together into figure and ground. A possible candidate for assembling local information already within the primary visual cortex is the plexus of long-range horizontal connections (e.g., Gilbert 1992). These connections are thought to establish connections between neighboring processing units, thereby aiding the segmentation process. The mechanism by which local interactions combine and boundaries of a visual object form is, however, unknown (Figure 14.2). Can all of this be based on local interactions?

The Problem of Accumulating Local Errors: The Puzzle of the Dome

To illustrate the problem of relying on local operators, consider an example from the history of architecture. Construction of the Florence Cathedral started in 1296, and its magnificent dome was completed in 1436. Driven by the urge to surpass Pisa and Siena in the size and decoration of the cathedral, the Florentine cupola is still the largest masonry dome in the world. Unfortunately,

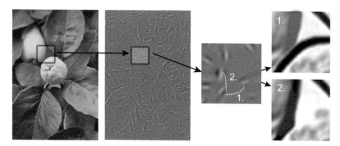

Figure 14.2 The problem of visual segmentation. The primary visual cortex provides a neural description of oriented edge primitives and their locations at a number of spatial scales. The natural image shown in the left panel activates a very large number of cortical filters; however, those that receive input according to their selectivities will be more active. The next panel shows the most activated filters for each location within the image. Neural interactions within the primary visual cortex are assumed to connect the most active filters in an orientation selective, facilitatory manner (e.g., similar-to-similar orientations). However, when viewed locally within the inset, it seems that these connections might be ambiguous. Which one is the better connection? Number 1 is a good choice, as the central object will be well segmented. Number 2 would be a bad choice, as it connects the boundaries of two independent objects. Is there some global reference guiding these local decisions?

only sparse information exists on how this wonderful three-dimensional shape was constructed from bricks and mortar. The dome was built using approximately 4 million specially designed bricks, collectively weighing about 37,000 tons. Filippo Brunelleschi's masonry techniques made a great contribution to architecture (King 2000), and his work is probably the best example of Renaissance engineering. However, he was also very secretive and never revealed his methods in detail. Although the dome is probably the most studied building on the planet, it is still uncertain how it was built. It is not known exactly, for example, what instruction was given to the, assumedly, eight groups of bricklayers as they raised the cupola's eight sides. How were they to know the exact positioning of each brick? Walls are easy to construct vertically, although some quasi-global reference (e.g., a masons' level or a plumb line) is needed from time to time. However, when the wall is curved, there are three dimensions to control: (a) the longitudinal curvature of the dome, (b) the circumferential curvature, and (c) the course by course, precise, and nonuniform change of the inward tilt of the bricks. A simple plumb line, even if it is combined with a level, cannot obviously control these three dimensions.

Brunelleschi was a great geometer, and the plan of the dome was probably perfect. He even took care of designing the shapes of the bricks for each new course himself. However, even with the best planning, and greatest care, bricklayers make slight errors, and if they only use local references (e.g., the previous course of bricks, or the ribs of the dome), these small errors will accumulate. Even an error of a hundredth of a degree in any of the three dimensions would result in a catastrophe in the case of 4 million bricks. In fact, such a catastrophe was observed in Siena after a massive addition to the existing cathedral and dome was undertaken in 1339. Although Sienese ambition was at least as great as Florentine ambition, Brunelleschi's ingenuity only served Florence. Just how did Brunelleschi manage to control the global shape of the dome and achieve this unprecedented and still unequaled construction?

With respect to the global shape of the dome, it may have seemed a good idea to use some central reference during the construction. However, wooden centering or scaffolding, which could have served to support as well as guide the overall arches, was not employed. How was the shape of the dome preserved without any scaffolds to guide it? How does the complicated pattern of bricks fill the spaces between the corners of the dome? Imagine that eight bricklayer groups are working on the wall, and other than along the circumference of the wall, they cannot compare notes. If there is no central pole of any kind (and there was certainly none because the dome is over 80 meters high) to use as a reference, and the masons cannot communicate with each other directly, how are the eight sides going to meet at the top? Perhaps there was some central reference after all, and it was removed after the dome was completed.

Massimo Ricci, a contemporary Italian architect, spent almost as much time trying to figure out the secret of the dome as Brunelleschi did building it. It took Ricci fifteen perplexing years to come up with an idea that might explain the riddle posed by the dome's construction. According to Ricci, the real secret of Santa Maria del Fiore lies in its extremely simple, although, in this context, surprising shape: a flower (Figure 14.3). The eight petals of the flower grew out of a circle, centered within the octagonal base, with a diameter three-fifths of the octagon diameter (the dome is based on a quinto-acuto, four-fifths measure). The flower was probably made of metal, and long ropes were attached to it, traversing the internal space of the dome. The shape of the petal controlled the circumferential curvature (the second dimension); the length of the ropes attached to the petals controlled the longitudinal curvature (the first dimension); and the tilt angle of the ropes controlled the inward tilt of the bricks (the third dimension). Each rope, connecting a certain location of the wall to the petal across the base of the dome, was adjusted to cross the central axis of the cupola. This was achieved by "centering" ropes between the corners of the cupola and the vertices of the flower. When a bricklayer wanted to align a new brick, he would move his rope to the new position, and his apprentice would shift the other end of the rope along the petal until the rope crossed the central axis again in a straight line. Perhaps the procedure was not repeated for each individual brick, but whenever it was done, the brick adjusted in this way would fall into place in accordance to the global reference point, and earlier local errors would not accumulate. This sounds like the solution indeed!

Even if historical evidence for the flower theory is missing, Ricci's scale model of the dome attests to its feasibility. The idea is simple: *when only local operators are given, use axis-based global reference to achieve a global shape and avoid the accumulation of local errors*. According to Ricci, this is the only

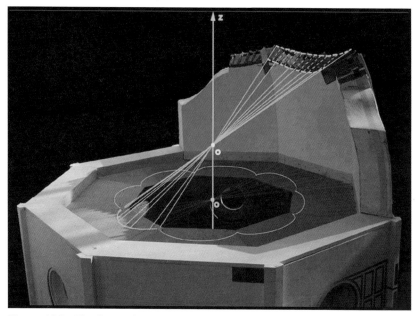

Figure 14.3 The flower theory based on Massimo Ricci Theory (used with kind permission from Luciana Burdi). The flower-shaped "skeleton" and many ropes between the flower and the wall serve to adjust the spatial position of the bricks according to the requirements of longitudinal, circumferential, and inward tilts. It is the symmetry axis of the shape that is employed as a global reference; however, the axis does not have to be there. It is determined with the help of ropes.

way of constructing such a shape. This might appear too ambitious; however, the usefulness of symmetry axes in avoiding the accumulation of local errors is very clearly illustrated with this example.

What Are the "Strings"? The Puzzle of "Closure"[1]

To challenge the local filters and local interactions of the primary visual cortex, and to investigate the type of integration that might be carried out at this "early" cortical level, Kovács and Julesz (1993) designed a psychophysical paradigm. The stimulus used in this paradigm consisted of a closed chain of co-linearly aligned Gabor signals (contour) and a background of randomly oriented and positioned Gabor signals (noise) (Figure 14.4). Gabor signals roughly model the receptive field properties of orientation selective simple cells in the primary visual cortex. Therefore, they are appropriate stimuli for the examination of these small spatial filters and their interactions. Notice that

---

[1] Closure is an old Gestaltist term used, e.g., by Kurt Koffka (1935:150) in *Principles of Gestalt Psychology*. According to Koffka, the term refers to the superiority of closed contours over open ones.

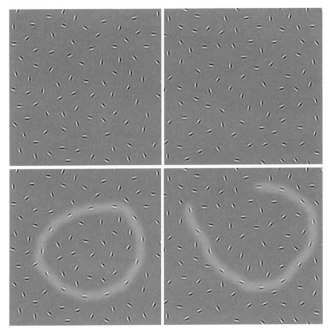

Figure 14.4 The closure superiority effect. The top two panels both have a contour embedded in noise. The solutions are presented in the bottom panels. Most observers find the closed contour in the upper left panel very easy to see, while it is difficult to trace the open contour even in the presence of the solution. (Because of individual variability in noise-tolerance, certain individuals might need different noise levels for the closure effect to appear.)

the contours cannot be detected by purely local filters or by neurons with large receptive field sizes corresponding to the size of the contour. The long-range orientation correlations along the path of the contour can only be found by the integration of local orientation measurements. The noise forces the observer to do these local measurements at the scale of the individual Gabor signals and to rely solely on long-range interactions between local filters while connecting the signals perceptually.

Since the "contour in noise" stimulus was designed to isolate long-range interactions that subserve spatial integration of orientation information in the primary visual cortex, all we expected was that human observers would be able to detect the contours even if noise density is greater than contour density (in other words, when noise elements are closer to each other than contour elements). This, indeed, was observed, but there was another surprising observation: closed contours in these images were much easier to see than open ones (Kovács and Julesz 1993; Mathes and Fahle 2007). We called this a "closure superiority" effect. As described in Kovács and Julesz (1993), closure superiority can be measured at perceptual thresholds. Threshold effects are difficult to illustrate; however, Figure 14.4 is an attempt to show the results of the experiments in an instant demonstration. According to the results, in spite of the locally equivalent parameters (same elements and same spacing parameters), closed contours are perceived differently: they seem to get a kick during the process of segmentation.

If local features are detected by local filters, and their interactions are also local (between neighbors), what causes the elements of a closed contour to jump out, while the same types of elements along an open one blend in with noise? Closure is a global shape property (similar to the shape of the dome). Local filters and local interactions—even if they form long chains—cannot deal with global shape properties. Local errors will accumulate along the chains of local interactions, and the result will be uncertain. Top-down instructions that arrive from higher levels of the cortical hierarchy may also not help. The higher-level "hypotheses" about the shape will meet unsegmented local orientation signals (contour + noise) and, in such a dense field of elements, any global suggestion can take shape, and the result will be uncertain again. Are there more than just local interactions during segmentation in the primary visual cortex? Does some kind of global reference serve segmentation much the same way as the Brunelleschi–Ricci flower served the construction of the Florence Cathedral dome? If there is such a reference system, what are the "strings" (taking the analogy further) in the brain that connect the flower and the bricks?

How Does Neural Activity Signal Gestalts?

The answer to this question might be found in computational models (e.g., Li 2005; Mundhenk and Itti 2005) or in the careful investigation of cortical microcircuits (e.g., Angelucci et al. 2002). It seems to be clear that long-range lateral interactions between neighboring neurons in the primary visual cortex are relevant in co-linearity-based contour grouping; however, these might be insufficient to account for integration beyond that of neighboring neurons (e.g., Loffler 2008). The closure superiority effect might share the underlying neural mechanisms with perceptual phenomena such as surface perception, surface interpolation, or "filling-in" (an excellent review is provided by Komatsu 2006). This Forum has provided a platform to explore these issues: in particular, whether the binding problem as defined by von der Malsburg (1981/1994), temporal correlations in the activity patterns of neurons (e.g., Engel et al. 1992; Phillips and Singer 1997), or a flexible assembly of spatial patterns of coordination (Haken et al. 1990; Kelso 1995) might explain global effects, such as closure superiority. I suggest that it might be essential to consider representational constraints before pointing out actual neural mechanisms. I will define and illustrate one of these constraints in the next section.

The Problem of Space–Time Resolution: Representing Living Things

Parsing an image into figure and ground is still far removed from the goal(s) vision may have evolved to accomplish. In addition to the role vision plays in guiding locomotion across space, perceiving the complex movements of living things is also essential. Although our visual environment is dominated by artificial objects, the human visual system appears to be fine-tuned to extract effortlessly socially relevant information from the movements of another person—a task that is essential for interpersonal interactions. Considering the nonrigid movements of the body, this requires an efficient and coupled coding of visual shape and motion information.

To illustrate the representational constraints on coding for biological movement information, let us turn to another historical example, this time from photography. A French medical doctor, Etienne-Jules Marey (1830–1904), attempted to capture time and to make all movements of the human body visible and measurable. He invented several devices to track circulation, respiration, and muscle function. In his studies on locomotion, his goal was to generate a description of complex human motion and depict the relationships, both in time and space, between various body parts (e.g., during a walk) using a single representation, within a single image. Marey realized that the two-dimensional graphs, which he used earlier to record the changes of a single parameter in either time or space, would not be useful in this case. While thinking about the appropriate space–time representation, he realized that photography might be an appropriate tool to capture and characterize human movement in time. Marey invented a camera with a fixed photographic plate and a rotating, slotted-disk shutter, which allowed him to overlay multiple exposures on the same plate and to reduce blur that would result when trying to take a shot of a moving subject. However, even with the fixed-plate camera, the problem of spatial blur was not completely solved. In fact, there was a trade-off between acuity in time and in space. If Marey increased the number of exposures (number of slots in the shutter), there were more pictures, and the resulting temporal resolution was better. However, due to contour overlap, spatial resolution was poor, and the images were blurred. Conversely, spatial resolution could be improved by decreasing the number of slots, but only at the expense of temporal resolution. Marey explicitly recognized the trade-off between spatial and temporal resolution (cited in Braun 1995:83):

> In this method of photographic analysis the two elements of movement, time and space, cannot both be estimated in a perfect manner. Knowledge of positions the body occupies in space presumes that complete and distinct images are possessed; yet to have such images, a relatively long temporal interval must be had between two successive photographs. But if it is the notion of time one desires to bring to perfection, the only way of doing so is to augment greatly the frequency of images, and this forces each of them to be reduced to lines.

Chronophotography provided the final solution for capturing time. To prepare for his chronophotographs, Marey dressed his subject in a black costume and marked the joints with shiny buttons connected by metal bands (Figure 14.5a). The subject moved around in the dark, such that only the movements of the buttons and wires were recorded in the picture. By selecting what he considered the most informative points and lines, he was able to read the successive postures of the body in his plates and follow the important trajectories of motion (Figure 14.5b). The relevance of this solution for human vision was later confirmed by Gunnar Johansson's work on the perception of biological motion in the point-light walker displays (Johansson 1973), and by the motion-capture method employed in psychology (Troje 2002) and in modern animation techniques.

The Role of Symmetry in Optimizing Space–Time Resolution

I began discussion of the closure superiority effect by referring to Kurt Koffka, who emphasized the relevance of surface regions enclosed by closed contours. The conclusions of the "dome" and the "closure" stories might be according to the taste of Gestalt psychologists. Indeed, the wonderful shape of the dome is not simply a collection of arches, but rather a three-dimensional volume; the closed line is not simply a collection of line segments, but rather a two dimensional surface. Ricci's flower theory suggests how the Gestalt of the dome can

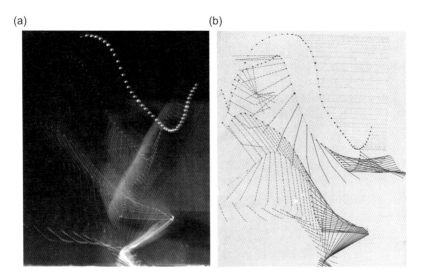

Figure 14.5 Geometric chronophotography by Marey. (a) The subject, dressed in black and photographed against a dark background, has been recorded jumping from a chair. (b) The chronophotograph of a jump is analyzed with graphics. The original glass plates (from 1883) are held by the College de France Archives.

be assembled from millions of bricks using an ingeniously simple line of reference. This line of reference is the main symmetry axis of the dome.

Marey's space–time diagrams are wonderful examples of the representational issues inherent in capturing the movement of complex bodies. In addition to their beauty, the chronophotographs demonstrate the pertinence of symmetry axes. The question arises whether the brain, when processing similarly complex information, uses similar representational solutions. Do symmetry axes play a crucial role in vision? In a series of experiments (Kovács and Julesz 1994), this question was posed with respect to the representation of simple shapes within the primary visual cortex. How is an extended, closed circle represented in the activity pattern of orientation-tuned neurons that have small receptive fields?

Simultaneous activity of a large number of interacting neural elements can be revealed by tracking the activity of several units simultaneously in search of their higher-order correlations, such as in electrophysiological cross correlation and multiunit studies. An alternative way is to estimate how the activity of one unit is affected in the context of the activity of other units. The latter approach was used in a psychophysical reverse mapping technique, where the activity of one unit is measured as a function of the changing context. Psychophysically measured local contrast sensitivity reflects the local activity of neurons, and the context of the interaction pattern can be manipulated by changing the overall stimulus design. Closed contours (illustrated in Figure 14.4) were employed as context, and single Gabor signals were used to obtain local contrast sensitivity of human observers (Kovács and Julesz 1994). The position of the local signals varied within the closed contours, and by measuring sensitivity for many locations within these contours, a map of contrast sensitivity was obtained for each investigated shape: circle, ellipse, cardioid, and triangle (detailed in Kovács et al. 1996, 1998).

To be able to see if any sensitivity change within these contours is related to the symmetry axes of the shape, we used a medial-axis type transformation (Blum 1967; Siddiqi and Pizer 2007). The D_ε function, as shown in Figure 14.6, is based on an equidistance metric, where the D_ε value of each internal point represents the degree to which this point can be considered as the center of the local boundary segment around it. The transformation provides a nonuniform skeleton of the shape, with one or more peak values. The peaks are very important, and are equidistant from the longest segments of the boundary. In other words, these are the most informative points, and long contour segments can be traded for them. We used the maxima of the D_ε function, which we called the medial-point representation, to predict potential sensitivity changes within the simple shapes mentioned above. If there is any change related to symmetry axes in local contrast sensitivity, it should be around these maxima!

To our great surprise, the D_ε function was a wonderful predictor of psychophysical performance. The prediction worked for simple shapes, circles, and ellipses as well as for shapes with curved and branching symmetry axes

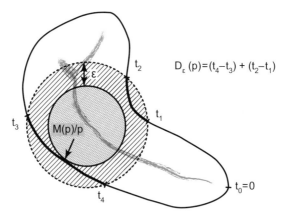

Figure 14.6 The D_ε function: D_ε is defined for each internal point by the percentage of the boundary points that are equidistant from the internal point within a tolerance of ε.

(Kovács et al. 1998). The contrast sensitivity changes were extremely specific, not simply some inside-specific enhancements. The maxima of the sensitivity changes corresponded to the maxima of the D_ε function. Neural correlates of these results have also been found in the modulation profiles of single-cell activity in the primary visual cortex (Lee et al. 1998; Lee 2003), although further confirmation of these would be useful. The psychophysical and neurophysiological data indicate that in addition to being sensitive to global shape properties (e.g., closure, and figure–ground relationships), the primary visual cortex is sensitive to specific shape properties and can host a medial-point type representation. The most provocative possibility is that this early cortical area provides a sparse skeletal code of shape!

Notice the similarities between the chronophotographs (Figure 14.5) and the medial-point representation (Figures 14.6, 14.7). Both provide a very small number of local points that can replace an "infinite" number of points composing a shape. The compactness does not preclude the representation from reflecting global shape properties. Such a sparse shape-coding would be a very desirable tool of communication between low- and high-level cortical visual areas, such as between the primary visual cortex and the inferotemporal area. Using only a handful of "hot-spots" to send information of segmentation in a bottom-up manner, and to send knowledge-based expectations top-down,

Figure 14.7 D_ε for sequential frames of the movements of an animal. The maxima of the function are good candidates as primitives for biological motion computations.

might provide the brain with a channel that works better than any current artificial image-compressing tool.

The physical world seems to operate along basic laws that exhibit known symmetries (e.g., translation in space, rotation through a fixed angle, etc.). The near-symmetries and broken symmetries might be even more interesting for physicists. A powerful example of a broken symmetry involves the phase change of water with decreasing temperature. As temperature goes down to 0°, liquid begins to solidify. Interestingly, at the same time, rotational symmetry disappears from the structure that H_2O molecules line up into. This might, in fact, be generalized to phase transitions in the domain of visual shape, and breaking those simple symmetries might be more interesting than the symmetries themselves.

Are the above-mentioned contrast sensitivity maps purely epiphenomenal, or are they proof of intelligent image compression in the visual cortex? If the latter, how is this implemented precisely by the cortex? Perhaps (not necessarily oscillatory) synchronous firing of orientation-tuned neurons mediates the compression and provides "hot-spots" in the neural representation of the segmented visual input. In what neural language would these "hot-spots" then be transmitted to more abstract levels of processing to meet with linguistic representations and semantic memory? In addition, how would the enriched information advance thereafter through the feedback pathways to enhance segmentation?

Acknowledgments

The author was supported by the ETOCOM project (TAMOP-4.2.2-08/1/KMR-2008-0007) through the Hungarian National Development Agency in the framework of the Social Renewal Operative Programme supported by the EU and co-financed by the European Social Fund.

15

Coordination in Sensory Integration

Jochen Triesch, Constantin Rothkopf,
and Thomas Weisswange

Abstract

Effective perception requires the integration of many noisy and ambiguous sensory signals across different modalities (e.g., vision, audition) into stable percepts. This chapter discusses some of the core questions related to sensory integration: Why does the brain integrate sensory signals, and how does it do so? How does it learn this ability? How does it know when to integrate signals and when to treat them separately? How dynamic is the process of sensory integration?

Introduction

Merriam-Webster defines coordination as "the harmonious functioning of parts for effective results" and as such it is omnipresent in brain and behavior. During perception, the brain has to make sense of sensory signals from a number of different modalities (e.g., vision, audition, olfaction, touch, proprioception). These signals need to be processed and integrated to compute an (usually correct) interpretation of the environment. How this happens is a fundamental problem. In this chapter we raise a number of central questions regarding how sensory integration pays special attention to dynamic coordination in the brain. A more detailed review of sensory integration is provided elsewhere (Rothkopf et al. 2010).

Why Integrate?

Perception is a difficult computational problem. The state of the world must be inferred from noisy and ambiguous sensory signals. To reach a solution, the brain must rely on making use of all available sources of information—from the different sensory modalities mentioned above (Stein and Meredith 1993), to different so-called *cues* within one modality. There are, for example, many so-called *depth cues* in visual perception that are thought to contribute to the perception of an object's distance from the observer (Landy et al. 1995). Next to the integration of various sources of evidence, perception utilizes, to a large extent, previously acquired knowledge about the world. Such prior information will be particularly important when the sensory data are very ambiguous (Weiss et al. 2002) and may be either innate or the result of learning, and thus subject to constant adaptation.

The benefit of combining several sources of information is twofold. First, our estimates of the state of the world will become more accurate as we integrate several noisy sources of information. To date, most work on sensory integration has focused on this aspect, and has been demonstrated amply in humans and various animal species. Second, we may be able to respond more quickly; that is, processing time is reduced when stimuli are presented in more than one modality. Both aspects are of obvious relevance for an organism's survival and well-being and can be closely related. Thus, when several sources of noisy evidence are available, under certain assumptions we can obtain the same amount of information from these sources if we observe a single source for a long time or several of them for a correspondingly shorter time.

How to Integrate?

A natural starting point for determining how the brain might integrate sensory information from different cues or modalities is to ask: What is the *optimal solution* to the problem? Such questions can be answered in the popular framework of *Bayesian inference* (Pearl 1988), for which many reviews are available (Kersten et al. 2004; Kersten and Yuille 2003; Yuille and Kersten 2006). In this framework one can construct so-called "ideal observers" that use all the available sensory information in an optimal fashion according to the laws of probability and statistics. After the ideal observer has been constructed and its behavior has been analyzed, it can be compared to that of human subjects or animals. In many (simple) situations, human behavior has been well-modeled by an appropriate ideal observer model, and this is usually taken as evidence that the brain performs Bayesian inference.

Unfortunately, however, solving the Bayesian inference problem and constructing appropriate ideal observer models can be a very difficult task. In the most general setting, Bayesian inference belongs to a class of computational problems that requires an exponentially increasing amount of processing as the problem size gets bigger (e.g., the more sensory variables are involved). In these situations, it may be infeasible to construct the ideal observer, and thus approximations have to be made. As a consequence, it is impossible to judge whether human behavior is optimal. However, since the brain will also have to use approximations to solve the Bayesian inference problem, it is important to ask what kinds of approximations it is using. In principle, this question can be answered within the Bayesian inference framework but we do not know of specific examples where this has been demonstrated.

Finally, viewing cue integration (and more generally perception) as Bayesian inference happens entirely on a computational level. It is still unclear how the neural implementation of Bayesian inference (or approximations of it) would look at the level of groups of neurons exchanging action potentials. This constitutes one of the most important and pressing questions in the field of computational neuroscience (Deneve 2005; Ma et al. 2006). A promising view is put forward by Phillips, von der Malsburg, and Singer (this volume), who argue that dynamic coordination is relevant to Bayesian inference because the distinction between driving and modulatory interactions is implied by the way in which posterior probabilities are computed from current data and prior probabilities.

How Does the Brain Learn How to Integrate?

The concept of Bayesian inference provides a powerful framework for studying sensory integration, but it does not address how the brain acquires the necessary probabilistic models: How does it decide what sensory variables to represent? How does it learn their statistical relationships? In the machine learning and statistics communities, progress has been in understanding how such models and their parameters can be learned, but optimal Bayesian learners are even harder to construct than ideal observers, and human learning can deviate strongly from the ideal case.

Experimental evidence regarding the acquisition of sensory integration abilities stems from developmental studies with children and learning experiments with adults. Interestingly, recent experiments with children suggest that it may take many years before children exhibit appropriate sensory integration abilities consistent with ideal observer models (Gori et al. 2008; Nardini et al. 2008; Neil et al. 2006). Initially, they may not be integrating different modalities at all (Gori et al. 2008).

In adult learning experiments, a relatively simple case is the one where the set of different cues is fixed and only their relative weighting changes. In visual cue integration, for example, Ernst et al. (2000) and Atkins et al. (2001) showed that when two conflicting visual cues are paired with a haptic cue, subjects will, over the course of a few days, learn to increase the weight of the visual cue that is consistent with the haptic cue and decrease the weight of the inconsistent cue. Thus, it appears that the haptic cue serves as a reference model for adjusting the visual cues.

When to Integrate?

Another fundamental issue concerns the timing of signals from different modalities or cues: when should they be combined or when should they be considered separately (Koerding et al. 2007)? An interesting problem in this context is that of audiovisual source localization. Imagine your task is to estimate the location of one or two target objects that are presented simultaneously in the auditory and/or visual domain through brief light flashes and sounds. When one auditory and one visual signal are received, but they seem to be very far apart, then it is prudent to assume that they did not originate from the same object and thus should not be integrated into a single percept. In contrast, if the two sources are sufficiently close, we assume that there is only a single object giving rise to both the auditory and the visual signal. In this case, it may be better to integrate the position estimates to arrive at a single, more precise estimate.

In Bayesian terms, the brain considers two different models to explain observed sensory signals. The first posits that there are two distinct objects: one producing the visual signal; the other producing the auditory signal. The second model posits that there is only one object giving rise to both the visual and auditory stimulus. How, then, might the brain make the appropriate determination? One obvious strategy is to evaluate both models and choose the one judged most probable, a technique called *model selection*. A plausible alternative is to evaluate both models and average their interpretations; different weights are given to both models in the averaging process, depending on how likely they appear. This technique is called *model averaging*. Recent research has started to address which strategy human subjects use (Shams and Beierholm 2009). However, thus far, evidence has been mixed. People appear to behave differently in different tasks, and large individual differences between subjects have been observed.

How does the brain *learn* when to integrate signals from different modalities and when to treat them separately? Recently, Weisswange et al. (2009) demonstrated that this ability could be acquired through generic reinforcement learning mechanisms (Sutton and Barto 1998). In their model, an agent needs to make orienting movements toward objects and is rewarded for localizing them precisely. In the situation where a visual and an auditory input are close together, the model will integrate them into a single position estimate. When they are far apart, the model will orient toward either the visual or the auditory stimulus without trying to integrate them. Although this model cannot prove that the brain acquires the ability to select the appropriate model in a similar way, it shows that the underlying reinforcement learning mechanism is sufficient to produce this behavior.

How Dynamic Is Cue Integration?

Whether or not different pieces of sensory information are integrated depends on the current situation (e.g., stimuli, context, behavioral goals). For a decision to be reached, different modalities and cues need to be *dynamically coordinated*.

Though the need to determine dynamically what aspects of the current sensory input should be integrated and what should be segregated has been well studied within submodalities such as motion perception (Braddick 1993), very little work has addressed to date this issue in relation to multicue or multimodal integration. Instead, most work has assumed a fixed situation, with a fixed set of sensory cues, and with fixed reliabilities that are known to the subject. In situations where different cues are in conflict with each other, subjects are known to suppress and/or recalibrate discordant cues flexibly (Murphy 1996). How exactly this occurs is an issue that has received little attention.

One exception can be found in research on self-organized cue integration. In many real-world situations, the usefulness of different sensory cues or modalities for a certain task will change over time. Cues may sometimes be conflicting or may require recalibration. Unfortunately, it is usually not clear a priori which cues are to be trusted and which ones should be suppressed or recalibrated. Triesch and von der Malsburg (2001) proposed the idea of *democratic cue integration*. In democratic integration, several cues are merged into an estimate of the state of the world, and the result is fed back to all individual cues to drive quick adaptations. Cues that conflict with the agreed-upon result are suppressed and/or recalibrated. The system is simply driven to maintain agreement among the different cues.

While initial work on democratic integration explored the benefits of such a scheme in the context of a computer vision problem of tracking people in video sequences, more recent work has studied the topic psychophysically. Triesch et al. (2002) showed that human subjects who track objects among distractors quickly reweight different cues (e.g., size, color, and shape of the tracked object), depending on the reliability of the cues. This reweighting occurs within one second. The neural basis of this flexible reweighting of different information sources remains a promising topic for future studies.

Conclusion

The integration of different sensory modalities and cues poses a central problem in perception. Although the Bayesian framework has proved very useful in understanding subjects' behavior on a range of tasks, more research is needed to understand the neural implementation of Bayesian inference processes and the approximations that the brain may be using. Furthermore, the learning mechanisms that set up the system to perform in a near-optimal fashion require investigation. Since much of this learning takes place in the context of goal-directed actions, the concept of reinforcement learning can be used to frame enquiry into these issues.

16

Neural Coordination and Human Cognition

Catherine Tallon-Baudry

Abstract

Understanding how thoughts can arise from a hundred billion or so interconnected neurons is the ultimate but still unreached goal of human cognitive neuroscience. The huge advances in human brain imaging over the last twenty years have led to a parcellation of the brain into a multitude of functional regions, yet understanding how activity is coordinated in the activated network remains a great challenge. This chapter summarizes the different modes of neural coordination that have been considered in the human literature, both theoretically and experimentally. It underlines the distinction between well-learned behavior, which can take advantage of prespecified neural routes, and dynamic neural coordination in flexibly defined neural ensembles, which can generate new percepts and/or creative behaviors. Discussion is devoted to the role that brain rhythms play in human cognition, with the underlying assumption that brain rhythms are a signature of dynamic coordination. A hypothetical but comprehensive schema is delineated to explain why different frequency bands coexist and interact.

Modes of Coordination

Coordination has distinct meanings. It can refer to the way information is transmitted from one region to the other. For instance, during attentional orienting, do the frontal areas drive the parietal regions, or is it the other way round (Buschman and Miller 2007; Grent-'t-Jong and Woldorff 2007; Bressler et al. 2008; Green and McDonald 2008)? Coordination may also refer to the way something new is created by the interaction process itself, in agreement with the aphorism "the whole is larger than the sum of its parts."

Information Flow and Its Limits

Characterizing the information flow between brain regions has been a very active field over the last fifty years. Roughly, two main streams can be distinguished. The historically older one is the attempt to organize the ever-increasing number of functional maps of the visual system (Van Essen 1979; Hadjikhani et al. 1998) into a global coherent model (Felleman and Van Essen 1991; Young 1992). This approach is fraught with a number of problems, in particular the choice of the criterion used to place one area on top of the other. Using the pattern of feedforward, feedback, and lateral connections generates too many plausible solutions (Hilgetag et al. 1996), whereas using the sequence of response latency yields a quite different organization of the visual system (Schmolesky et al. 1998).

The other stream is more recent and has been developed more specifically in humans, probably because the noninvasive imaging techniques used in humans (fMRI as well as EEG or MEG) sample the whole brain. Models based on such methods often lead to a rather sequential description of brain regions "lighting up" one after the other. This "boxology" approach has a strong descriptive power and fits well with the models derived from experimental psychology.

Pushing either of these two approaches to its limits raises some difficult issues. The diagrams of brain organization that summarize how information flows from one area to the other are oriented. They usually begin with a sensory input and go through a number of processing stages, which, depending on the authors, end up in supramodal regions or the hippocampal formation (the convergence zones of Damasio 1989), or in the fronto-parietal network as in global workspace theories (Baars 1997; Dehaene et al. 1998). Pushing this line of reasoning to its extreme reveals a number of similarities between convergence zones and grandmother cells. Indeed, most the objections raised against grandmother cells can be transposed to the description level of convergence zones (Singer and Gray 1995). In other words, the general principles of population coding may be relevant no matter whether the code unit is a cell or a functional module. In particular, both convergence zones and grandmother cells integrate across many inputs and, as a result, are faced with similar combinatorial problems. While integration by convergence may take place in dedicated neural circuits for well-learned stimuli (Li et al. 2004), more flexible mechanisms would be needed for new objects and situations.

A particularly crucial issue concerns the endpoint of any process: recognizing an object, a situation, or a mental state for what it is. In other words, when is the readout from the preceding stage sufficient to reach a decision? How does the system know it has reached a state that corresponds to a solution? If one considers a simple stimulus-response association, the process ends with the production of the motor response. However, in the case of mental states that do not end up with a movement, the answer is much less clear. Daniel Dennett is one of the most famous opponents of what he called the "Cartesian Theater" hypothesis (Dennett 1991). In this view, the mind, or the ultimate convergence zone, sits in front of a theater where percepts and thoughts are displayed. The Cartesian Theater hypothesis thus posits the existence of a "person inside" who acts as the ultimate witness (or convergence zone) of everything that occurs in the conscious mind. Is there, however, another person inside the homunculus, since the sentence, "I am aware that I am aware that I am late delivering my paper," makes sense? Any model of the brain that uses convergence as an intrinsic mechanism confronts, at some point, the "person inside" or homunculus issue. Some posit that activity in the prefrontal cortex is a necessary condition for awareness (Gaillard et al. 2009; cf. Goldberg et al. 2006). Others consider that the neural correlates of perceptual awareness are found in sensory cortices and do not necessarily require a frontal involvement (e.g., Lamme and Roelfsema 2000; Ress and Heeger 2003; Kouider et al. 2007; Wyart and Tallon-Baudry 2009). Recent models such as global workspace theories (Baars 1997; Dehaene et al. 1998) try to circumvent the difficult issue of the homunculus by mapping awareness onto a combination of sensory modules communicating with a large parieto-frontal network.

Emergence

Emergence is a central concept in complex systems analysis and can be roughly summarized as "the whole is larger than the sum of its parts." The idea is that some new information or knowledge is created at the system level, through simple interactions between lower-level components. As a result, some properties that do not exist in any constitutive elements of the system can emerge at the population level. Emergence is typically observed in flocking or herding behavior. A well-known example of such behavior can be found in the field of artificial intelligence (Reynolds 1987): "boids" are moving objects following simple local rules (avoiding bumping into their closest neighbors, moving roughly in the same direction and with the same speed as their closest neighbors, staying close to other boids). These three simple local rules are sufficient to produce a group behavior similar to that of a flock of birds, including the V-shaped flight of ducks. Coherent behavior can thus emerge from local rules, without a need for either an explicit global schema or for a group leader. Let us also consider what happens at the end of a theater performance: clapping usually begins in a loud and disorganized manner, but after several curtain calls applause becomes rhythmic. A temporal structure spontaneously emerges, simply because people tend to listen to each other. This property—emergence of a global coherent behavior without the need of conductor—is particularly interesting when related to the search for the neural correlates of awareness because of the commonly admitted view that there is not a single anatomical module responsible for awareness (Crick and Koch 1990; Engel and Singer 2001; Alkire et al. 2008). In this view, the adequate description level of some cognitive processes would be the whole brain or a substantial part of it (Lashley 1931; Haxby et al. 2001), rather than a functional module, and new cognitive knowledge would be created by between-area interactions, in addition to specific local computations.

Emergence versus Information Transfer?

There are some fundamental differences between the concepts of emergence and of information flow. Let us consider that Mary was in A at t and in B at $t + 1$. If Mary is a commercial traveler and has to meet a client in B, then knowing that Mary arrived on time in B is sufficient. However, if Mary is a ballet dancer, this description does not help: what matters is whether the other dancers also moved from A to B or not; in other words, it is the dynamics of the global picture that are relevant. In practice, one or the other strategy may be preferentially used depending on the task to perform. One of the crucial factors is probably the amount of learning and degree of automaticity required, versus novelty and flexibility: a key idea about emergence (or dynamic coordination) is its flexibility, its ability at signaling new relationships.

Feedforward, Feedback, and Recurrent Processing

Emergence and information, as defined here, are concepts rather than experimentally tractable neural mechanisms. How do they relate to the classical typology of neural coordination (feedforward, feedback, and recurrent processing)? The initial wave of feedforward activity has been mainly associated with fast and automatic analysis of the visual scene (Thorpe et al. 1996; Hochstein and Ahissar 2002) and may reflect unconscious processing (Lamme and Roelfsema 2000). Both feedback (Bullier 2001) and recurrent processing (Lamme and Roelfsema 2000) have been related to more sustained levels activity and could be involved in conscious processes. As a first approximation, one could consider that fast feedforward processing is fully automated and, therefore, would not reflect *dynamic coordination* but rather information flowing along a fully prespecified neural route. Conversely, recurrent processing would be more flexible and a signature of emergence. However, there is again another level of approximation, since assessing whether a neural process occurs in a feedforward or feedback manner is always difficult and almost impossible with the noninvasive tools used with humans.

In the human literature, feedforward processing is often linked to the rapid cascade of early-evoked potentials that is essentially maintained during anesthesia (Alkire et al. 2008). In opposition to the idea that this initial volley of activity is rather automatic, there is growing evidence that top-down expectancies can meet the feedforward stream at very early latencies, before 100 ms (Chaumon et al. 2008; Kelly et al. 2008; Poghosyan and Ioannides 2008; Dambacher et al. 2009). Feedback and recurrent processing are usually associated with longer latencies and with more sustained states. The frequency content of the signal is most often used to describe such sustained states, but late slow waves which can show up in event-related potentials (ERPs), such as the contingent negative variation or the P300, can also be interesting indexes.

If we accept the simplifying assumption that early ERPs reflect mainly feedforward processing and that oscillations reflect mainly dynamic coordination, a crucial issue is how the two interact. This is only beginning to be addressed experimentally. For instance, several groups showed that the amplitude of the response evoked by the stimulus depends on the phase and/or amplitude of ongoing alpha (Jansen and Brandt 1991; Becker et al. 2008) or gamma rhythms (Fries, Neuenschwander et al. 2001). The mechanisms responsible for these interactions could be a general phase resetting of ongoing oscillations by stimulus onset (Makeig et al. 2002), a controversial proposal (Shah et al. 2004; Mazaheri and Jensen 2006; Risner et al. 2009), or an asymmetry in ongoing alpha oscillations (Nikulin et al. 2007; Mazaheri and Jensen 2008). Last but not least, it seems that the nature and strength of the interaction could depend on the resting state frequency characteristics of each subject (Koch et al. 2008).

Distinct Roles for Different Frequency Bands

In this section, discussion is devoted to brain rhythms, with the underlying assumption that brain rhythms are one of the signatures of dynamic coordination. More precisely, I will focus on how brain rhythms are related to cognitive processing.

Mapping Cognitive Functions onto Frequency Bands

Historically, each frequency band has been associated preferentially to a given type of cognitive or physiological process. Delta waves were associated with sleep, theta activity with memory, and alpha rhythms with vigilance fluctuations, whereas the beta and gamma ranges were associated initially with active awake stages and more recently to feature binding, attention, and memory. It is beyond the scope of this chapter to review all of the literature on all frequency bands (for recent reviews, see Jensen et al. 2007; Klimesch et al. 2007; Palva and Palva 2007; Schroeder and Lakatos 2009; Tallon-Baudry 2009), but a few examples are sufficient to demonstrate the absence of a strict correspondence between a frequency band and a cognitive process.

There is a large and converging body of evidence that grouping features into a coherent percept is accompanied by changes in the gamma range (reviewed by Jensen et al. 2007; Tallon-Baudry 2009). However, the formation of coherent percepts can also be accompanied by modulations of oscillatory synchrony in the alpha range (Mima et al. 2001; Freunberger et al. 2008). The alerting, orienting, and executive attentional networks engaged in many attentional tasks affect oscillatory synchrony in different frequency ranges, from theta to gamma frequencies (Thut et al. 2006; Fan et al. 2007; Siegel et al. 2008). Episodic memory encoding and retrieval typically affects both theta and gamma oscillatory synchrony (Sederberg et al. 2003; Osipova et al. 2006), but some also report modifications in the alpha band (Klimesch, Doppelmayr, Schimke et al. 1997; Klimesch et al. 1999; Sauseng et al. 2002). Visual short-term memory retention is associated with sustained gamma and beta oscillations that originate from distinct areas (Tallon-Baudry et al. 1998, 2001; Tallon-Baudry 2004), but also with modulations in the alpha range (Jensen et al. 2002; Jokisch and Jensen 2007; Grimault et al. 2009). The historical associations between a frequency band and a cognitive process should therefore be reconsidered: cognitive functions do not map directly onto frequency bands.

It is also important to underline a fact that sounds trivial: the functional role of oscillatory synchrony in distinct frequency bands may simply depend on the functional specialization of the area that generates these oscillations (Tallon-Baudry et al. 2005), much as the functional significance of ERPs depends on the areas that generate them. This might seem like a statement of the obvious, but this simple statement had surprisingly disappeared from human literature on brain rhythms. For instance, local gamma oscillations are observed in a wide range of areas in human intracranial recordings, from visual (Lachaux et al. 2000, 2005; Tallon-Baudry et al. 2005), to frontal (Howard et al. 2003; Mainy et al. 2007), and medial temporal lobe structures (Tanji et al. 2005; Sederberg et al. 2007). It would indeed seem quite unlikely that those gamma oscillations should all reflect the same cognitive function, given the variety of their anatomical location. Intracranial data in humans remain scarce, but fortunately increasingly more MEG/EEG studies include a source reconstruction approach that provides quite a precise localization of increase and decrease of oscillatory synchrony (Grimault et al. 2009; Hillebrand et al. 2005; Hoogenboom et al. 2006; Medendorp et al. 2007; Siegel et al. 2008; Gross et al. 2004; Wyart and Tallon-Baudry 2009).

Subdivisions within Frequency Ranges

Another source of complexity stems from the existence of functional subdivisions within a given frequency range. It has long been known, for instance, that the upper and lower alpha ranges could display distinct functional variations (Klimesch, Doppelmayr, Pachinger et al. 1997; Petsche et al. 1997). More recently, we showed that distinct cognitive processes can elicit gamma oscillations in different locations: grouping and selective attention simultaneously affect gamma-band oscillations, but in distinct subfrequency bands and at distinct locations (Vidal et al. 2006). Similarly, learning and conscious perception are associated with oscillations in the gamma range, but in different subbands and in distinct areas (Chaumon et al. 2008). Finally, varying attention and awareness simultaneously revealed that distinct frequency bands within the gamma range varied separately with visual awareness and spatial attention (Wyart and Tallon-Baudry 2009). It has also been suggested that the detailed frequency content of gamma-band oscillations could encode specific physical features, such as spatial frequency (Hadjipapas et al. 2007) or sound lateralization (Kaiser et al. 2009). There is no doubt that gamma-band oscillations are influenced by stimulus low-level features in sensory regions (Hall et al. 2005; Adjamian et al. 2008), but whether this still holds true for higher-level areas remains an open issue. In any case, oscillatory synchrony in a given frequency band should not be considered as a single phenomenon, functionally and anatomically homogenous (Tallon-Baudry 2009).

What Are the Relevant Criteria of Frequency-band Selection?

The absence of a direct correspondence between a frequency range and a cognitive function raises a fundamental issue: what determines the preferential use of a given frequency? There appears to be a large flexibility: the same fronto-parietal network (Buschman and Miller 2007) or the same visual (Wyart and Tallon-Baudry 2009), olfactory (Cenier et al. 2009), or audio-visual (Chandrasekaran and Ghazanfar 2009) region can engage into oscillatory synchrony at distinct frequencies, involved in distinct cognitive functions. This flexibility may be subtended by distinct local networks and cellular types: *in vitro* experiments reveal that there are distinct frequencies (20–30 Hz vs. 30–70 Hz) in the infra- and supra-granular layers, respectively (Cunningham et al. 2004; Roopun et al. 2006). It is tempting to suggest that each frequency band corresponds to a specific microcircuitry; for instance, gamma-band oscillations would critically depend on $GABA_A$ interneurons in upper layers. This basic cellular equipment, present in each and every cortical area but in varying proportions, would define natural frequency domains corresponding to the typical frequency ranges, from theta to alpha and gamma. In a given cognitive task, the use of a frequency band would depend on two sets of factors.

The first group of factors relates to the task's physiological requirements. It was initially suggested that frequency depends on the network's size and geometry (Kopell et al. 2000; von Stein and Sarnthein 2000): because conduction delays increase in large network, synchronization takes place at lower frequencies (Buzsáki and Draguhn 2004). Another potentially important factor is the time constant of the biological mechanisms involved (Koch et al. 1996) and the coding precision required (Desbordes et al. 2008). If, for instance, time-dependent synaptic plasticity is required, then a precision of 10–20 ms is necessary (Bi and Rubin 2005; Markram et al. 1997) and the whole network might shift to the gamma frequency range. Finally, the metabolic costs of establishing sustained oscillations may vary between frequency bands. It has been suggested, in particular, that there may be a stronger relationship between the BOLD response and gamma-band activity (Mukamel et al. 2005; Niessing et al. 2005), although the relationship at rest may be more complex (Mantini et al. 2007; Nir et al. 2008). The preferential use of a frequency band might therefore also be influenced by metabolic demands.

A second group of factors can be found in cognitive constraints. First, the time constant of the task is likely to influence the pace of the system: if there are only 500 ms to complete a visual search, for instance, frequencies below 5–10 Hz are unlikely to be relevant, whereas if there is no time constraint to perform a task, then one might shift to lower frequencies. Similarly, if there is any regularity in the temporal structure of the task, subjects are likely to form windows of temporal expectancies (Tallon-Baudry 2004; Praamstra and Pope 2007; Schroeder and Lakatos 2009). Second, oscillations could be used to define chunks of processing, in which data will be grouped and isolated from those of the preceding and following period. Sensory or cognitive chunks can potentially be created at many timescales; for instance, arbitrary associations can be learned through a wide range of time intervals (Balsam and Gallistel 2009). Examples of chunking can be found in vision and olfaction and have been related to beta-range oscillations (Uchida et al. 2006; VanRullen et al. 2006). Along the same line of reasoning, it would be tempting, although premature, to relate very slow (< 0.3 Hz) oscillations (Monto et al. 2008; Nir et al. 2008) to the "psychological present," the few seconds during which successive events form a perceptual unity and can be apprehended without voluntary recall. Third, another interesting potential constraint is the number of cognitive processes to be multiplexed. Searching for someone in a crowd typically involves retrieval of information about the person from long-term memory, attentional suppression of nonmatching faces, and bottom-up feature-binding processes. One possibility to coordinate these three cognitive processes would be to use distinct frequencies, as detailed in the next section. In this view, the frequency tuning of each process is likely to depend on the total number of processes required by a task.

Integrating between Frequencies

Multiplexing, Integration/Segregation

Analyzing a situation and reacting in an appropriate manner requires the coordination of a number of cognitive processes. As shown above, searching for someone in a crowd, for example, involves a sensory analysis of the visual scene, the recall of a face template from memory, and the attentional scanning of all the faces potentially matching that template. All these cognitive processes have to be integrated into the general search task, but because they reflect distinct operations they should nevertheless remain segregated. The neural correlates of these distinct cognitive processes are likely to show up in different frequency bands, and one way to coordinate them without fusing them is to coordinate activity between frequency bands. Coordination of one frequency band with another would be an elegant solution to multiplex information while keeping a reasonable trade-off between integration and segregation. Besides, because the number of simultaneous frequencies as well as the number of multiplexing patterns is limited, the number of concurrent processes is naturally restricted to a finite number, in line with the idea that simultaneous tasks might tap into shared and limited resources.

Different Types of Coupling Can Be Considered

Coupling between frequency and/or areas can appear in many different ways. The first candidate that was considered was between-area phase coupling in the same frequency band (Lachaux et al. 1999; Varela et al. 2001). In this view, two or more distant areas oscillate in the same frequency range with relatively constant phase relationships, and there is indeed experimental evidence that such phase coupling between distant sites can occur and play a cognitive role (see, e.g., Tallon-Baudry 2004; Tallon-Baudry et al. 2001; Uhlhaas, Linden et al. 2006; Melloni et al. 2007; Doesburg et al. 2008). Between-area coupling can also appear as amplitude covariation of the signal at the same frequency (Bruns and Eckhorn 2004). Both phase- and amplitude-coupling can also occur between frequency bands, recorded at the same site or at different sites (Bullock et al. 1997; Palva et al. 2005; Meltzer et al. 2008). Finally, a mixed version of phase- and amplitude-coupling occurs when high frequency oscillations occur preferentially during peaks (or troughs) of a lower frequency rhythm. Such coupling patterns have been observed, for example, in rat entorhinal (Chrobak and Buzsáki 1998a) and neocortex (Sirota et al. 2008), as well as in cat visual cortex (Grenier et al. 2001). So-called nested oscillations attracted a great deal of interest because of an influential model of memory storage that would account for the limits of human memory capacity by an interplay between theta and gamma oscillations (Lisman and Idiart 1995), and there is growing evidence in humans for such theta/gamma relationships (Canolty et al. 2006; Sauseng et al. 2008).

Conclusions and Prospects

A picture emerges in which activity in the brain can be coordinated at multiple spatial and temporal scales using different frequency bands. This overall picture is much more complex than the initial attempts at matching a cognitive process with a type of neural coordination, but much more flexible and deemed to have a bright future. Indeed, it suggests that cross frequency coupling could be used to integrate as well as segregate information over behavioral time and cognitive space, a feature that is necessary to obtain a high level of flexibility. In addition, because the biologically constrained circuits that generate oscillations in different frequency bands are limited in numbers, the numbers of different frequencies, and therefore the number of cross frequency coupling schemata, is limited. This inbuilt constraint of the brain would be reflected in the well-known cognitive capacity limitations.

17

Failures of Dynamic Coordination in Disease States and Their Implications for Normal Brain Function

Steven M. Silverstein

Abstract

Dynamic coordination may become compromised due to several nonindependent factors, including a reduced ability to generate oscillations and synchrony, neurotransmitter and receptor excesses and reductions, anatomical and cellular changes that impair connectivity at local and global scales, and changes in gene expression that stem from primary genetic or environmental causes. This chapter presents a review of disorders in which dynamic coordination failures have been identified. It compares and contrasts the forms and severity levels of the impairments in each disorder and links these to biological causes, as far as is currently known. Questions are posed regarding the implications of data on coordination failures for increasing our understanding of normal coordination and the possibility of its enhancement.

Overview and Issues

Dynamic coordination refers to the ongoing combination and recombination of neural signals to form higher-level, adaptive patterns of activity that retain the identity of the original signals yet are nonlinear due to feedback and recurrent processing. This can be contrasted with driving (or feedforward) input, which refers to the signals whose timing, salience, and relative contributions to emergent patterns of activation are modulated by coordinating processes. Dynamic coordination of neural activity allows for rapid and effective adaptation to changing conditions. However, the nearly infinite combination of potential networks that can be formed by local and long-range connectivity and the complex timing requirements—both within and across frequency bands—that are needed to support them result in a complex system in which even small errors can have significant, large-scale consequences. In this chapter, I address the nature of these errors and their consequences.

Because dynamic coordination at the neural level—via synchronization of firing activity or other proposed mechanisms—is thought to support a range of cognitive functions—including visual perception, working memory, language, and self-representation—it is reasonable to ask whether reduced coordination is the mechanism underlying the dysfunctions in these processes found in certain neurological and psychiatric conditions. If this were to be the case, information on dynamic coordination could provide a unifying framework for understanding the core pathology underlying specific disease states. This knowledge could also foster cross-disorder comparisons, thereby clarifying genetic and neurodevelopmental contributions common to multiple diseases. By increasing our understanding of how disordered coordination leads to abnormal mental activity, we may be able to shed light on how normal coordination supports effective mental functioning and how coordination can be enhanced.

I begin by reviewing the evidence for failures of dynamic coordination in several illnesses: amblyopia, schizophrenia, Williams syndrome, autism, epilepsy, and Alzheimer's disease. Here the focus is on paradigmatic examples of coordination failures as demonstrated in studies of visual perceptual organization. Behavioral, functional magnetic resonance imaging (fMRI), and electrophysiological data are presented to clarify the similarities and differences in the severity, valence, and scope of coordination failures in these disorders. Discussion in the next section is more speculative, focusing on evidence for other forms of cognitive impairment in these and other disorders and addressing the question of whether these phenomena can also be seen as evidence of reduced coordination capacity. Thereafter, I address the issue of whether coordination failures are due primarily to reductions in a coordinating function, versus an increase in the strength of driving input, or a local attentional bias, or whether evidence for all of these exist. Following this discussion, I address neurotransmitter systems and their interactions, as well as neuronal and genetic abnormalities that may be involved in coordination changes. The final section poses questions regarding what abnormal dynamic coordination and its effects can teach us about normal brain and cognitive functioning.

Paradigmatic Evidence for Impaired Dynamic Coordination in Brain Diseases

Perceptual organization in vision is a paradigmatic example of dynamic coordination in that features are grouped together into an emergent holistic representation based upon their relationships to each other, while the signal for each individual feature remains intact. This process can be seen clearly in the case of contour integration (Figure 17.1), which has been studied extensively in healthy and clinical populations. In a typical contour integration task, the conditions under which integration can occur, and the mechanisms responsible for it, are determined by employing stimuli with a continuous path of Gabor signals embedded in noise. Gabor signals closely model the receptive field properties of orientation-selective simple cells in primary visual cortex (V1) and are therefore ideal for the examination of these small spatial filters and their integration and interactions. Embedded contours cannot be detected by purely local filters or by the known types of orientation-tuned neurons with large receptive fields. The long-range orientation correlations along the path of the contour can only be found by the integration of local orientation measurements, and this can be seen as a classic example of dynamic coordination (Figure 17.1). Therefore, evidence of contour integration impairment associated with a brain disorder would be *prima facie* evidence of abnormal dynamic coordination in that condition. In the following subsections, this evidence is reviewed. Evidence for dynamic coordination failures using other measures of perceptual organization will also be noted.

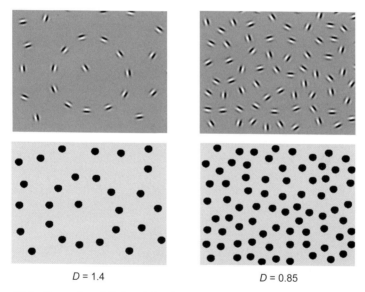

Figure 17.1 Examples of Gabor-defined contours with different D values (left: D = 1.4, right: D = 0.85). D is the ratio of the average distance between adjacent background elements to the average distance between adjacent contour elements; this is equivalent to the actual signal-to-noise ratio. In the bottom panels, Gabor elements were replaced by disks. Without orientation cues, the contour remains invisible at $D < 1$. This is the range where perceptual organization depends on long-range, horizontal, excitatory interactions between feature detectors based on relationships present in the input (in this case, correlations between orientations of adjacent contour elements) and represents a paradigmatic example of dynamic coordination. At $D > 1$ the contour can be perceived simply via density cues. Figure designed by Ilona Kovács; reprinted from Silverstein et al. (2000), with permission from Elsevier.

Amblyopia

Amblyopia is a condition characterized by abnormal binocular input due to problems with a single eye, leading to suppression of input from that eye. This can occur because of muscle weakness in one eye (e.g., "lazy eye"; strabismic amblyopia), or to one eye having significantly greater refractive error than the other (anisometropic amblyopia). Amblyopia has been associated with reduced contour-integration performance (Kovács et al. 2000), especially when contours must be linked within background noise and when there is positional uncertainty. These impairments have been found in both strabismic and anisometropic amblyopia, although less consistently in the latter condition. Findings of amblyopia-related reduced contour integration have been replicated in animal studies (Kiorpes 2006). The findings cannot be attributed to reduced contrast sensitivity or other low-level visual factors. In humans, contour-integration performance varies as a function of the degree to which binocular input was restored in childhood via treatment (Kovács et al. 2000).

Because all studies to date of contour integration in amblyopia have solely used behavioral measures, there is a relative lack of direct physiological evidence on biological mechanisms involved in the impairment. Research in cats indicates that visual processing of high frequency gratings by the amblyopic eye is associated with significantly reduced neural synchrony compared to that observed in the fellow eye, suggesting that altered synchrony and its associated excitatory and inhibitory mechanisms may be involved in human amblyopia (Roelfsema et al. 1994). A recent structural MRI study found reduced gray matter in visual processing areas (Mendola et al. 2005), suggesting a link between abnormal contour integration and gray matter loss. Interestingly, of all the disorders reviewed in this section, amblyopia has the most limited form of dynamic coordination failure (limited only to vision), and has been associated with only gray matter reduction. In contrast, other disorders have more widespread coordination failures, and this greater degree of impairment has often been associated with both gray and white matter reductions (see below). This supports the hypothesis that white matter tracts are critical for inter-regional coordination, whereas they are less critical for intra-regional coordination, of which contour integration may be an example.

Williams Syndrome

Williams syndrome is a genetic and neurodevelopmental disorder characterized by cardiac anomalies, overdeveloped expressive language skills, normal to superior memory abilities, a strong desire to talk but often poor social understanding, and poor motor and visuospatial skills. Many, but not all, studies demonstrate that perceptual organization, including contour integration, is impaired in Williams syndrome (Martens et al. 2008). Some of the discrepancies may be accounted for by evidence that grouping by some features (e.g., luminance, closure, alignment) is intact, whereas grouping by others (e.g., shape, orientation, proximity) is impaired (Farran 2005). A recent fMRI study indicated reduced activation in visual and parietal cortices in Williams syndrome patients during a processing of global forms made up of local features (Mobbs et al. 2007). Evidence of reduced functional connectivity between cortical regions during object processing has also been found (Martens et al. 2008). Evidence of both gray and white matter reduction has been found in Williams syndrome. To a large extent, these findings parallel those found in schizophrenia (see below), although the clinical features of the syndromes are very different. This is an important consideration as attempts are made to determine the role of coordination failures in the overall pathophysiology of each disorder.

Schizophrenia

In contrast to amblyopia, schizophrenia is a condition in which dynamic coordination appears to be impaired throughout the cortex, as evidenced by multiple cognitive impairments. All studies of contour integration in schizophrenia indicate impaired performance. Performance is also significantly, and inversely, correlated with scores on the Ebbinghaus illusion task (see Figure 17.2), in which schizophrenia patients are more accurate in size judgments secondary to reduced effects of contextual integration (Uhlhaas, Phillips et al. 2006). A recent fMRI study indicated that impaired contour integration was associated with reduced activation in V2–V4 in schizophrenia patients, areas where

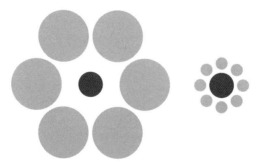

Figure 17.2 Examples of the Ebbinghaus illusion. The two inner circles are the same diameter. When the inner circle is surrounded by smaller circles, most observers perceive it as larger than its actual size. In contrast, when surrounded by larger circles, most observers perceive it as smaller than its actual size. People with schizophrenia have been found to be less susceptible to this illusion. In most laboratory tasks incorporating this phenomenon, the size of the inner circle is compared to a no-surround circle, or two circles with surrounds (one larger, one smaller) are shown and the subject is required to determine which inner circle is larger/smaller. By manipulating the actual sizes of the inner circles, the sizes of the outer circles, and/or the distance between inner and outer circles, a parametric determination of context sensitivity can be obtained.

integration of features into wholes occurs (Silverstein et al. 2009). These data are consistent with findings from human (nonclinical) and monkey fMRI studies that identified cortical regions involved in contour integration (e.g., Kourtzi et al. 2003).

Studies of perceptual organization using paradigms other than contour integration also consistently indicate that integration of noncontiguous elements is impaired in schizophrenia (for a review, see Uhlhaas and Silverstein 2005). In contrast, processing of textons, continuous contour, and features such as symmetry are intact. This suggests that in schizophrenia, coordination via prespecified feature hierarchies is not affected, whereas dynamic coordination is impaired. As further evidence of this, studies that specifically studied top-down effects on grouping indicate that performance is especially poor under these conditions.

A consistent problem in schizophrenia research is the generalized deficit, or the tendency of patients to perform poorly on nearly every measure, for reasons that may have nothing to do with the process purportedly being measured (e.g., sedation from medications, poor motivation) (Silverstein 2008). However, in ten of the studies of perceptual organization in schizophrenia, the reduced ability to integrate information and the subsequent reduced influence of visual context led to superior performance, compared to controls, in terms of either making decisions about individual features or reduced susceptibility to illusions. Therefore, evidence for impairments in visual integration has been convincingly demonstrated independent of a generalized deficit. These impairments also cannot be accounted for by medication, as they have been demonstrated in nonmedicated patients, and task performance is not correlated with medication dose in medicated patients. This extensive experimental literature is consistent with earlier clinical descriptions and first-hand subjective patient accounts of fragmented face, object, and scene perception in schizophrenia (see Carr and Wale 1986).

Electrophysiological studies of perceptual organization in schizophrenia have identified reduced P100 amplitude when patients viewed fragmented pictures (e.g., Foxe et al. 2001). A recent study found reduced N150 amplitude to global fragmented targets in a global-local task in schizophrenia (Johnson et al. 2005), and the source of that waveform has been localized to V3/V3a within the lateral occipital complex, an area in which feature integration occurs (Di Russo et al. 2001). Especially relevant to the issue of dynamic coordination are studies of synchronized neural activity during performance of perceptual organization tasks. In nonpatients, synchronization of oscillatory activity has been identified within the gamma (reflecting binding of activity at shorter cortical distances, often within-region) and beta (reflecting longer-distances, often between regions) bands (von Stein et al. 1999). Uhlhaas, Phillips et al. (2006) demonstrated smaller increases in beta-band synchrony among patients viewing degraded facial images, evidence consistent with earlier data on smaller increases in gamma-band synchrony during processing of illusory contours

(Spencer et al. 2004). Further evidence for abnormal dynamic coordination in schizophrenia comes from findings of reduced stimulus-locked power within the beta and gamma bands, reduced evoked stimulus-locked oscillatory activity within the gamma band, and reduced non-stimulus-locked oscillations within the gamma band during visual or auditory processing tasks (Uhlhaas, Haenschel et al. 2008; Uhlhaas and Singer 2006). Abnormal oscillatory activity is present at least as early as the first illness episode (Symond et al. 2005).

In contrast with findings of reduced synchrony in first episode patients, behavioral evidence for perceptual organization dysfunction has not been found among either high risk or first episode patients (Silverstein et al. 2006; Parnas et al. 2001), except in one study, in which it was only found among subjects with increased symptoms (Uhlhaas et al. 2004). In contrast, perceptual organization dysfunction has been consistently identified in older patients with a chronic disease course. This parallels diffusion tensor imaging data, which indicate smaller white matter changes at first episode compared to patients with chronic illness (Friedman et al. 2008), supporting a link between white matter integrity (and corticocortical connectivity) and dynamic coordination. Taken together, these data suggest both a core illness-related impairment and a progressive process. Further evidence for this is that perceptual organization impairments in schizophrenia are most reliably found in individuals with histories of poor premorbid social functioning (Knight and Silverstein 1998), suggesting that the impairment is associated with a core illness subtype with neurodevelopmental disturbances: the subgroup of patients with histories of good functioning before the initial psychotic episode generally demonstrate normal perceptual organization. Even among the poor premorbid functioning subgroup, however, the severity of the impairment correlates with degree of symptomatology, especially disorganized thinking.

Autism

Autism is a neurodevelopmental disorder characterized by severe cognitive and social functioning deficits. Studies have demonstrated that autism is associated with reduced context sensitivity and an increased ability to recognize or make decisions about individual elements embedded within visual displays (Happe and Frith 2006). Studies specifically investigating integration of spatially separated stimuli reveal conflicting findings; some show reduced integration whereas others show normal integration (e.g., Del Viva et al. 2006). As with schizophrenia, it appears as if a determinant of integration impairment in autism is illness severity. Deficits in autism are most reliably found in persons with comorbid mental retardation, a frequent aspect of the disorder. People with autism with normal IQs and people with autism spectrum disorders are more likely to demonstrate normal perceptual organization of both static and moving displays. As reviewed by Uhlhaas and Singer (2006), converging evidence from EEG and fMRI studies indicates reduced functional connectivity and smaller increases in gamma-band synchrony compared to controls during cognitive tasks. The biological bases of these abnormalities are less clear, but it has been proposed that autism may be characterized by excess excitation and unstable cortical networks. White matter abnormalities have also been identified in autism, although these have not been specifically linked to task performance.

The apparent similarities between schizophrenia and autism suggest that each should have significant comorbidity of features from the other disorder. Indeed, this has been found (Rapoport et al. 2009; Sprong et al. 2008), raising the possibility that genetic and neurobiological studies exploring the boundaries and overlap between these conditions may help identify the core pathology caused by coordination failures.

Alzheimer's Disease

In contrast to amblyopia, in which coordination failures are caused by abnormal sensory input, and schizophrenia, in which they are caused by an interaction between neurodevelopmental abnormalities and progressive illness effects, in Alzheimer's disease, dynamic coordination failures are due to late life neurodegeneration. At least six studies to date (reviewed in Uhlhaas, Pantel et al. 2008) have documented impaired perceptual organization in Alzheimer's disease. Patients with white matter atrophy demonstrated the poorest contour integration ability.

Parkinson's Disease

Parkinson's disease involves massive loss of dopaminergic neurons in the substantia nigra, leading to akinesia, tremor, and cognitive deficits. Unlike other disorders reviewed here, Parkinson's disease is characterized by increases in relative synchrony. This evidence is largely physiological and has thus far been seen only in the motor domain. It is not yet known whether Parkinson's disease is characterized by nonmotor manifestations of impaired coordination. Uhlhaas and Singer (2006) reviewed evidence that increased beta-band synchronization in cortical motor areas is related to akinesia in Parkinson's disease. There is also evidence for abnormal oscillatory activity that is coherent with the frequency of limb tremor, and data showing that tremor is associated with abnormal synchronization of motor neurons. Parkinson's disease offers an interesting contrast to amblyopia. Both involve focal abnormalities of synchronized activity: the former is characterized by increased patterned motor activity, the latter by a reduced ability to generate patterned visual representations. The focal activity in Parkinson's disease can also be viewed as an analog to the focal activity in schizophrenia (see next section) and epilepsy (discussed below). In all cases, there is heightened self-organized activity that is relatively impermeable to mental or environmental influences.

Epilepsy

Epilepsy is a heterogeneous category that includes conditions involving restricted seizure activity as well as seizures involving the spread of activity throughout the cortex. Because it is not possible to conduct behavioral testing during seizure activity, the evidence on dynamic coordination in epilepsy comes solely from physiological recording. Evidence for both increased and decreased synchronization of activity has been found. A synthesis of the evidence (Uhlhaas and Singer 2006) suggests that synchronization between distant cortical regions is reduced, whereas it is increased in the epileptic focus (i.e., the site of seizure origin). Moreover, reduction in synchrony may play a causal role in seizure formation by allowing for the formation of the epileptic focus, whose self-sustaining activity is relatively uninfluenced by other cortical activity. In general, the setting of reduced synchrony—along with isolated areas of increased synchrony—that is hypothesized to characterize seizure proneness has also been suggested to exist in schizophrenia (see next section). Interestingly, there is an increased risk of schizophrenia in people with epilepsy, and NMDA receptor antagonists (used to model schizophrenia; see below) lead to EEG changes similar to those seen in some forms of epilepsy (Lisman et al. 2008).

Putative Evidence for Impaired Dynamic Coordination in Brain Diseases

Phillips and Singer (1997) hypothesized that the cortical algorithm and cortical circuitry involved in dynamic coordination are implemented throughout the cortex and are operative for multiple cognitive functions, including perceptual organization, attention, working memory, long-term memory, language, and consciousness. Implementation of this algorithm modulates the timing and strength of driving input and creates the higher-order representations that emerge from it, based on contextual relationships that can be established in both the input itself and information about past experience with similar stimuli. The neural circuitry necessary to produce the contextual fields supporting dynamic coordination is thought to rely heavily on NMDA receptor (excitatory, pyramidal cell) and GABAergic (inhibitory, interneuron) activity. A similar proposal, in terms of an isomorphism in neural network activity underlying multiple forms of cognition, was advanced by Fuster (2003). Evidence in support of the Phillips and Singer (1997) and Fuster (2003) theories includes similar modulatory circuitry in all areas of the cortex, the role of interneurons and neural networks in generating synchronized oscillations, and the role of gamma-, beta-, and theta-band synchrony in implementing all of the cognitive functions hypothesized to be based on this circuitry. Models based on principles of perceptual organization have also been applied to binding of social information in social cognition (e.g., theory of mind; Blakemore and Decety 2001) as well as to segregation of information in observed behavior and relationships between unit size and attributions for behaviors (Baldwin et al. 2001). These functions, however, have not yet been linked to biological processes.

Regarding brain diseases, support for the above models would come from evidence of impairment in nonperceptual cognitive functions, as well as evidence that these impairments involve reduced coordination and abnormal synchronization. The latter is critical, because while there is much evidence for cognitive deficits in attention, memory, and language in psychiatric and neurologic disorders, this evidence has typically not been understood in terms of integrative and coordinating processes. A summary of the evidence, which suggests that disorders with multiple cognitive deficits can be understood within the framework of widespread dynamic coordination failure, is presented below. To date, this has largely come from studies of schizophrenia.

Amblyopia

As noted above, in amblyopia, deficits in dynamic coordination appear to be limited to visual processing, secondary to impaired input early in development.

Williams Syndrome

The hypothesis of multiple examples of dynamic coordination failure in Williams syndrome has not yet been investigated. Evidence of fine motor-sequencing difficulties, attentional problems (especially at younger ages), and impaired social inference are areas worthy of further study. It is also possible that the specific genetic factors involved in Williams syndrome produce a relatively circumscribed deficit in the area of visuospatial processing that is unrelated to other illness features.

Schizophrenia

Three sources of evidence support the hypothesis of widespread dynamic coordination failure: (a) behavioral studies indicating reduced binding in nonperceptual cognitive impairments, (b) involvement of abnormal oscillatory or synchronized activity in these impairments, and (c) significant correlations between indices of these impairments. Regarding the first, a recent study indicated that reduced perceptual organization was associated with attentional disengagement (i.e., reduced ability to maintain attentional focus within groups of stimuli) in an attention task (van Assche and Giersch 2009). Another recent study (Lefèbvre et al. 2009) indicated reduced binding of features during encoding of both spatial and temporal information in a working memory task, suggesting that the binding impairment is not limited to spatial information. These data are consistent with poor performance on tasks of coherent motion detection (Tschacher et al. 2008), which involve both spatial and temporal processing. It has also been shown that schizophrenia is characterized by reduced context-based binding of cues in episodic memory (Waters et al. 2004) and reduced relational memory organization (Titone et al. 2004). Regarding the second point, Uhlhaas, Haenschel et al. (2008) reviewed evidence for abnormal oscillatory activity and synchrony during a range of cognitive tasks in schizophrenia.

Evidence from several studies indicates that impairments on multiple indices of reduced dynamic coordination are related in schizophrenia. For example, severity of impairment on tests of perceptual organization is significantly correlated with the level of thought disorganization (Knight and Silverstein 1998; Uhlhaas and Silverstein 2005). Test scores are also significantly correlated with scores on theory of mind tasks, supporting views of similar integrative mechanisms underlying both perception and aspects of social cognition involving inferential or propositional reasoning (Uhlhaas, Linden et al. 2006). These data support the hypotheses of Carr and Wale (1986) and Phillips and Silverstein (2003) that schizophrenia is characterized by multiple variants of the same basic dysfunction, leading to widespread failures in binding related information together into coherent representations to support effective thought and behavior. This impairment is present in pre-attentive and post-attentive as well as spatial and temporal processing. It has also been observed in functions as high level as autobiographical memory, the disturbance of which has been attributed to a reduced binding of self-representations with observed actions during encoding of episodic memories (Danion et al. 1999).

Overall, evidence from studies of schizophrenia suggests failures of perceptual organization and associated widespread reductions in oscillatory and synchronized neural activity. However, there are also suggestions of excessive synchronization. For example, Hoffman and McGlashan (2001) suggested that symptoms such as delusional ideas or hallucinations might reflect "parasitic foci." This refers to self-perpetuating attractor states, which, although reflecting abnormally strong connectivity themselves, are thought to be formed within a context of reduced connectivity (i.e., functional fragmentation) in which nonreality-based combinations of representations are more likely to occur. This hypothesis is consistent with evidence of abnormal neural coactivation secondary to white matter abnormalities in patients with auditory hallucinations (Hubl et al. 2004). On a more global scale, there is evidence that the smaller post-stimulus increases in gamma-band synchrony in schizophrenia reflect an abnormally high baseline level of synchrony, against which only small relative increases are possible (Flynn et al. 2008). This could, as does the hypothesis of reduced coordination, also account for the reduced ability of emergent feature properties (e.g., gestalts) to take precedence in conscious awareness over background information. Such data are consistent with recent findings of network hyperactivity in patients who experience an exaggerated self-awareness during conditions when self-awareness is normally suppressed, and findings that this hyper-connectivity is present even at rest (Whitfield-Gabrieli et al. 2009). The latter may be the neural signature of the symptom of "hyper-reflexivity" discussed by Sass and Parnas (2003). To resolve the competing positions of hyper- and hypo-connectivity, it was suggested that for relatively inflexible networks, such as parasitic foci (and associated stable symptoms such as delusional ideas and hallucinations), to form, there must be a reduction in functional connectivity in surrounding areas (similar to that proposed for seizures, see below), whereas symptoms such as rapid shifts in perspective, and disorganized thinking and speech, are characterized by states of transient, hyper-plastic connectivity (Guterman 2007). An important unresolved issue here is the extent to which heterogeneity in coordination abnormalities in schizophrenia is related to heterogeneity in autonomic arousal abnormalities, which are also observed in schizophrenia.

Much research in schizophrenia has implicated dysfunction of the prefrontal cortex (PFC), and has noted the control functions of this region in domains such as temporal context processing, working memory, relational encoding in memory, and action planning. Recent evidence (Barbalat et al. 2009) suggests, however, that information is hierarchically organized in the PFC in part based on the temporal framing of action and events, that activity in the caudal lateral PFC varies as a function of episodic and contextual signals, and that activity is reduced in this region in schizophrenia. Therefore, rather than viewing the PFC as a monolithic control center that is responsible for imposing order on the output of operations from other cortical areas, it, as do other regions, may operate via dynamic coordination, with, in this case, the coordinating algorithm generating the typical "gestalts" of the PFC (e.g., action plans, anticipated behavior–consequence links, and temporal context). To what extent then can the evidence for PFC abnormalities in schizophrenia be accounted for by coordination failures within this region? Conversely, to what extent does the evidence that the frontal cortex is involved in grouping of distant, but not closely separated, visual elements (Ciaramelli et al. 2007) and that in schizophrenia, perceptual organization deficits are most pronounced when top-down feedback is required for effective performance, implicate a specific role for the PFC in dynamic coordination in this and other illnesses with coordination failures, and also in healthy individuals?

Autism

Uhlhaas and Singer (2006) review evidence that in autism, a reduced ability to group stimuli is found in auditory processing, linguistic context processing, and social cognition, in addition to vision. To date, there have been few studies of physiological processes associated with impairment on these tasks, but one study (Grice et al. 2001) found reduced gamma-band synchrony during a face perception task.

Alzheimer's Disease

Despite behavioral evidence for reduced perceptual organization and profound memory deficits, and resting physiological evidence for reduced neural synchrony, there has been only one task-related study, and this indicated reduced synchrony during a cognitive (working memory) task. Uhlhaas and Singer (2006) concluded that in addition to loss of neurons, phenomena found in this disease also reflect impairments in the coordination of distributed neural activity, which could be due to gray and white matter reduction.

Other Conditions

Studies in other conditions (e.g., epilepsy, Parkinson's disease) could clarify the extent to which *multiple* impairments reflect reduced dynamic coordination, but have yet to be done. Interestingly, recent findings of reduced synchrony in response to steady state auditory stimulation in multiple sclerosis (Arrondo et al. 2009), a disorder characterized by white matter degeneration, support the hypothesis that these tracts are involved in dynamic coordination, and therefore that abnormalities therein could produce a range of coordination failures in disorders with this feature.

Issues that Arise

Can Life Experience Cause Reduced Dynamic Coordination?

Evidence concerning the disorders reviewed above suggests that abnormal dynamic coordination can occur in the context of abnormal sensory input (e.g., amblyopia), neurodevelopment (e.g., Williams syndrome, autism), neurodevelopmental abnormalities interacting with stress and neurotransmitter/receptor changes (e.g., schizophrenia), degenerative processes (Alzheimer's disease, Parkinson's disease), or developmental, injury-related, or idiopathic causes (epilepsy). It remains to be determined whether, and to what extent, these different etiologies produce qualitatively different forms of coordination abnormalities. Another relatively unexplored issue is whether dynamic coordination failures can occur due primarily to neurobiological changes caused by abnormal life experience.

As an example of this, it has been hypothesized that auditory hallucinations in posttraumatic stress disorder and schizophrenia may consist of sensory components of memories of traumatic incidents (e.g., sexual abuse) that are decontextualized from the majority of the episodic memory trace and its associated affect (Read et al. 2005). This is similar to the model for dissociative symptoms postulated long ago by Breuer and Freud (1895), in which the affect associated with the traumatic experience is split off from ideation related to the experience. This is also similar to the ideas of Janet (1889), who proposed two core phenomena in mental functioning: one that preserves and recreates the past, and one that involves integration (van der Hart and Friedman 1989). The latter "reunites more or less numerous given phenomena into a new phenomenon different from its elements. At every moment of life, this activity effectuates new combinations which are necessary to maintain the organism in equilibrium with the changes of the surroundings" (Janet 1889, cited in van der Hart and Friedman 1989:5)—a view similar to Phillips and Singer's (1997) concept of dynamic coordination. In Janet's view, in schizophrenia and other mental disorders involving cognitive fragmentation there is reduced integration, such that components (e.g., memory traces) are unmodulated and appear magnified relative to ongoing events in the person's life.

Evidence in support of a link between trauma and reduced coordination of mental activity comes from a class of mental disorders known as dissociative disorders, which are characterized by losses of conscious awareness of aspects of experience. This can involve identity (psychogenic fugue states), aspects of remembered experience (psychogenic amnesia), or aspects of self (e.g., dissociative identity—multiple personality—disorder). Dissociative identity disorder is commonly associated with histories of childhood physical and/or sexual abuse, and psychogenic fugue and psychogenic amnesia are often associated with intolerable stress in adulthood. While no studies have yet examined dynamic coordination in these disorders, clarification of the extent to which dissociation involves reduced coordination is provided by studies of hypnosis, in which dissociation of consciousness (e.g., the non-experience of pain), and phenomena such as hallucinations can be temporarily induced, especially in highly hypnotizable subjects (Silverstein 1993). Preliminary evidence from hypnosis research indeed suggests that splitting apart of normally integrated representations does involve reduced coordination. For example, Fingelkurts et al. (2007), in a case study, demonstrated reduced functional connectivity, across multiple frequency bands, after hypnotic induction compared to baseline. The authors concluded that, in highly hypnotizable subjects, cognitive modules and subsystems may be temporarily incapable of communicating with each other. In a controlled study, Croft et al. (2002) found that prior to hypnosis, gamma synchrony predicted pain ratings for both high- and low-hypnotizable subjects. However, during hypnosis, while this relationship was again observed for low-hypnotizable subjects, it was eliminated for high-hypnotizable subjects. These data suggested that hypnosis involves a functional disconnection between the frontal cortex and other areas, and thus that the symptoms of dissociative disorders may reflect functional and/or anatomical disconnections.

Interestingly, increasing evidence suggests that people with schizophrenia have high rates of childhood trauma, and one study has found links between trauma history in schizophrenia, more severe illness, and more impaired contour integration (Schenkel et al. 2005). Trauma is rarely examined as a correlate of cognitive or biological functioning. However, it could be a common factor in dissociation and impaired dynamic coordination across a range of mental disorders that develop after childhood. Relatedly, factors such as being bullied in childhood, racial discrimination, and chronic social defeat have also been linked to the later development of psychosis, all raising the possibility that chronic profound stress alters dynamic coordination (the biological basis of this possibility will be discussed below).

Is Reduced Dynamic Coordination Involved in Emotion-processing Abnormalities?

To date, nearly all research on dynamic coordination in mental disorders has focused on cognitive phenomena (e.g., perceptual organization, working memory, hallucinations). However, a hallmark of several of the disorders considered is altered emotion processing and expression. To what extent is altered dynamic coordination involved in these abnormalities, and/or in the apparent decoupling between ideation and emotional experience that can be found in these disorders?

Intriguing evidence comes from studies of alexithymia, a personality trait characterized by a reduced ability to identify and verbally label emotional experiences. A recent study (Matsumoto et al. 2006) found that whereas in healthy controls gamma-band power and phase synchronization were increased when processing emotionally negative stimuli, individuals with alexithymia did not demonstrate either increase. This suggests that people with alexithymia may be characterized by a reduction in communication between brain regions and a related reduction in the integration of mnemonic and/or emotional information during processing of emotional stimuli. These data also support the hypothesis that altered gamma-band synchronization is involved in the splitting of ideation and affect that has been hypothesized to occur in people with histories of trauma, as noted above.

In contrast to alexithymia, a disorder that is characterized by excessive emotional activity (and often a history of childhood trauma) is borderline personality disorder (BPD). BPD is characterized by emotional dysregulation, including anger outbursts and intense sadness, feelings of loneliness and emptiness, transient psychotic symptoms, and an unstable sense of identity (and unstable relationships). In a recent study (Williams et al. 2006), during a tone discrimination task, patients with BPD demonstrated a delay in the generation of gamma synchrony over posterior cortical sites and a reduction in gamma synchrony over right hemisphere sites. Moreover, the delay in posterior synchrony was associated with ideational distortions involving self and others, and reduced right hemisphere synchrony was associated with behavioral impulsivity. Williams et al. (2006) suggest that the data indicate reduced functional connectivity between posterior and frontal networks, and that this is a mechanism in the abnormal evaluation of stimulus significance and dyscoordinated emotional responses seen in BPD.

Taken together, the preliminary data from studies of alexithymia and BPD—conditions at the extremes of emotional experience—suggest that there is an optimal degree of synchronization necessary for adaptive integration of emotional and cognitive experience. Abnormalities in networks involved in emotion–cognition integration can lead either to reduced emotional experience and expression or to unmodulated expression. This perspective on coordination and emotions can be seen as analogous to the difference between, for example, amblyopia (reduced coordination) and Parkinson's disease (increased coordination).

The research presented in the last two subsections suggests that manifestations of dynamic coordination failures may extend beyond the cognitive and motor phenomena that have been studied thus far. Specifically, it is suggested that dysregulated emotional experience and behavior may also be manifestations of impaired dynamic coordination. Neuroscience research techniques have only begun to be applied to these important aspects of brain disease. Nonetheless, theories have appeared in which (a) mood regulation; (b) coordination of feelings, thoughts, and behavior; and (c) balance between thought and instinctual drives are seen as three core dimensions of brain function, with all of these rooted in synchronized oscillations. In this view, a range of mental disorders (e.g., mood disorders, schizophrenia, obsessive-compulsive disorder, phobias, BPD) reflects brain region-specific variation in the neurodevelopment of the capacity for coordinated activity (Pediaditakis 2006).

Similarly, the extent to which the apparent splitting between mental contents, and/or between mental and emotional content following traumatic or other chronically stressful life experiences, involves reduced dynamic coordination requires further study. Preliminary evidence from a number of psychiatric conditions suggests that a history of trauma may lead to selective areas of reduced coordination, and that this can account for commonality of symptoms across a variety of disorders including dissociative, borderline personality, and psychotic disorders—most of which were once grouped together as either "hysteria" or "schizophrenia" where similarities in the core integrative psychic dysfunction were suggested long ago (e.g., Jung 1907). Questions arising are whether corrective emotional experiences (e.g., dynamic psychotherapy, cognitive behavior therapy, healing relationships) can reduce or eliminate manifestations of altered dynamic coordination, and if so, whether this is a mechanism by which psychotherapy is effective.

Overall, the research reviewed above suggests that dynamic coordination abnormalities can manifest in multiple ways, causing a variety of pathological phenomena depending on their scope (i.e., extent and loci of distribution in the cortex), valence (excessive or reduced), and severity. An unresolved issue is the extent to which differences on these three dimensions can parsimoniously account for the clinical presentations in each condition. What is needed, therefore, are (a) additional biomarkers of cognitive coordination that are separable from generalized performance difficulties and symptom severity, and (b) more sophisticated models that explain and clarify the mechanisms whereby differences in scope, valence, and severity of coordination failures specifically account for individual symptoms and syndromes, and their differences.

Is Apparent Dynamic Coordination Failure Caused by Too Little Integration or Too Much Feature Processing?

The data cited in the first section provide consistent evidence for reduced performance on tests of perceptual organization in specific brain diseases. This literature generally assumes that this reflects reduced integrative or modulatory ability (except where noted, e.g., Parkinson's disease) in the presence of normal intensity of driving input. However, it is also possible that these findings reflect (a) excessive feature processing even in the face of normal integrative functions or (b) an attentional bias toward local stimuli in the presence of normal global processing. Studies of perceptual organization typically have not been able to distinguish between these explanations; however, some evidence suggests that this issue is worth exploring. For example, studies of global and local processing with compound stimuli have demonstrated that adopting a focus on one level or the other (i.e., an attentional bias) can change which level appears to take precedence, raising the possibility that an illness-related attentional bias could produce what looks like a dynamic coordination failure. In Williams syndrome and autism, there is evidence of attentional bias toward local processing (Porter and Coltheart 2006), although not in all studies. This contrasts with the excessive global bias found in Down's syndrome, and thus it is possible that illness-related and illness-specific attentional biases exist, in addition to, or rather than, the hypothesized illness related integration deficit. In schizophrenia, a long history of findings of sensory gating disturbance and subjective reports of increased subjective intensity of sensory stimuli are consistent with the hypothesis of excessive feature processing. Further, in autism, there is some evidence for excessive processing of the meaning of words (Uhlhaas and Singer 2006). Questions that arise from these considerations include:

- Are the findings of increased baseline synchrony (Flynn et al. 2008) and hyperactivity in cortical networks (Whitfield-Gabrieli et al. 2009) in schizophrenia compatible with excessive levels of driving input?
- Can explanations involving excessive driving input or local attentional bias account for reduced organization in other cognitive functions (e.g., memory, language, social cognition) in schizophrenia, autism, and Williams syndrome?
- To what extent do attentional biases to local or global levels reflect coordination impairments (see Engel et al., this volume)?
- Is it possible to have both excessive processing of features and reduced integration? This is consistent with the conclusions of an fMRI study

demonstrating excessive left-right frontal connectivity and reduced anterior-posterior connectivity in schizophrenia (Foucher et al. 2005).
- Can the problem of apparent similarities in dynamic coordination failures in disorders with different clinical presentations be, at least in part, resolved by attributing the deficits to different causes (e.g., integration deficit, excessive feature processing, attentional bias)?
- Alternatively, to what extent are the differences in coordination failures and/or clinical presentations in the different disorders a function of variation in the spatial distances (and corresponding differences in frequency bands) over which coordination can and cannot be implemented?

Neurobiological Candidate Mechanisms for Abnormal Dynamic Coordination

Biological explanations for impaired dynamic coordination have generally focused on three levels of analysis: (a) coordination within and between brain systems (e.g., electrophysiology), (b) anatomy (e.g., gray and white matter reductions and other cellular abnormalities), and (c) neurotransmitters. Representative evidence for the first of these was presented above. Anatomical findings related to impaired dynamic coordination include:

- gray matter reductions in amblyopia, schizophrenia, and possibly in autism;
- white matter reductions in schizophrenia, Alzheimer's disease, Williams syndrome, and possibly in autism;
- increased neuronal cell packing density in visual cortical regions (consistent with reduced connectivity and decreased synaptic signalling) in schizophrenia (Selemon 2001); and
- decreased dendritic field size in schizophrenia.

Hypotheses and data regarding neurotransmitter-related abnormalities in dynamic coordination failures have generally focused on:

- NMDA-receptor hypofunction as a basis for impaired excitatory binding of relevant features (Phillips and Silverstein 2003), as well as for changes in GABA-related inhibitory function (Roopun, Cunningham et al. 2008);
- abnormal regulation of NMDA receptor-dependent synaptic plasticity by neurotransmitters such as dopamine, acetylcholine, and serotonin (Stephan et al. 2009);
- a primary impairment in GABAergic (inhibitory) activity including reduced generation of oscillations via inhibitory interneurons (Gonzalez-Burgos and Lewis 2008);

- alterations in acetylcholine receptors, which must be active for cortical networks to engage in synchronized, high frequency oscillations (Uhlhaas, Haenschel et al. 2008); and
- excessive activity at cannabinoid receptors, which leads to both diminished oscillatory activity and impaired sensory gating (Hajós et al. 2008).

Genetic factors related to neurotransmitters and neuroplasticity may also play a role in illness-related dynamic coordination failures (Lisman et al. 2008). There is little research on this specific issue to date, although evidence is accumulating regarding genetic contributions to (a) neurotransmitter function and expression of different types within the same class of (e.g., NMDA) receptor, (b) neurotransmitter abnormalities in schizophrenia, and (c) the development of autism, Williams syndrome, epilepsy, and other relevant disorders. For example, decreased expression of genes related to synaptic function has been found in schizophrenia (Mirnics et al. 2001). Recently, an increase in NOS1AP has been found in postmortem samples in schizophrenia, and this has been hypothesized to lead to decreased signaling at NMDA receptors and reduced dendritic field size (Brzustowicz 2008). A question suggested by these findings is whether—if disorders differ in their type of dynamic coordination failure—these differences can be accounted for by differences in neuroanatomical, neurotransmitter, or genetic factors.

A rarely addressed issue concerns the extent to which the biological basis of dynamic coordination can be affected by factors such as reduced social interaction, poor diet, lack of exercise, poor maternal care, or physical or sexual abuse. For example, it was recently proposed that reduced gray matter in schizophrenia may be secondary to reduced cardiovascular functioning and related reduction in neuronal growth, which in turn may be related to lack of physical activity and related reduction in brain-derived neurotrophic factor (BDNF) expression, overweight status and poor diet (Ward 2009). Diet has been demonstrated to affect gene expression related to brain plasticity (McGowan et al. 2008). It has also been demonstrated that, in rats, poor maternal care is associated with reduced NMDA receptor levels, reduced BDNF expression, and impaired spatial learning (Liu et al. 2000). Moreover, in humans, child abuse alters expression of genes that control stress responsiveness later in life; this can lead to the sequence of chronically increased cortisol, hypothalamus–pituitary–adrenal axis dysregulation, and hippocampal cell death, thereby reducing the hippocampus's ability to integrate contextual features during recall of episodic memories (McGowan et al. 2009). As noted above, this phenomenon has been suggested to be a cause of auditory hallucinations. All of these findings are relevant for models of reduced dynamic coordination in disorders associated with trauma and/or unhealthy lifestyles (e.g., schizophrenia, BPD, dissociative disorders, as reviewed above).

Questions Regarding Normal Brain Functioning

To what extent can data on dynamic coordination abnormalities inform our understanding of normality? For example, to what extent can findings of changes in disorders with genetic contributions allow us eventually to understand the genetic factors associated with normal coordinating processes? How would this add significantly to our understanding of coordinating functions beyond what is known about neurotransmitters and neuroanatomy? Also, to what extent can increases or decreases in coordination result from self-generated changes in mood or thought? The extent of downward causation from mental activity has yet to be explored.

Phillips and Singer (1997) suggest a distinction between prespecified feature hierarchies and dynamic coordination. This is supported by evidence from schizophrenia, in which grouping based on principles consistent with the former is intact, whereas performance deficits appear on tasks involving the latter. Does this suggest that these two processes are separable? Relatedly, do the relationships between neuronal loss, abnormal synchrony, and binding disturbances, as well as the correlations between indices of dynamic coordination failure in disorders such as schizophrenia, support Phillips and Singer's (1997) hypothesis that "the cortical algorithm everywhere is the same?" Do such data provide disconfirmatory evidence for Rolls' (2006) hypothesis that synchronization is most relevant for binding in feature hierarchy networks in early cortical areas (e.g., V2–V4) whereas information representation in higher areas is conveyed nearly completely by spike rates?

Data from schizophrenia and epilepsy suggest that reduced synchrony is the mechanism by which attractor states form that are relatively isolated from other cortical functioning. The data overall agree with Uhlhaas and Singer's (2006) contention that a trade-off between correlated and decorrelated activity is critical for normal brain function. Is it possible to quantify this trade-off and, if so (e.g., via neurofeedback), to enhance performance at various job functions or in life in general?

Phillips et al. (this volume) suggest that in addition to dynamic grouping and contextual disambiguation, concepts such as dynamic embedding, dynamic linking, and dynamic routing are relevant to understanding cognition. For example, they suggest that thoughts about thoughts (i.e., metacognition) may require dynamically embedded groupings. This suggests that disorders such as schizophrenia and autism, both characterized by disturbances in metacognition and theory of mind, may involve failures in dynamic embedding. In addition, data on autobiographical memory support a form of dynamic embedding involving grouping by causation, temporal proximity, and similarity in content (Brown and Schopflocher 1998). Can the concept of dynamic embedding guide the development of techniques to improve metacognition, social cognition, and memory function in both healthy and ill persons?

A potential criticism of the construct of dynamic coordination is that if it can explain nearly all aspects of normal and abnormal mental functioning, then it is too broad and simply too general a term for what the brain does. What are the implications of the data from illness states for determining how the construct of dynamic coordination adds to or conflicts with views such as Hebb's (1949) seminal theory of cell assemblies, Hemsley's (2005) theory of the role of the hippocampus in integrating sensory input with memory traces, and Andreasen's (2008) view that cerebellar activity is a primary determinant of cognitive coordination? Engel et al. (this volume) have made a first pass at clarifying which aspects of cognition do, and do not, involve cognitive coordination. In addition, they highlight the theoretical perspectives on cognitive coordination that can help differentiate it from other theories.

Do the differences in type and severity of dynamic coordination changes across disorders have implications for understanding individual differences? For example, can the tendency to bind normally uncorrelated representations, as found to an extreme in schizophrenia, be the basis for creativity? Is context sensitivity, which is reduced to an extreme degree in schizophrenia and autism—two conditions with social functioning deficits—the basis for social skill in the "neurotypical" population?

The questions posed here are but a sample of those that could be generated from a reading of the data laid out in this chapter. We may be far from an answer to these and other questions at present, but the relative ease with which they can be generated suggests that understanding failures of dynamic coordination has the potential to increase our understanding of normal brain function, and possibly to lead to the development of techniques to improve the cognitive and emotional functioning of healthy people as well.

Acknowledgments

I thank Angus MacDonald III and Bill Phillips for helpful comments on an earlier version of this chapter.

Overleaf (left to right, top to bottom):
Karl Friston, Andreas Engel, Ilona Kovács
Peter Uhlhaas, Angus MacDonald, Bill Phillips
Peter König, group discussion
Earl Miller, Catherine Tallon-Baudry, Jochen Triesch
Scott Kelso, Steve Silverstein

18

Coordination in Behavior and Cognition

Andreas K. Engel, Karl Friston, J. A. Scott Kelso,
Peter König, Ilona Kovács, Angus MacDonald III,
Earl K. Miller, William A. Phillips, Steven M. Silverstein,
Catherine Tallon-Baudry, Jochen Triesch, and Peter Uhlhaas

Abstract

What is coordination and how is it achieved? This chapter begins with a discussion of the concept and key features of dynamic coordination. Next, its relation to cognitive functions and learning processes are explored, as is the role of neural oscillations in different frequency bands for dynamic coordination. Thereafter, modulation of coordination at the systems level is reviewed, and the relation of the mechanisms discussed to neuropsychiatric disorders is pursued. The purpose of this chapter is not to delineate all properties of coordination or all of its different manifestations. Instead, our intent is to portray the multifaceted problem that stands before us.

Introduction

It is a truism that we know coordination when we see it. At the same time, coordination may be subject to precise measurement and observation. How well we can measure and observe depends on how accessible coordination is and at which level we choose to observe it.

We say that someone like Tiger Woods is coordinated because his golf swing (a) is dynamic, evolving over time in a well-defined sequence directed toward achieving a goal; (b) involves the orchestration of many different sensory, motor, and cognitive processes at many different levels; (c) involves a reduction in dimensionality (i.e., despite the enormous number of degrees of freedom, coordination is coherent and low dimensional); (d) adapts to the perceived environmental conditions as well as the sensed state of the body; (e) is a flexible, creative process that involves decision making and planning; (f) is stable (i.e., resistant to perturbations over the timescale of the behavior); (g) involves learning and is subject to modification by a number of factors, such as level of attention, stress, or others.

The problem of coordination involves understanding how component parts and processes relate in an orderly fashion to produce a recognizable function. Coordination may thus be defined as a functional ordering among interacting components in space and time. Coming in many guises, coordination represents one of the most striking features of living organisms. Some of the basic phenomena that seem to be of particular relevance to understanding dynamic coordination in the brain and cognition are:

- Patterned states of coordination remain stable in time despite perturbations.
- Component parts and processes (dis)engage in a flexible fashion depending on functional demands and/or changes in environmental conditions.
- Multiple coordination states may exist rendering living things multifunctional, effectively satisfying the same (or different) sets of circumstances.
- Switching from partially to fully coordinated states and vice versa is commonplace.
- Selection of coordination patterns is tailored to suit the current needs of the organism.
- Coordination patterns adapt to changing internal and external contingencies.
- Depending on a balance between competitive and cooperative processes, learning may take the form of abrupt transitions from one coordinated pattern to another.
- The system may remain in the current pattern of coordination even when conditions change, thus exhibiting memory.

We begin by reviewing the concept and key features of dynamic coordination.

How Can Dynamic Coordination Underlying Behavior and Cognitive Processing Be Conceptually and Formally Specified?

Dynamic Coordination as a Result of Self-organization

A key concept for understanding dynamic coordination in complex systems is self-organization. Self-organization refers to the spontaneous formation of patterns and pattern change in systems that are open to exchanges of information with the environment and whose elements adapt to the very patterns of behavior they create. Inevitably, when interacting elements form a coupled system with the environment, coordinated patterns of behavior arise. Naturally occurring environmental conditions or intrinsic, endogenous factors may take the form of control parameters in a dynamical system. For example, candidate control parameters in neural circuits include neuromodulators and synaptic drive. A circuit may be capable of operating in distinctly different stable modes and switching between them depending on the level of synaptic drive and the degree of neuromodulation (e.g., Briggman and Kristan 2008).

When a control parameter crosses a critical value, instability occurs and leads to the formation of new (or different) patterns. In self-organizing dynamical systems, such as fluids, lasers, and chemical reactions, the enormous compression of degrees of freedom near critical points arises because events occur on different timescales: the faster individual elements in the system become "enslaved" to the slower, "emergent" collective variables which now constitute the relevant information for the system's dynamic behavior. Collective variables are relational quantities, spanning or enfolding different domains that reflect the coupling among component parts and processes.

Alternatively, and perhaps more in line with how nervous systems are coordinated, one may conceive of a hierarchy of timescales for various processes involved in coordination. On a given level of the hierarchy, dynamic coordination may be subject to constraints (e.g., of the task) that act as boundary conditions on lower-level processes. At the next level down are component processes and events that typically operate on faster timescales. Thus a complete description of coordination on a chosen level of description would seem to require identification of (a) the boundary conditions and control parameters that establish the context for particular coordination phenomena to occur, (b) the relevant collective variables and their dynamics, and (c) the component level and its dynamics including the nonlinear coupling between components.

Self-organized pattern formation in the brain—a subject of much active investigation in the neurosciences—expresses itself in various forms, including brain oscillations (e.g., Basar et al. 2000; Buzsáki 2006), transient phase synchrony among neural populations (e.g., Singer and Gray 1995; Varela et al. 2001; Engel et al. 2001; Bressler and Kelso 2001), multistability, abrupt phase transitions ("switches") in cortical activity patterns, and so forth. Long ago Katchalsky et al. (1974:58) noted: "The possibility of waves, oscillation, macrostates emerging out of cooperative processes, sudden transitions, prepatterning, etc., seems made to order to assist in the understanding of integrative processes…particularly in advancing questions of higher order functions that remain unexplained in terms of contemporary neurophysiology." Here we will discuss numerous studies that explicitly address such manifestations of dynamic coordination in the brain and relate these to cognitive and behavioral functions.

Dimensionality Reduction

The key feature of coordination is that a very large number of heterogeneous elements characterized by mutual interaction "live" in a subspace or manifold whose dynamics are low dimensional. Near instability, the individual elements must order themselves in new or different ways to accommodate current conditions. The patterns that emerge may be defined as attractor states of the collective variable dynamics; that is, the collective variable may converge in time to a certain limit set or attractor solution. Mathematically, systems composed of many interacting elements are described in terms of a (large) number of time-dependent states that trace out a trajectory in a high-dimensional state-space (so that the current state is represented by a point in state-space). The long-term evolution of many systems can then be characterized in terms of those parts of state-space to which all trajectories are attracted (the attractor manifold). Crucially, in instances of dynamic coordination, this manifold usually has a low dimension and supports wandering (itinerant) trajectories; this means that the states are not fixed but revisit a subset of states in a flexible but reproducible way (e.g., metastability and bistable perception).

The fact that the manifold is low dimensional is key for understanding how order can be synthesized from multiple interacting systems, like neurons or macro columns, and may explain the emergence of percepts with a unitary nature. The mathematical analysis of coupled systems suggests that these manifolds enforce synchronization of the coupled systems (in fact they are referred to as synchronization manifolds). This is the key to understanding the central role of oscillations and synchrony in binding the dynamics of distributed populations in the brain. Furthermore, it speaks to synchronization as a pragmatic measure that can be used to infer the presence of dynamic coordination.

Importantly, the reduction in dimensionality associated with dynamic coordination implies the creation of new knowledge. Let us consider the example of an image that contains a large number of Gabor patches. A full description of the image would require, for each Gabor patch, its location and orientation. However, if those Gabor patches can be integrated in a contour, then the image is fully described by the contour itself: this corresponds to a reduction of the number of state variables necessary to describe the image or, in other words, to the fact that additional knowledge has been created by coordinating the elements into a whole: the contour. This idea of creation of new knowledge is also captured by the axiom "the whole is different from the sum of its parts" and thus dynamic coordination can be viewed as those processes which foster the emergence of the whole from the parts.

The Theory of Coherent Infomax

The concept of dynamic coordination can be specified in informational theoretical terms within the theory of Coherent Infomax (Kay et al. 1998; Phillips et al., this volume). In short, contextual modulation affects the transmission of the information that it modulates, while, in contrast to the signal that it modulates, transmitting little or no conditional mutual information about itself. Alternatively, another way to see what is meant by the phrase "dynamic coordination" is to think of it as a cover term for at least three fundamental neurocomputational functions: multiplicative gain modulation, dynamic grouping or "binding," and dynamic routing (cf. Phillips et al., this volume). All three functions can be viewed as involving interactions that affect neural activity but without changing the information transmitted by the cells producing that activity. Tiesinga et al. (2008:106) state that "multiplicative gain modulation is important because it increases or decreases the overall strength of the neuron's response while preserving the stimulus preference of the neuron." Multiplicative gain modulation has been closely related to attention, coordinate transformations, the perceptual constancies, and other cases of contextual modulation, which shows the breadth of its range of potential application.

Our discussions, however, also reflected somewhat different views on the degree to which coordinating interactions can change the local "meaning" or representational contents. The Coherent Infomax Theory suggests that coordinating interactions are essentially modulatory in nature and, thus, have only weak effects on the information carried by neural responses, which are considered to result mainly from bottom-up inputs into the respective circuit. However, there may also be cases of coordination where the coordinating interactions are actually constitutive for the meaning (or functional role) of the local neural signals. Recording of a single neuron supplies a high amount of mutual information on the activities of other neurons. Therefore, the information on the activity pattern of one neuron cannot be inferred from the stimulus as such, but only through additional knowledge regarding the activity pattern of other neurons. The interaction of the neurons constitutes a free variable that is not directly affected by the stimulus. Hence, in information theoretic terms, the activity pattern is not determined completely by the evidence (stimulus); the prior (here the conditional probability given that other neurons fire in a specific pattern) has to be taken into account.

Put in neural terms, this latter view implies that the activity of individual neurons, taken on their own, does not carry completely invariant information; it is the context of the neural population (the assembly) that actually determines functional impact and "meaning" for the individual neural responses. Obvious examples can be found at the level of perceptual grouping, where it is known that perception of a complex object is mediated by coordination of massive neural populations; in this case, representational contents are established only at the population level, and coordinating interactions (e.g., synchronization of the respective neural signals) are a necessary condition for generating "meaning" and transmitting information on the "whole" rather than on its "parts."

Dynamic Pattern Theory

An alternative theoretical approach that also describes the phenomena associated with dynamic coordination is known as Dynamic Pattern Theory. This approach has been formally defined and studied by Kelso (1995) and others, drawing on Haken's work on synergetics. Important insights into principles of coordination have come from studies of motor coordination, such as that within and between hands, arms, and legs. The mathematical formulations of synergetics and Dynamic Pattern Theory have been more extensively developed than that of Coherent Infomax, but the two approaches seem to have much in common. A number of general conclusions have emerged from studies of motor coordination that are very similar to what has been proposed as applying to "dynamic coordination":

1. Dynamic coordination in sensorimotor control is highly distributed. Swinnen (2002:350), reviewing work on bimanual coordination concludes: "Although the prevailing viewpoint has been to assign bimanual coordination to a single brain locus, more recent evidence points to a distributed network that governs the processes of neural synchronization and desynchronization that underlie the rich variety of coordinated functions."
2. Population coding is common. Most movements are not driven by a single cell but by the combined activity of a population of cells. This has been well illustrated by Georgopoulos (1995), who has shown that the precise direction of movements is more closely related to the activity of cell populations than to that of single cells.
3. The precise timing of neural signals is crucial. In-phase or counterphase synchronizations between rhythmically contracted homologous muscle groups are particularly common (e.g., in walking, running, and swimming), but other phase relationships can be learned, and constraints on this learning are the subject of much research.
4. The system falls easily into certain preferred patterns of activity (Kelso 1995). Modes of coordination in the motor system are patterns of movement that are easily performed and resistant to perturbation. They are usually well rehearsed and highly automated. They can be thought of as attractors in an energy landscape and are analogous to the idea of attractors in population codes for concepts and memories. An advantage of applying the notion of attractors to the motor system is that the process of attraction can be made concretely visible. This can be done, for example, by making clockwise circling motions with both hands, but with one leading the other by a small proportion of the cycle frequency. As cycle frequency increases there will be a tendency for the two movements to become in phase such that the homologous muscle groups are activated synchronously. As this is a highly stable movement pattern, it attracts similar but unfamiliar movement patterns toward it, thus making the idea of an attractor visible. Specific patterns of interaction occur between command streams at various stages of planning and execution. These can be either mutually interfering or

mutually supportive. This is the basic premise of the "neural crosstalk" approach developed by Swinnen (2002) and others. Well-coordinated actions are therefore those in which mutual support is maximized and mutual interference is minimized.

Understanding Dynamic Coordination

Several key concepts mentioned in the introduction will reemerge throughout the remainder of this chapter. These include dimension reduction (i.e., the notion that dynamic coordination reduces a high number of dimensions or degrees of freedom to a low-dimensional space or manifold); the notion of infomax and the importance of maintaining a high mutual information between the sensorium and its internal representation; and the functional role of dynamic coordination, which can be defined in terms of optimization for both perception and action. It is helpful to see these three constructs as intimately related facets of the same basic process: Put simply, we suppose that the purpose of the brain is to represent the world in a parsimonious and accurate fashion. This entails a mapping from sensory input to an internal representation, so that the representation provides a parsimonious account or explanation of sensory input. This parsimony corresponds to dimension reduction; namely, a collapse of a high-dimensional input space into a low-dimensional representation. The accuracy of this representation means that the sensory inputs can be predicted with minimal error and suggests that dynamic coordination optimizes prediction error. The fact that sensory inputs can be predicted implies a high degree of mutual information between those inputs and their representation. This is the essence of infomax. Coherent infomax addresses the fact that a better representation can be constructed by selectively preserving information that is predictably related across local processors that operate upon different parts of the input data. By so doing it might be possible to discover distal causes in the proximal data. We will return to the theme of optimized mappings between high-dimensional input spaces and low-dimensional representations later, when we consider the relationship between cognitive processes and dynamic coordination.

Is Dynamic Coordination the Basis of Specific Cognitive Functions or Is It a General Optimizer of Cognitive Functioning?

Cases of Dynamic Coordination

Dynamic coordination was originally conceived as a process that was mediated by synchronous interactions among neurons or neuronal populations during perceptual synthesis. It emphasized the contextual disambiguation of cause and content during the processing of sensory information in neuronal circuits, both local and distributed, and the mergence of unitary representations of the external causes of sensations. Its ubiquitous role in neuronal and biological processes has led to a more inclusive use of the concept to cover the coordination of distributed dynamics, not just among sensory neurons but in the coordination of behavior and, more generally, the coupling of dynamical systems from the electrophysiology level to ethology.

It is useful to seek examples of the various forms of dynamic coordination at different levels of analysis from local circuits to behavior. Examples of contextual gain modulation at the level of single cells, or local groups of cells, include much evidence for the modulatory effects of attention and concurrent stimulus inputs from beyond the classical receptive field, as reviewed, for example, by Reynolds and Desimone (1999).

Examples at the level of cognition and behavior include endless demonstrations of contextual disambiguation, both within and between modalities. Rigorous examples from perception range from the effects of task irrelevant collinear or non-collinear surrounds on contrast detection, as studied in humans by Polat and Sagi (1994) and in awake-behaving monkeys by Kapadia et al. (1995), to effects on speech, face and scene perception. Many other examples are given by Edelman (2008a) and include paintings by Magritte designed to increase our awareness of the way in which context modulates interpretation of the local data.

Examples of dynamic grouping at the behavioral level include the contour integration task (reviewed by Kovács and Julesz 1993; Field et al. 1993; Hess et al. 2003). This task is designed so that explanations in terms of prespecified feature detectors are unlikely. It has also been used to test dynamic grouping in awake-behaving monkeys by Kreiter and Singer (1996) as well as in experiments combining behavioral and physiological measures (e.g., Müller et al. 1997).

Taking the example of visual perception, there is an obvious need for back-and-forth interaction among several levels of processing. Let us consider the perception of moving bodies: the nonrigid movements of the different body parts require an efficient and coupled coding of visual shape and motion information. For the efficient transfer of this coupled information, a representation (or code) is needed that optimizes space-time resolution (e.g., preserving a sufficient amount of spatial information in the presence of good temporal resolution). The point-light displays used by Johansson (1973) and the motion capture techniques more recently employed in studies of biological motion perception (Troje 2002) exemplify the effectiveness of dimensionality reduction that bring about "meaning" in the course of dynamic coordination.

Cognitive Processes and Coordination

One of the challenges encountered in our discussions of the relation between coordination and cognition was that the concept of "dynamic coordination" seems to relate to an enormous variety of neural and cognitive functions. In the extreme, "dynamic coordination" could even become a synonym for "cognition." One way to address the problem of over-inclusiveness is to assume a heuristic rather than formal approach. This can be accomplished by conceptualizing the framework along two axes (Figure 18.1): the number of elements to be coordinated during the process and the degree of flexibility associated with the process under study. For clarity, only a few neural and cognitive processes are shown in this diagram, highlighting typical examples used during our discussions.

In Figure 18.1, the horizontal axis ("number of coordinated elements") represents intuitions about the extent to which a task is complex, as defined by the combination of the number of elements required. As we shall see below, this dimension seems to covary with the extent to which synchronous oscillatory activity is evoked, the number of neurons or neural modules involved, as well as the extent to which intra- or inter-regional feedback is required. The vertical axis ("flexibility") refers to the number of choices in the given task and represents intuitions about the extent to which a task involves selecting an interpretation (from an ambiguous input) or action (when performing a task) from an increasingly broad range of options. In addition, this dimension maps

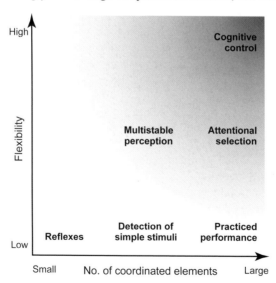

Figure 18.1 Two dimensions of dynamic coordination. The horizontal axis represents the number of coordinated elements; the vertical axis depicts "flexibility." Gray shading indicates degree of dynamic coordination.

the extent to which a task is more novel or requires more control. Accordingly, a high degree of dynamic coordination would be associated with both coordination of large numbers of elements and high flexibility, as indicated by the gray shading of the figure background. The key question is whether the same parameter ranges are typically required for cognitive processes. In our discussions, consideration of numerous examples suggests that this may generally be the case, although exceptions seem possible.

Clearly, the relation between both dimensions is not symmetric. Relevant examples suggest that increasing the demands on flexibility (e.g., by allowing for selection between a wide range of choices) imposes a necessity for coordinating increasingly large numbers of elements in the system. Suitable examples include, at an intermediate level of both parameters, multistable perception or, at a very high level in both axes, the process of cognitive control. We note, however, that the converse is not true; coordination of large numbers of elements can take place without large flexibility on the task. For example, skilled performance in an overlearned motor task may represent a high level of coordination even when the flexibility is highly constrained. Thus, a proficient piano player touching a specific key at a specific time with a specific force clearly requires a high level of coordination even when no other outputs are under consideration. Importantly, for the current purposes, dynamic coordination is less evident, or absent, in a number of familiar functions (e.g., simple perceptions, reflexes, and prepotent responses). These processes share the property of being capable of being accomplished with first order statistics.

Silverstein (this volume) has detailed the nature of perceptual organization and how it places demands on dynamic coordination. The example of attentional gain control is also illustrative in this instance. One well-studied aspect of this process is the selection of one of a number of sensory stimuli for further processing from an array of different sensory stimuli. Here, the selection of one object in a visual scene for detailed processing by higher visual cortical areas requires the harmonious coordination of a number of lower and higher visual-processing areas. Such a selection process can be seen as a special case of the more general notion of dynamically coordinating the flow of information through the brain. Consider the example of driving a car while engaging in a conversation. Visual information about the location of the car, with respect to the road and the current traffic situation, is processed and ultimately routed to the muscles controlling our hands on the steering wheel and our feet on the accelerator or brake pedal. At the same time, acoustic information associated with the words of our conversation partner will be processed and ultimately routed to our own vocal apparatus as we respond. However, this pattern of information flow can be quickly and flexibly altered if, for example, our conversation partner asks us to stop the car—an acoustic stimulus prompting us to step on the brake.

Dynamic coordination, by definition, is required when the output of a given level of processors cannot be specified in advance on the basis of the input.

That is, multiple outputs are possible, and the output is determined in part by contextual factors. In this way, dynamic coordination can be seen as a basis of any process that requires such coordination; as noted above, this includes a wide range of perceptual and cognitive functions. What differentiates cognitive functions is, of course, the nature of the representations that must be created and transformed: visual vs. auditory input, codes for muscle movement, memory representations, sensorimotor integration, semantic information, etc. Moreover, the nature of the coordinating process appears to depend on a number of factors, including the speed at which coordination is needed (which can affect whether the primary coordinating mechanism is an initial neural spike vs. a slower or faster rhythm) and the number of processes that must be carried out simultaneously. In the latter case, multiple frequency bands will be "opened" and oscillatory activity in each will subserve a different process, although these may interact (Tallon-Baudry, this volume).

Dynamic Coordination as Optimization

It may be useful to understand some essential characteristics of dynamic coordination in terms of optimization. This rests on reducing the function of the brain to the optimization of specific quantities and thinking about what this entails for neuronal dynamics.

For nearly every aspect of brain function, from elementary perceptual categorization to optimal decision making under uncertainty, one can frame the problem (objective) faced by the brain in terms of an objective function. For perception, this objective function is the evidence for an internal (generative) model of how its sensory inputs were caused. Under some simplifying assumptions, this reduces to the amount of prediction error—the mismatch between sensory inputs and the predicted inputs under an optimized model. Perception, therefore, reduces to the suppression or explanation of (bottom-up) prediction error by (top-down) predictions; this is known as predictive coding and is a special case of more general Bayesian formulations in terms of free energy (Friston and Stephan 2007).

In action and motor control, several objective functions have been proposed, most in the service of finessing forward models of motor control. Recent formulations suggest that the same sort of prediction error minimized by predictive coding is minimized in optimal motor control (i.e., sensory prediction errors on proprioceptive channels reflect the mismatch between sensed and anticipated, or desired, consequences of movement). In procedural and reinforcement learning, the role of (reward) prediction error is again central to many formulations, such as temporal difference models and simpler versions in psychology (e.g., the Rescorla-Wagner model). In game theory and optimal decision theory, the objective function comprises expected loss under uncertainty or (in behavioral economics) its complement: expected utility or value. In fact, the value or expected reward can be linked to the surprise or amount of prediction error minimized by action and perception. The basic idea here is that we can think of brain function as optimizing something (usually minimizing prediction error). So what does this entail for the optimization process?

Biophysical optimization schemes generally use some type of stochastic search (e.g., natural selection) or gradient descent. Gradient descent is inherently a dynamic process, most often formulated in terms of differential or difference equations. This means that any optimization in the brain (that uses gradient decent) must be a dynamic process (if it involves distributed neuronal states) and must involve dynamic coordination. The signature of gradient descent is that the dynamics move current states of the brain towards an attractor that represents the (fixed-point) optimal solution; where the objective function is brought to an extreme. For example, in perception, this would be the maximum a posteriori estimate of the causes of sensory input, at which point the prediction error is usually minimized. This means that dynamic coordination must attract brain states to an invariant set (the desired or optimal solution). Happily, this is the hallmark of dynamic coordination: the organization of the degrees of freedom in large numbers of distributed neuronal systems, so that they are contained in a low-dimensional space.

Can we be more precise about the sorts of dynamic coordination this might evoke? To do so, we need to consider dynamic coordination at different scales. A fair amount of evidence suggests that the reciprocal message passing between different levels or cortical areas in visual cortex is a key determinant of coordinated dynamics and, counterintuitively, may be faster than local lateral interactions within an area. This recursive and self-organized message passing is mandated by biological formulations of (predictive coding) perceptual inference and calls on dynamic coordination at a timescale of tens to hundred of milliseconds over a spatial scale of millimeters to centimeters. Most importantly, it must be self-limiting because its function is to suppress or explain away prediction errors (cf. the self-limiting transients observed electrophysiologically). This means there must be some variant of a feedback loop that ensures convergence to the optimal state. Thus, we might expect (functionally) suppressive effects of top-down extrinsic (between area) connections.

Synchrony and Dynamic Coordination

How does the preceding discussion relate to synchronous interactions between neurons or neuronal populations? A large number of studies carried out over the past two decades suggests that temporal correlations in neural activity play a key role for dynamic coordination in various sensory modalities (von der Malsburg and Schneider 1986; Singer and Gray 1995; Singer 1999; Tallon-Baudry and Bertrand 1999; Engel et al. 1992, 2001; Herrmann, Munk et al. 2004; Fries 2005). As shown by numerous studies in both animals and humans, synchronized oscillatory activity, in particular at gamma-band frequencies (> 30 Hz), is related to a large variety of cognitive and sensorimotor functions. Fast synchronization may play many essential roles. They all rely on a key mechanistic aspect of fast (e.g., gamma) synchronization that enhances the effective coupling between neurons—synchronous gain—in a flexible and context-dependent fashion. The majority of the available studies were conducted in the visual modality, relating gamma-band coherence of neural assemblies to processes such as feature integration over short and long distances (Engel, König, and Singer 1991; Engel, König, Kreiter et al. 1991; Tallon-Baudry et al. 1996), surface segregation (Gray et al. 1989; Castelo-Branco et al. 2000), perceptual stimulus selection (Fries et al. 1997; Siegel et al. 2007), and attention (Müller et al. 2000; Fries, Reynolds et al. 2001; Siegel et al. 2008).

Beyond the visual modality, gamma-band synchrony has also been observed in the auditory (Brosch et al. 2002; Debener et al. 2003), somatosensory (Bauer et al. 2006), and olfactory (Wehr and Laurent 1996) systems. Moreover, gamma-band synchrony has been implicated in processes such as sensorimotor integration (Roelfsema et al. 1997; Womelsdorf et al. 2006), movement preparation (Sanes and Donoghue 1993; Farmer 1998), and memory formation (Fell et al. 2001; Csicsvari et al. 2003; Gruber and Müller 2005; Herrmann, Lenz et al. 2004). Collectively, these data provide strong support for the hypothesis that synchronization of neural signals is a key mechanism for integrating and selecting information in distributed networks (Singer and Gray 1995; Singer 1999; Engel et al. 2001). What they suggest is that coherence of neural signals allows for the establishment of highly specific patterns of effective neuronal coupling, thus enabling flexible and context-dependent binding, the selection of relevant information, and the efficient routing of signals through processing pathways (Salinas and Sejnowski 2001a; Fries 2005; Womelsdorf et al. 2007).

Consideration of numerous examples for processes requiring dynamic coordination shows that these are generally associated with task- or context-specific changes in oscillatory activity and/or coherence. Supportive evidence for this includes studies on Gestalt perception, attention, long-term memory encoding and retrieval, working memory, choice and cognitive sequencing, multimodal integration, language comprehension, and even awareness. Table 18.1 presents an overview of key examples for such studies. As the table shows, oscillatory processes covary with cognitive functions in multiple frequency bands (discussed further below).

In summary, dynamic coordination at a timescale of tens to hundreds of milliseconds may be essential for optimization of distributed representations in the brain. The dynamic nature of this optimization is shaped by the underlying connectivity, which is subject to plasticity on a slower timescale. Emerging from anatomical connectivity, patterns of synchronous interactions may be coordinated dynamically on a faster timescale. Processes occurring at these two timescales can mutually constrain each other through mechanisms

Table 18.1 Cognitive processes and oscillatory neural activity.

Cognitive process	Evidence for oscillations	Evidence for coherence	Frequency band(s)	Species	Brain region	Method	Reference
Grouping	+	−	Gamma	Human	Posterior cortex	EEG	Tallon-Baudry et al. (1996)
Grouping	−	+	Alpha	Human	Posterior cortex	EEG	Mima et al. (2001)
Spatial attention	+	−	Alpha	Human	Posterior cortex	EEG	Thut et al. (2006)
Spatial attention	+	+	Gamma	Human	FEF, IPS, MT+	MEG	Siegel et al. (2008)
Attention (gain)	+	+	Gamma	Monkey	V4, LIP, PFC	Microelectrode recording	Fries, Reynolds et al. (2001); Buschman and Miller (2007)
Attention (timing)	+	−	Beta	Monkey	FEF	Microelectrode recording	Buschman and Miller (2009)
Working memory	+	+	Beta	Human	Posterior cortex	intracranial EEG	Tallon-Baudry et al. (2001)
Working memory	−	+	Beta	Monkey	IT	ECoG	Tallon-Baudry et al. (2004)
Working memory	+	−	Beta, gamma	Monkey	PFC	Microelectrode recording	Siegel et al. (2009)
Long-term memory encoding	−	+	Gamma	Human	Hippocampus	Intracranial EEG	Fell et al. (2001)
Long-term memory retrieval	+	−	Gamma	Human	Posterior cortex	EEG	Herrmann, Lenz et al. (2004)

Cognitive process	Evidence for oscillations	Evidence for coherence	Frequency band(s)	Species	Brain region	Method	Reference
Awareness	+	−	Gamma	Human	Sensory areas	MEG	Wyart and Tallon-Baudry (2008)
Awareness	+	−	Alpha, beta	Human	Sensory areas	MEG	Linkenkaer et al. (2004)
Multisensory integration	+	−	Gamma	Human	Temporal cortex	EEG	Schneider et al. (2008)
Multisensory integration	+	+	Alpha, beta	Monkey	Auditory cortex, temporal cortex	Microelectrode recording	Kayser and Logothetis (2009)
Visuomotor integration	+	−	Alpha, beta, gamma	Human	Sensorimotor cortex	MEG	Chen et al. (2003)
Language comprehension	+	−	Theta, alpha, beta	Human	Temporal cortex	EEG	Bastiaansen et al. (2005)
Social coordination	+	−	"Phi" (alpha range)	Human	Centroparietal cortex	EEG	Tognoli et al. (2007)

FEF: frontal eye fields
IPS: intraparietal sulcus
MT+: middle temporal region
LIP: lateral intraparietal

PFC: prefrontal cortex
IT: inferotemporal cortex
EcoG: electrocorticography

of time-dependent plasticity and experience-dependent consolidation of architectures selected by synchronization of the sort indexed by gamma oscillations.

Is Dynamic Coordination Learned and Does Dynamic Coordination Modify Learning?

Relation between Dynamic Coordination and Learning

Dynamic coordination and learning are likely to have a profound influence on each other, but understanding this relationship is far from trivial. On one hand, dynamic coordination may be the result of learning processes and/or subject to developmental change. On the other, dynamic coordination may shape learning by selecting flexibly created relationships between internal representations that should be laid down in synaptic weight patterns.

Essentially all behaviors or complex competencies associated with dynamic coordination discussed above improve during ontogenetic development. This could mean that the competencies and neural structures being coordinated are improving, but it also suggests that coordination may be improving through learning. What may be the underlying mechanisms?

Learning processes are associated with a range of synaptic (and other) plasticity mechanisms. Dynamic coordination may also be realized through a number of mechanisms among which the synchronization of neuronal responses and oscillations in the gamma frequency band are prominent candidates. The question then is: How do the known plasticity mechanisms shape the (putative) mechanisms for coordination? For example, are there learning processes that will tend to improve neuronal synchronization? Or should the ability to synchronize be viewed instead as a generic circuit property, which learning processes may exploit for, say, the dynamic routing of information? To what extent do learning mechanisms shape dynamic coordination in a task-dependent manner that is influenced by reward signals to improve behavioral performance? Do learning processes ultimately transfer dynamically coordinated states into efficient but inflexible special purpose circuits?

Importantly, dynamic coordination may also affect learning. In particular, it may select from a multitude of possible association patterns those that are meaningful and should be remembered. As a specific example, consider the situation where dynamic coordination takes the form of synchronizing the firing of a population of neurons at the level of a few milliseconds. Interestingly, work on spike-timing-dependent plasticity (STDP) shows that whether a synapse between two neurons is strengthened or weakened can depend on the millisecond-scale precise timing of their action potentials (Markram et al. 1997). If the presynaptic neuron fires shortly before the postsynaptic neuron, the connection is strengthened. If, however, the presynaptic neuron fires shortly after the postsynaptic neuron, the synapse is depressed. Thus, dynamic coordination processes that control the spike timing of groups of neurons (e.g., by synchronizing them) will have a huge impact on what synapses are strengthened or weakened. Dynamic coordination processes, together with the action of neuromodulators, may in fact be controlling the expression of plasticity.

Neural Synchrony during Human Ontogeny

The development and maturation of cortical networks critically depends on neuronal activity, whereby synchronized oscillatory activity plays an important role in the stabilization and pruning of connections. In STDP, pre- and postsynaptic spiking within a critical window of tens of milliseconds has profound functional implications (Markram et al. 1997). Stimulation at the depolarizing peak of the theta cycle in the hippocampus favors long-term potentiation (LTP), whereas stimulation in the trough causes depotentiation (LTD) (Huerta and Lisman 1993). The same relationship holds for oscillations in the beta and gamma frequency range (Wespatat et al. 2004).

Furthermore, synchronization of oscillatory activity is an important index of the maturity and efficiency of cortical networks. Neural oscillations are energy-efficient mechanisms for the coordination of distributed neural activity that are dependent upon anatomical and physiological parameters (Buzsáki and Draguhn 2004) which undergo significant changes during development. Thus, synchronization of oscillatory activity in the beta and gamma frequency range is dependent upon corticocortical connections that reciprocally link cells situated in the same cortical area, across different areas or even across the two hemispheres (Engel, König, and Singer 1991; Engel, König, Kreiter et al. 1991). Furthermore, GABAergic interneurons play a pivotal role in establishing neural synchrony in local circuits as indicated by research that shows that a single GABAergic neuron may be sufficient to synchronize the firing of a large population of pyramidal neurons (Cobb et al. 1995) and the duration of the inhibitory postsynaptic potential (IPSP) can determine the dominant frequency of oscillations within a network (Wang and Buzsáki 1996). As brain maturation involves changes in both GABAergic neurotransmission (Hashimoto et al. 2009; Doischer et al. 2008) and the myelination of long axonal tracts (Ashtari et al. 2007; Perrin et al. 2009), changes can be expected in the frequency and amplitude of oscillations as well as in the precision with which rhythmic activity can be synchronized over longer distances at different developmental stages.

During development of resting state activity, there is a reduction in the amplitude of slow-wave (delta, theta, alpha activity) rhythms, while fast (beta- and gamma-band) rhythms increase during childhood and adolescence. This is accompanied by increases in the coherence of oscillatory activity (for a review, see Niedermeyer and Silva 2005). Development of task-related activity in the gamma band coincides with the emergence of cognitive functions during early childhood (Csibra et al. 2000), suggesting that the maturation of high frequency activity could be related to cognitive development. Following infancy, continued development of neural synchrony is observed whereby oscillations shift to higher frequencies and synchronization becomes more precise. Specifically, this is not complete until early adulthood; neural synchrony continues to mature throughout the adolescent period, which represents a critical phase of brain maturation (Uhlhaas et al. 2009).

Coordination in Perceptual and Motor Development

The human development of perceptual organization as measured behaviorally (Kovács 2000) seems to follow a similar maturational course, as suggested above, with respect to neural synchrony, continuing into adolescence and early adulthood. Perceptual integration can be taken as an example. Sensitivity to contour closure—a Gestalt property, definitely requiring dynamic perceptual organization—has been shown to be a measurable skill in adult human observers, enhancing the segmentation of noisy images (Kovács and Julesz 1993). Closure sensitivity is missing in three-month-old human infants (Gerhardtstein et al. 2004), and the underlying ability to integrate spatial information across the visual field develops until the end of adolescence in humans (Kovács et al. 1999). The normal course of development in perceptual organization is affected by the nature of input to the visual system. Abnormal visual input in, for example, amblyopia leads to a severe deficit in perceptual organization related to the amblyopic eye (Kovács et al. 2000). The contour integration stimuli used in the amblyopic study have been designed mainly to involve primary visual cortex processing, and a neuropsychological study confirmed the sufficiency of the primary visual cortex in this task (Giersch et al. 2000). It has been shown in the cat that amblyopia is associated with altered intracortical processing in V1 (Schmidt et al. 2004), and reduced synchronization of population responses has been suggested as a neurophysiological correlate of strabismic amblyopia (König et al. 1993; Roelfsema et al. 1994). It remains to be seen, however, whether synchronization in V1 underlies both intact contour integration in humans and deficient processing in human amblyopes.

Is reduced perceptual performance in children—either in terms of precision or timing—due to less efficient or slower synchrony, or to the fact that basic visual skills can become overlearned and automated over the course of development, leading to more efficient/faster processing? Alternatively, can all dynamically coordinated activities be trained to an "automatic" level? Wonderful examples in both perceptual (e.g., Karni and Sagi 1993) and motor learning (Karni et al. 1998) demonstrate that performance is improved over time, both in terms of precision and in terms of the time taken in individual trials. In both cases, there seems to be an initial phase of learning that involves the activity of a number of cortical areas; later on, plastic changes will be specific to the primary sensory or motor cortices. It might be argued that dynamic coordination is only involved in the first phase of learning, while during the second stage a more inflexible structure carries out the task. Perceptual and motor learning might be excellent grounds for testing the idea of reduced dynamic coordination following practice as both seem to show a very high level of specificity in terms of learning. This specificity is graded, however, and the amount of transfer varies across tasks. Transfer properties of different skill learning cases might tell us about the involved cortical structures and the flexibility of coordination.

Another important aspect of basic skill learning in the perceptual and motor domains is that learning is sleep-dependent. It seems that sleep actively contributes to performance improvements in procedural learning (Stickgold et al. 2000; Walker et al. 2002, 2003). In addition, it contributes to learning in the contour integration task—a task designed to rely heavily on dynamic integration (Gervan and Kovács 2010).

Learning and Dynamic Coordination

One of the roles of dynamic coordination is to encode new relationships. However, if those relationships turn out to be stable ones, encountered repeatedly, then it is probably worth creating a neural route dedicated to those items that are bound by stable links. This new route would result from the modification of synaptic efficiency in repeated jointly activated assemblies. It seems reasonable to assume that this new route is fast and recruits a smaller number of neurons. Thus it does not necessarily require flexible, dynamic coordination any longer: the new route would correspond to a prespecified spatiotemporal pattern of neural activity.

Recent data in humans are compatible with this schema. Subjects performed a typical visual search task in which they had to detect a target at different locations and report its orientation. Two types of images were interspersed: in predictive images, the layout of the distractors predicted accurately the location of the target, while in nonpredictive images, the target could appear anywhere. Subjects learned these regularities: after five or six presentations, they were faster at reporting the orientation of the target in predictive images than in nonpredictive ones. During the learning phase, before any behavioral advantage to predictive images occurred, oscillations in the low-gamma range appeared specifically in response to predictive images, suggesting that the brain was detecting the stability of the relationship between distractors and target (Chaumon, Hasboun et al. 2009). As soon as those relations were learned and the behavioral advantage to predictive images appeared, those low frequency gamma oscillations stopped. Predictive images were processed now in a very different way, with evoked responses specific to predictive images in the temporal lobe and orbitofrontal cortex occurring before 100 ms (Chaumon et al. 2008; Chaumon, Schwartz et al. 2009). The result of learning thus seems to be a modification of the early volley of feedforward processing in response to predictive images.

What Is the Role of Different Frequency Bands for Dynamic Coordination?

Oscillations in Different Frequency Bands

Ongoing intrinsic and event-related oscillations are usually categorized into five frequency bands: delta (0.5–3.5 Hz), theta (4–7 Hz), alpha (8–12 Hz), beta (13–30 Hz), and gamma (> 30 Hz). A large body of evidence suggests that oscillatory activity in these frequency bands is linked to a broad variety of perceptual, sensorimotor, and cognitive operations (Engel et al. 1992, 2001; Singer and Gray 1995; Basar et al. 2000; Klimesch et al. 2006; Palva and Palva 2007; see also Table 18.1). Oscillatory activity in the delta band has been related to motivational processes, the brain reward system, and is the predominant frequency during deep sleep phases (Basar et al. 2000; Knyazev 2007). Activity in the theta band has been linked to working memory functions, emotional arousal, and fear conditioning (Knyazev 2007; Jensen and Lisman 2005). The prominent alpha-band responses, discovered in the human EEG by Hans Berger in the late 1920s, have been suggested to reflect cortical operations during the awake-resting state in the absence of sensory inputs. More recent theories have proposed that alpha-band oscillations may also relate to disengagement of task-irrelevant brain areas (Klimesch et al. 2006), as well as working memory function and short-term memory retention (Palva and Palva 2007). Neuronal responses in the beta band have been frequently linked to sensorimotor processing (e.g., Roelfsema et al. 1997; Brovelli et al. 2004) as well as many other functions including working memory and multisensory integration (see Table 18.1). As discussed above, the putative functions of synchronization in the gamma band seem to be particularly diverse, ranging from feature integration, stimulus selection, attention, and awareness to sensorimotor integration, movement preparation, and memory formation. This striking diversity indicates that it may be difficult to associate cognitive functions or even classes of functions in any unique and direct way with oscillatory dynamic coupling in specific frequency bands. The examples discussed below yield a rather complex picture. While they provide clear cases of task- or context-related modulation of frequencies or even switching between different frequency ranges, they do not yet suggest generalizable conclusions.

Currently, another unresolved issue concerns the interaction of multiple frequency bands. Phase synchrony and phase modulation of oscillations across different frequency bands has recently been suggested to play a key role for the organization of networks engaged in speech processing (Schroeder et al. 2008) and memory encoding (Palva and Palva 2007; Jensen and Lisman 2005). This clearly adds to the complexity of the picture already presented by the findings on individual frequency bands. The possibility of multifrequency coupling has also been addressed in the framework of Dynamical Systems Theory (Kelso 1995). As mentioned earlier, the effective degrees of freedom of complex, dynamical systems are often reduced to the space of coupled nonlinear oscillators where a rich variety of behaviors is possible. In particular, a system's ability to generate multifrequency behavior is governed by the differential stability of mode-lockings as seen through so-called "Arnold Tongues" (named after the Russian mathematician Vladamir Arnold). In this dynamic scenario, pattern complexity is related to a hierarchy of frequency ratios.

Role of Different Frequency Bands in Sensory Processing

Recording in primary visual cortex of alert cats, Siegel and König (2003) demonstrated that neuronal activity, as characterized by the local field potential, is optimally orientation-tuned when the gamma band in the frequency range between 40–100 Hz is considered. Synchronization in a lower frequency band with different properties suggested distinct functional roles of low and high frequency synchronization. Subsequently, it has been shown that natural visual stimuli induce robust responses in the gamma frequency band (Kayser et al. 2003). A second frequency band, located at the classical alpha and low beta bands (8–23 Hz), showed reliable tuning to stimulus features (Kayser and König 2004). In marked contrast, tight locking to temporal properties of the stimulus was found in the remaining frequency bands. This locking is independent of the spatial structure of the stimulus. Together these four frequency bands cover the whole frequency range investigated. These studies demonstrate that the entire frequency range of the local field potential can be assigned a role in visual processing, but presumably these roles differ profoundly.

Another study investigated interareal interactions during processing of expected and novel stimuli in the cat visual system (von Stein et al. 2000). Processing of expected stimuli was characterized by high alpha-band activity, and phase relationships and laminar distribution suggested an influence of higher onto lower areas. In contrast, new and surprising stimuli induced high gamma-band activity. These data could be accounted for in a detailed simulation assigning gamma activity to an iterative bottom-up directed processing mode and alpha activity to a top-down directed processing mode. These data offer a new perspective to the classical view that alpha activity is an idling rhythm (i.e., expressing properties of the visual system at rest), whereas gamma activity is thought to be involved in Gestalt perception and figure–ground segmentation (see above). The alpha rhythm might be better described as reflecting visual processing guided by expectations, whereas gamma activity may arise as new stimulus configurations are freshly interpreted in light of previous experience.

Recent evidence suggests that multisensory integration may also relate to neuronal interactions in different frequency bands. Schall et al. (2009) investigated audiovisual binding by presenting continuously changing, temporally congruent and incongruent stimuli. Spectrotemporal analysis of EEG signals revealed locking to visual stimulus dynamics in both a broad alpha band and the lower beta band. This matches results on the role of different frequency bands during processing of natural visual stimuli observed in alert cats (Kayser and König 2004).

Role of Different Frequency Bands in Attention and Awareness

Some evidence for different roles of frequency bands comes from recent observations that different forms of visual attention result in increased coherence in different frequency bands (Buschman and Miller 2007, 2009). When monkeys shifted attention in a bottom-up fashion to a salient "pop-out" stimulus, there was a greater increase in coherence between the frontal and parietal cortices in an upper frequency band (35–55 Hz). By contrast, when attention was shifted in a top-down fashion (to a target that matched one held in short-term memory), there was a greater increase in a lower frequency band (22–34 Hz). Higher frequency oscillations may result from superficial pyramidal cells, which form feedforward connections. They show stronger gamma frequencies than the deep pyramidal cells that originate feedback connections. This also fits well with more recent observations that lower frequency coherence may play a role in controlling the timing of the shifts of attention in the top-down attention condition. Buschman and Miller (2009) found that monkeys shifted the location of their attention every 40 ms as they searched for the visual target. This was reflected in frontal eye fields (FEF) spiking activity and was correlated with the lower frequency band oscillations, suggesting that the lower frequency oscillations can provide a "clocking" signal that helps coordinate when different brain areas contribute to a shift of attention.

Indeed, one role for oscillations may be to coordinate complex, multistep computations. An oscillating wave of inhibition would allow computations to be temporally constrained and information to be released in a "packet" at a time when a downstream area is ready to receive it. This would cause a "discretizing" of events and explain psychophysical observations of a periodic allotment of attention (VanRullen et al. 2007). Attention appears to be allocated in discrete chunks of time and not as a continuous function that smoothly shifts from location to location.

This predicts that oscillation frequencies would vary with the nature of the computation. Highly localized computations may be able to oscillate at higher frequencies while more complex, integrative, or inherently slower computations may result in slower oscillations. For example, consider covert versus overt attention. A purely covert search task without eye movements (as in Buschman and Miller 2007, 2009) allows faster shifts of attention and thus locking to relatively higher (beta) frequency oscillations. By contrast, in overt attention there is a slower time constant because of the increased "overhead" of moving the eyes with each attentional shift. Under these conditions, there are stronger theta oscillations that are time-locked to the eye movements (Desimone, pers. comm.). Computations that have even less temporal overhead and that occur within more local networks might lock to even higher frequencies. For example, gamma-band oscillations have been associated with working memory, surface segregation, perceptual stimulus selection, and focused attention not requiring serial shifts of attention (see above). Computations might use the closest inherent "eigenfrequency" or resonance of the cortical network given the constraints of the computation and the demands of the task at hand.

Evidence from human studies suggests that different subranges in the gamma frequency band can relate to distinct cognitive functions (Tallon-Baudry, this volume). In one experiment (Wyart and Tallon-Baudry 2008, 2009), subjects were cued to attend to the left or right hemifield, and were then presented with a faint oriented grating, either on the attended or unattended side. At each trial, subjects were asked whether they had experienced the stimulus consciously. Each stimulus can therefore be classified as (a) attended or unattended and (b) consciously perceived or not. These two cognitive functions were expressed separately in distinct subfrequency ranges within the gamma range. Gamma-band oscillations related to awareness originated in lateral occipital cortex and were centered around 60 Hz. They were not influenced by spatial attention. Attention-related gamma oscillations were observed at more parietal locations and around 80 Hz, without any influence of whether the subject had consciously perceived the stimulus or not. This could suggest that whenever two distinct cognitive processes have to remain segregated, gamma-band oscillations appear in a narrower frequency band, dedicated to that process, leaving other "slots" available for other concomitant processes to be implemented.

Role of Different Frequency Bands in Motor Circuits

Neurophysiological evidence on the complementary role of oscillations in different frequency bands also comes from recordings in human subcortical structures that are carried out during stereotactic operations for the treatment of movement disorders, such as Parkinson's disease (Brown and Marsden 1998; Brown 2003; Engel et al. 2005). This surgical approach opens up the possibility of recording both unit and field potential signals from the target structures and testing the presence of oscillatory activity and its coherence with EEG and EMG signals during motor tasks in the patients. In a series of studies, Brown and coworkers investigated task- and dopamine-dependent changes of neural coherence between cortex and basal ganglia structures (Brown et al. 2001; Marsden et al. 2001; Cassidy et al. 2002). They investigated shifts in the frequency range of coherence when the patient was under different states of medication or in different behavioral states. Measurements without medication showed that in the akinetic "OFF" state, coherence between the basal ganglia and cortex is dominated by tremor frequencies and frequencies in the beta band. Interestingly, treatment with the dopamine precursor levo-dopa reduced low frequency activity and resulted in a new coherence peak at 70 Hz in the gamma band (Brown et al. 2001). Importantly, electrical stimulation at those sites where beta-band coherence was highest with the EEG and the contralateral EMG yielded the best amelioration of Parkinsonian symptoms (Marsden et al. 2001). In another study, the functional significance of high frequency activity was investigated by testing the modulation of coherence before and during voluntary movement. In the OFF state, beta activity was suppressed during movement preparation and execution, whereas in the ON state (i.e., after levo-dopa treatment), gamma coherence was enhanced in relation to the movement (Cassidy et al. 2002). These findings are compatible with a model in which interactions between the basal ganglia, thalamus, and cortex in different frequency bands modulate basal ganglia functions in a task- and state-dependent way (Brown 2003; Brown and Marsden 1998). Slow oscillations at tremor frequencies or in the beta band, resulting from dopamine depletion, seem to disrupt normal motor function. By contrast, gamma-band rhythms seem to be important for the organization of normal voluntary movement, as indicated by the emergence of these fast oscillations in the ON state, and by the prokinetic effects of deep brain stimulation at these frequencies or higher harmonics (Limousin et al. 1995).

Which Processes Modulate Dynamic Coordination at the Systems Level?

The importance of the emergence of dynamic coordination as a concept in cognitive neuroscience is that, among other things, it provides a balance to strictly localizationalist views of cognitive function. Based on data, computational models, the sociohistorical context of this intellectual development, and the position of this construct within the matrix of ideas in neuroscience, the focus of much research in this area has been on within-region coordination, with an emphasis on self-organization based on processing within regions. When discussed within the broader field of brain function, we can ask whether large-scale coordination operates according to similar mechanisms as more local or intra-regional coordination. Another important, but relatively unexplored issue, is to what extent the speed and strength of dynamic coordination is affected by more global modulatory influences in the brain. As noted above, data on development and learning indicate that there is plasticity in dynamic coordination. In addition, the well-known effects of psychopathology (e.g., schizophrenia; see Silverstein, this volume) and NMDA antagonists (e.g., ketamine, phencyclidine; Phillips and Silverstein 2003) on dynamic coordination indicate that these processes can operate within a wide range of efficiency.

However, relatively little attention has been paid to factors such as emotion, arousal, or fatigue.

There is both positive and negative evidence regarding the effects of emotion on coordination. For example, it has been demonstrated that the emotional content of pictures had little effect on early event-related potentials reflecting perceptual organization (Bradley et al. 2007). A more complex picture was revealed in a study by Colzato et al. (2007) in which the affective valence of pictures did not affect binding of visual features, but strongly affected binding of visual and response codes. Because it has been shown that sensory encoding in the visual cortex can be facilitated by affective cue-driven "natural selective attention" (Schupp et al. 2003), this suggests that affect modulates signals at a level beyond intra-sensory coordination, although it affects higher-level sensorimotor coordination. To date little is known about the effects of arousal or fatigue on dynamic coordination. However, the well-known effects of these factors on cognition, in general (e.g., Yerkes–Dodson law), suggest that it is worth exploring whether arousal effects occur at the level of coordinating interactions, and whether any such effects overlap completely with those of attention and occur at later stages.

We turn now to two closely related questions of particular interest: Are certain brain regions of particular importance for modulating dynamic coordination? To what extent is dynamic coordination constrained by top-down influences?

Prefrontal Cortex Modulates Dynamic Coordination

The ability to conceptualize and describe dynamic coordination in terms of formal models, in combination with the known similarity of local circuitry throughout the cortex (Phillips and Singer 1997), suggests that coordinating processes can occur within any brain region and, for coordination within a single sensory domain, no guidance from outside that region is necessary. The extent to which this is true needs further exploration. For example, while contour integration is typically seen as involving interactions only between neurons within the visual cortex, a recent study indicated that patients with traumatic brain injury to the frontal cortex were deficient in binding sparsely arranged, but still orientation-correlated, Gabor patches (Ciaramelli et al. 2007). This is consistent with evidence from a recent study of schizophrenia, in which reduced contour integration was associated with less frontal activity compared to healthy controls (Silverstein et al. 2009). In the latter case, while the largest and most consistent differences between groups were in visual cortex areas known to involve integration and to subserve perceptual grouping (e.g., V2, V3, V4), data on differences in frontal activation suggest that a larger network may be involved in normal dynamic coordination in vision than previously believed.

Given that the frontal lobe, and especially the prefrontal cortex (PFC), has been traditionally construed as a "central executive," it is important to clarify the extent to which activity in this region affects coordination in other areas, the conditions under which it occurs, and the specific types of coordination that it affects and provides. As we hope is evident in this chapter, dynamic coordination occurs throughout the brain at many levels of processing. Dynamic coordination is also required at a higher, meta level. Complex, goal-directed behavior would be impossible without brain systems and mechanisms that coordinate other systems to organize their processing and keep them on task and directed toward goals. Without it, thought and action would be determined solely by whatever sensory inputs happen to be most salient and the well-learned or reflexive responses associated with them. This ability is called cognitive control. It no doubt involves neural circuitry that extends over much of the brain, but it is commonly held that the PFC is particularly important.

The PFC occupies a far greater proportion of the human cerebral cortex than in other animals, suggesting that it might contribute to distinctively human cognitive capacities. Humans and monkeys with PFC damage seem to lack cognitive control: they become "stimulus-bound," their behavior dominated by reflexive reactions to the environment. Miller and Cohen (2001) proposed a guided activation model in which this cognitive control stems from the PFC's active maintenance of patterns of activity that represent goals and the means to achieve them (i.e., rules). This is thought to provide bias signals to other brain structures whose net effect is to guide the flow of activity along neural pathways that establish the proper mappings between inputs, internal states, and outputs needed to perform a given task, dynamically coordinating cortical processing to meet the demands of the task at hand.

Much of the evidence for this is, at the moment, indirect. One line holds that the PFC has complex multimodal properties that encode the type of information needed for goal direction: after training, many of its neurons reflect task rules (Miller 2000; Miller and Cohen 2001). Another line maintains that the PFC is anatomically well suited to coordinate cortical processing. It is interconnected with virtually the entire cerebral cortex, with the exception of primary sensory and motor areas. It is also interconnected with other major brain systems, including the basal ganglia, hippocampus, and anterior cingulate. Thus, it is well situated to integrate information about the external world and the animal's internal state and to send back signals that modulate processing in widespread brain areas.

More direct evidence for a role of the PFC in top-down modulation of posterior cortex has been mounting. Naya et al. (1996) showed that the PFC is needed for the recall of visual information in the inferior temporal cortex. By cutting the corpus callosum and the anterior commissure in two stages, they showed that activity in the inferior temporal cortex (reflecting the recall of an object from long-term memory) depended on top-down signals from the PFC. Moore and colleagues showed that microstimulation of the FEF causes attention-like modulations of neural responses in the visual cortex (Moore and Armstrong 2003). They electrically stimulated sites within the FEF and measured its effect on the activity of neurons in area V4. V4 neuron responses to a visual stimulus were enhanced after brief stimulation of retinotopically corresponding sites within the FEF below that needed to evoke saccades. Further, stimulation of noncorresponding FEF representations suppressed V4 responses. Buschman and Miller (2007) also found direct evidence that top-down attention signals arise from the frontal cortex and act on the posterior cortex. When monkeys shifted attention in a bottom-up fashion (to a salient, pop-out, stimulus), neurons in the parietal cortex reflected the shift of attention to the target before neurons in the PFC and FEF. By contrast, when attention was shifted in a top-down fashion (to a target that matched one held in short-term memory), the opposite was true: neurons in the frontal cortex showed a shorter latency to reflect the attention shift than those in the parietal cortex. This suggested that top-down and bottom-up attention signals arise from frontal and sensory cortex, respectively. Taken together, these considerations suggest that PFC is one of the dominant sources of modulatory signals that have an impact on dynamic coordination within and across other brain regions.

In addition to the frontal cortex, it has also been proposed that the cerebellum exerts a strong influence on cognitive coordination and that coordination impairments such as those occurring in schizophrenia can be attributed to cerebellar abnormalities (e.g., Andreasen and Pierson 2008). To date, however, data linking cerebellar function to a primary role in coordination of cognitive activity at the cortical level are lacking. However, it is still possible that the cerebellum contributes timing signals which serve a general coordinating function.

Top-down Processing and Neural Coherence

Most of the models considering the functional importance of top-down mechanisms make use of the *anatomical* notion of top-down processing: they assume that predictions or hypotheses about the features of environmental stimuli are expressed by signals travelling along feedback connections from "higher" to "lower" areas in a processing hierarchy. One of the earliest examples of such a model is the Adaptive Resonance Theory (Grossberg 1980). The theory assumes complementarity between ascending and descending pathways among sensory areas, the former allowing adaptive filtering of the input signals and the latter carrying predictive signals (templates of expected patterns that need to be matched by the current input). Related models that also postulate a key role of feedback influences in pattern recognition have been suggested by Mumford (1992) and Ullman (1995). These models also suggest that the comparison of sensory input with existing knowledge is essential for perception. Current top-down models of attentional selection and cognitive control (Frith and Dolan 1997; Fuster 1989; Miller 2000) assume that top-down influences originate in prefrontal and parietal cortical areas. As discussed above, a crucial idea is that assemblies of neurons that represent action goals in the PFC provide modulatory "bias signals" (Miller 2000) to sensorimotor circuits that have to carry out response selection. Thus, prefrontal signals are assumed to exert top-down control over the routing of information flow through specific sensorimotor loops.

A different idea of how top-down influences might be implemented neurally may lead to what could be called a dynamicist view (Engel et al. 2001). This view is motivated by the evidence that synchrony can be intrinsically generated (not imposed on the system by external stimuli) and modulated by intrinsic signals that reflect experience, contextual influences, and action goals (reviewed by Singer 1999; Engel et al. 2001). In this context, the search for the mechanisms of top-down control becomes equivalent to the investigation of the influence of ongoing patterns of activity on the processing of sensory signals and, in particular, on their selection and grouping through oscillatory patterning and synchronization. In contrast to the top-down models discussed above, the patterns relevant to the dynamic selection of input signals would be generated not only by assemblies in association cortices that carry more abstract, invariant representations, but as the result of continuous large-scale interactions between higher- and lower-order cortical areas. The patterns of coherent activity emanating from such large-scale interactions could bias the synchronization of input signals, leading to a selective enhancement of temporal correlations in subsets of the activated populations. This would result in enhanced saliency and a competitive advantage for the selected populations of neurons.

Evidence for top-down control by changes in the dynamics of intrinsically active networks has been obtained in a recent study of spatial attention shifts in humans (Siegel et al. 2008). In this study, MEG was combined in a spatially cued motion discrimination task with source-reconstruction techniques to characterize attentional effects on neuronal synchronization across key stages of the human dorsal visual pathway. The results demonstrate that visuospatial attention modulates oscillatory synchronization between visual, parietal cortex, and PFC in a spatially selective fashion. In particular, analysis of phase coherence in source space showed that during attentive processing of a visual stimulus, gamma-band coherence increases between regions corresponding to FEF, intraparietal sulcus, and middle temporal region. This suggests that attentional selection is mediated by frequency-specific synchronization between prefrontal, parietal, and visual cortex and that the intrinsic dynamics of frontoparietal networks is important for controlling attentional shifts (Engel et al. 2001; Corbetta and Shulman 2002). An interesting finding in this context is that this selection network did not involve prefrontal, but premotor regions, supporting what has been called the "premotor theory of attention" (Rizzolatti et al. 1987). Recently, similar observations have been made in studies on large-scale interactions accompanying shifts of attention in the monkey brain, showing that attention is associated with enhanced coherence between FEF and visual cortical areas (Gregoriou et al. 2009).

How Is Dynamic Coordination Related to Brain Disorders?

Disturbance of Dynamic Coordination in Schizophrenia

Research into psychiatric disorders may not only lead to insights into the mechanisms underlying abnormal mental functioning but may also be an ideal testing ground for examining the validity of current theories of healthy brain functioning. Schizophrenia is of particular relevance for testing the concept of dynamic coordination, and much of our discussion on the pathophysiological alteration of coordination mechanisms centered on this clinical condition.

From its earliest beginnings, the pathophysiology of schizophrenia has been described as a disorder involving a deficit in the integration and coordination of neural activity that leads to dysfunctions in cognition. Symptoms can involve false perceptual inference (e.g., hallucinations; failure to integrate subtle or distracting cues during perceptual grouping), false conceptual inference and contextual disambiguation (e.g., delusions and passivity phenomena), and a secondary failure of learning (psychomotor poverty). Bleuler (1911) chose the word "schizophrenia" to highlight the fragmentation of mental functions. Indeed, current theories of schizophrenia (Friston 1999; Phillips and Silverstein 2003) converge on the notion that core aspects of the pathophysiology are due to deficits in the coordination of distributed neural processes that involve multiple cortical areas. This perspective, which considers schizophrenia as a functional dysconnection syndrome, contrasts earlier views which emphasized a regionally specific pathophysiological process as the underlying cause for the signs and symptoms of schizophrenia. This view, which considers the symptoms experienced by the patient as a product of ensuing dysfunctional dynamic coordination, places abnormal synaptic function at the heart of the etiology. It emphasizes the primary role of synaptic plasticity and, in particular, its modulation by neuronal dynamics and neurotransmitter systems.

Support for the notion that dynamic coordination may be central to the pathophysiology of schizophrenia stems from the cognitive deficits central to the condition. These involve functions that are paradigmatic examples of dynamic coordination, such as working memory, attention, and perceptual organization (for a review, see Phillips and Silverstein 2003). In addition, if dynamic coordination is impaired in schizophrenia, one of the physiological manifestations should be impaired synchronous oscillatory activity. Evidence suggests that this is so. A substantial body of EEG/MEG studies support the hypothesis that cognitive deficits are related to impaired neural synchrony. Examination of auditory and visual steady-state responses to repetitive stimulation in patients with schizophrenia has revealed a specific reduction in the power of the stimulus-locked response in the beta and gamma frequency range, but not in the lower frequencies (Krishnan et al. 2005; Kwon et al. 1999). Reductions in evoked oscillatory activity have been reported for tasks involving visual binding (Spencer et al. 2003, 2004), for backward masking (Wynn et al. 2005), in auditory oddball paradigms (Gallinat et al. 2004), and during TMS-evoked activity over frontal regions (Ferrarelli et al. 2008). These results suggest selective deficiencies in the ability of cortical networks or cortico-thalamo-cortical loops to engage in precisely synchronized high frequency oscillations.

In addition to analyses of spectral power, several studies have also examined phase synchrony between distributed neuronal populations while patients performed cognitive tasks (Slewa-Younan et al. 2004; Spencer et al. 2003; Uhlhaas, Linden et al. 2006). Overall, these studies conclude that patients with schizophrenia are characterized by reduced phase locking of oscillations in the beta- and lower gamma-band range; this underscores that, in addition to abnormalities in local circuits, large-scale integration of neural activity is impaired. It is currently unclear, however, to what extent impairments in local circuits contribute to long-range synchronization impairments or whether these represent two independent phenomena.

Significant correlations, found in multiple studies across different laboratories, between reduced perceptual organization and reduced conceptual organization (i.e., the presence of formal thought disorder) (Uhlhaas and Silverstein 2005), and covariation of changes in both with treatment (Uhlhaas, Linden et al. 2006) provide evidence that different forms of dynamic coordination may be supported by a single mechanism. Correlations between cognitive dysfunctions and alterations in neural synchrony are furthermore suggested by relationships between the positive symptoms of schizophrenia and changes in the amplitude of beta- and gamma-band oscillations. Thus, patients with auditory hallucinations show an increase in oscillatory activity in temporal regions compared to patients without hallucinations (Lee et al. 2006; Spencer et al. 2008).

Further evidence for the role of neural synchrony in the pathophysiology of schizophrenia is the coincidence of symptom expression during the transition from adolescence to adulthood and developmental changes in power and synchronization of oscillations in the theta, beta, and gamma frequency range during normal brain development. Recently, Uhlhaas et al. (2009) showed that these parameters undergo profound changes during late adolescence, reflecting an increase in the temporal precision of cortical networks. This suggests the possibility that abnormal brain development in schizophrenia during the late adolescent period is unable to support precise temporal coding, which then leads to the decompensation of the network and the accompanying emergence of psychotic symptoms.

Impaired dynamic coordination in schizophrenia is consistent with deficits at the anatomical and physiological level in schizophrenia. One prominent candidate mechanism for the changes in neural synchrony is a dysfunction in GABAergic interneurons (for a review, see Lewis et al. 2005). For example, there is consistent evidence for reduced GABA synthesis in the parvalbumin-containing subpopulation of inhibitory GABA neurons in schizophrenia, which are critically involved in the generation of cortical and hippocampal oscillatory activity. Furthermore, impairments in long-range synchronization in schizophrenia can be related to changes in white matter volume and organization, as long distance synchronization of oscillatory responses is mediated by reciprocal corticocortical connections (Löwel and Singer 1992; König et al. 1993). This possibility is supported by *in vivo* anatomical examinations with diffusion tensor imaging that have revealed white matter anomalies throughout cortical and subcortical structures (for a review, see Kubicki et al. 2007).

These data suggest that dynamic coordination is a useful construct to understand the pathophysiology of schizophrenia. Yet, several questions remain open that are crucial for progress in this field of research. One intriguing phenomenon is the fact that cognitive and physiological dysfunctions are present throughout the cortex in schizophrenia. This raises the problem of which mechanisms can account for such a distributed impairment. One possible implication of this finding could be that core deficits in schizophrenia arise out of the altered global dynamics, which then lead to widespread impairments in local circuits. Accordingly, one strategy is to identify global coordination dynamics failures in schizophrenia. Furthermore, alterations in neural synchrony have been identified in several brain disorders and thus raise the question of diagnostic specificity. One possible assumption is that different syndromes are related to distinct but overlapping pathologies in the coordination of distributed neuronal activity patterns that are revealed by the systematic investigation of the temporal and spatial organization of neural synchrony across different frequency bands. This is undoubtedly a challenging task, but such a research program would ultimately result in better diagnostic tools for early diagnosis and intervention.

What Can Be Learned from This with Respect to Normal Brain Function?

Schizophrenia can be seen as a paradigmatic example of impaired dynamic coordination (Silverstein and Schenkel 1997; Phillips and Silverstein 2003) because there are several impairments which prima facie indicate reduced organization of elements into coherent wholes (e.g., in visual and auditory perception, working memory, selective attention, language, theory of mind, and binding of self-representation with action representation during ongoing behavior). Moreover, unlike much classic neuropsychological research, evidence of reduced dynamic coordination in schizophrenia is sometimes revealed by superior task performance (e.g., in terms of faster processing of single elements due to reduced contextual sensitivity), and these findings cannot be accounted for by medication effects.

To the extent that schizophrenia is seen as a model of relatively context-deficient cortical computation, the functions that are preserved in schizophrenia inform us about those functions that possibly do not require a high degree of dynamic coordination. Such functions include:

- understanding the meaning of individual words (as opposed to the reduced contextual constraint of words on later words in sentences),
- basic color perception (as opposed to color constancy or assimilation which rely on contextual cues),
- visual acuity,
- overlearned social behaviors (as opposed to being able to function in novel social contexts),
- basic motor functions,
- procedural memory, and
- understanding of basic cause-effect relationships (outside of interpretation of phenomena and events that affect self-esteem or sense of vulnerability).

Also, data on schizophrenia demonstrate the importance not only of NMDA receptor function for dynamic coordination, but also of GABA-mediated inhibitory interneuron function. However, schizophrenia is not the only disorder where impairments in dynamic coordination are evident. A comparison across disorders could assist our understanding of both the causes of dynamic coordination impairments and the bases of normal dynamic coordination (Silverstein, this volume).

Comparing schizophrenia with other developmental disorders, are there types of coordination impairments that generalize across disorders? For example, reductions in visual perceptual organization and theory of mind have been noted in developmental (autism spectrum) disorders (e.g., Silverstein and Palumbo 1995), and there is genetic and symptom overlap between schizophrenia and these disorders (Silverstein, this volume). Another interesting case is provided by Williams syndrome, a genetic condition in which a relatively consistent neurobehavioral phenotype is produced by a small deletion on chromosome 7 (e.g., Bellugi et al. 2000). One of the intriguing aspects of this syndrome is a characteristic abnormality of neural connectivity in a single cortical area, the primary visual cortex (Reiss et al. 2000; Galaburda and Bellugi 2000). Presumably due to these structural abnormalities in the primary visual cortex, patients with Williams syndrome present with a specific disruption of visual spatial integration (Kovács 2004). The example of Williams syndrome is particularly interesting because a well-defined structural abnormality is behind a perceptual deficit that is usually attributed to dynamic coordination. Strabismic amblyopia represents another interesting developmental abnormality, where a clear relation between structural changes in the cortical network, impairments of dynamic coordination, and disturbed perceptual and behavioral function has been established. As shown in studies in cat visual cortex, the impaired feature integration observed in amblyopic animals can be related to diminished intra- and interareal synchrony (König et al. 1993; Roelfsema et al. 1994; Schmidt et al. 2004).

These studies provide evidence for a common mechanism underlying contextual modulation in multiple functions, and evidence that neurodevelopmental changes can produce specific and profound changes in more than one of these domains. Identifiable neurobiological factors, such as the extent of white matter reduction, the number of cortical regions affected by suppression of sensory input during development, and the developmental onset of coordination failures, can account for variance in the cognitive and clinical manifestations in different disease states (Silverstein, this volume). The links established by those studies between behavioral evidence for impaired dynamic coordination (e.g., visual binding) and abnormal oscillatory activity (e.g., in gamma and beta bands) strongly support the notion that there is functional significance of coordinated neural activity, that multiple forms of cognitive functioning rely on dynamic coordination, and that coherence of neural oscillations in multiple frequency bands may indeed constitute one of the key mechanisms underlying dynamic coordination in the nervous system.

Acknowledgments

We thank Maurizio Corbetta, Shimon Edelman, Pascal Fries, Wolf Singer, and Matthew Wilson for contributing to our group discussions.

Bibliography

Note: Numbers in square brackets denote the chapter in which an entry is cited.

Abeles, M. 1982. Local Cortical Circuits. An Electrophysiological Study. New York: Springer-Verlag. [12]

———. 1991. Corticotronics: Neural Circuits of the Cerebral Cortex. Cambridge: Cambridge Univ. Press. [6, 13]

Adjamian, P., A. Hadjipapas, G. R. Barnes, A. Hillebrand, and I. E. Holliday. 2008. Induced gamma activity in primary visual cortex is related to luminance and not color contrast: An MEG study. *J. Vision* **8(7)**:41–47. [16]

Aertsen, A. M., G. L. Gerstein, M. K. Habib, and G. Palm. 1989. Dynamics of neuronal firing correlation: Modulation of "effective connectivity." *J. Neurophysiol.* **61**:900–917. [12, 13]

Albright, T. D., and G. R. Stoner. 2002. Contextual influences on visual processing. *Ann. Rev. Neurosci.* **25**:339–379. [1]

Albus, K. 1975. A quantitative study of the projection area of the central and the paracentral visual field in area 17 of the cat. I. The precision of the topography. *Exp. Brain Res.* **24**:159–179. [12]

Aldridge, J. W., and K. C. Berridge. 1998. Coding of serial order by neostriatal neurons: A "natural action" approach to movement sequence. *J. Neurosci.* **18**:2777–2787. [5]

Alkire, M. T., A. G. Hudetz, and G. Tononi. 2008. Consciousness and anesthesia. *Science* **322(5903)**:876–880. [16]

Allman, J., A. Hakeem, and K. Watson. 2002. Two phylogenetic specializations in the human brain. *Neuroscientist* **8**:335–346. [2]

Amit, D. 1989. Modeling Brain Function: The World of Attractor Neural Networks. New York: Cambridge Univ. Press. [6]

Amit, D., H. Gutfreund, and H. Sompolinsky. 1987. Information storage in neural networks with low levels of activity. *Phys. Rev. A* **35**:2293–2303. [6]

Amzica, F., and M. Steriade. 1997. The K-complex: Its slow (<1-Hz) rhythmicity and relation to delta waves. *Neurology* **49**:952–959. [7]

———. 1998. Electrophysiological correlates of sleep delta waves. *Electroenceph. Clin. Neurophysiol.* **107**:69–83. [7]

Andreasen, N. C., and R. Pierson. 2008. The role of the cerebellum in schizophrenia. *Biol. Psychiatry* **64**:81–88. [17, 18]

Angelucci, A., J. B. Levitt, E. J. S. Walton, et al. 2002. Circuits for local and global signal integration in primary visual cortex. *J. Neurosci.* **22**:8633–8646. [14]

Anstis, S. M., F. A. J. Verstraten, and G. Mather. 1998. The motion after-effect: A review. *Trends Cogn. Sci.* **2**:111–117. [12]

Arcelli, P., C. Frassoni, M. C. Regondi, S. de Biasi, and R. Spreafico. 1997. GABAergic neurons in mammalian thalamus: A marker of thalamic complexity? *Brain Res. Bull.* **42**:27–37. [2]

Ardid, S., X.-J. Wang, and A. Compte. 2007. An integrated microcircuit model of attentional processing in the neocortex. *J. Neurosci.* **27**:8486–8495. [6]

Ariav, G., A. Polsky, and J. Schiller. 2003. Submillisecond precision of the input-output transformation function mediated by fast sodium dendritic spikes in basal dendrites of CA1 pyramidal neurons. *J. Neurosci.* **23**:7750–7758. [11]

Ariëns-Kappers, C. E., G. C. Huber, and E. C. Crosby. 1936. The Comparative Anatomy of the Nervous System of Vertebrates, Including Man. New York: Hafner. [5]

Arrondo, G., M. Alegre, J. Sepulcre, et al. 2009. Abnormalities in brain synchronization are correlated with cognitive impairment in multiple sclerosis. *Mult. Sclerosis* **15**:509–516. [17]

Ashtari, M., K. L. Cervellione, K. M. Hasan, et al. 2007. White matter development during late adolescence in healthy males: A cross-sectional diffusion tensor imaging study. *NeuroImage* **35**:501–510. [18]

Atkins, J. E., J. Fiser, and R. A. Jacobs. 2001. Experience-dependent visual cue integration based on consistencies between visual and haptic percepts. *Vision Res.* **41(4)**:449–461. [15]

Azouz, R., and C. M. Gray. 2003. Adaptive coincidence detection and dynamic gain control in visual cortical neurons *in vivo*. *Neuron* **37**:513–523. [11]

Baars, B. J. 1997. In the theatre of consciousness: Global workspace theory, a rigorous scientific theory of consciousness. *J. Conscious. Stud.* **4(4)**:292–309. [16]

Baker, S. N. 2007. Oscillatory interactions between sensorimotor cortex and periphery. *Curr. Opin. Neurobiol.* **17**:649–655. [8]

Balaban, E. 1997. Changes in multiple brain regions underlie species differences in a complex, congenital behavior. *PNAS* **94**:2001–2006. [5]

Balaban, E., M.-A. Teillet, and N. Le Douarin. 1988. Application of the quail-chick chimeric system to the study of brain development and behavior. *Science* **241**:1339–1342. [5]

Baldwin, D. A., J. A. Baird, M. M. Saylor, and M. A. Clark. 2001. Infants parse dynamic action. *Child Devel.* **72**:708–717. [17]

Balsam, P. D., and C. R. Gallistel. 2009. Temporal maps and informativeness in associative learning. *Trends Neurosci.* **32(2)**:73–78. [16]

Banks, M. I., J. A. White, and R. A. Pearce. 2000. Interactions between distinct GABA(A) circuits in hippocampus. *Neuron* **25**:449–457. [8]

Barbalat, G., V. Chambon, N. Franck, E. Koechlin, and C. Farrer. 2009. Organization of cognitive control within the lateral prefrontal cortex in schizophrenia. *Arch. Gen. Psychiatry* **66(4)**:377–386. [17]

Bargmann, C. I. 1998. Neurobiology of the *Caenorhabditis elegans* genome. *Science* **282**:2028–2033. [5]

Basar, E., C. Basar-Eroglu, S. Karakas, and M. Schürmann. 2000. Brain oscillations in perception and memory. *Int. J. Psychophysiol.* **35**:95–124. [18]

Bastiaansen, M. C., M. v. d. Linden, M. T. Keurs, T. Dijkstra, and P. Hagoort. 2005. Theta responses are involved in lexical-semantic retrieval during language processing. *J. Cogn. Neurosci.* **17**:530–541. [18]

Battaglia, F. P., G. R. Sutherland, and B. L. McNaughton. 2004. Hippocampal sharp wave bursts coincide with neocortical "up-state" transitions. *Learn. Mem.* **11**:697–704. [7]

Bauer, M., R. Oostenveld, M. Peeters, and P. Fries. 2006. Tactile spatial attention enhances gamma-band activity in somatosensory cortex and reduces low-frequency activity in parieto-occipital areas. *J. Neurosci.* **26**:490–501. [18]

Beck, P. D., M. W. Pospichal, and J. H. Kaas. 1996. Topography, architecture, and connections of somatosensory cortex in opossums: Evidence for five somatosensory areas. *J. Comp. Neurol.* **366**:109–133. [2]

Becker, R., P. Ritter, and A. Villringer. 2008. Influence of ongoing alpha rhythm on the visual evoked potential. *NeuroImage* **39(2)**:707–716. [16]

Becker, S., and G. E. Hinton. 1992. A self-organizing neural network that discovers surfaces in random-dot stereograms. *Nature* **335**:161–163. [1]

Bedard, C., and A. Destexhe. 2008. A modified cable formalism for modeling neuronal membranes at high frequencies. *Biophys. J.* **94**:1133–1143. [12]

Beggs, J. M., and D. Plenz. 2003. Neuronal avalanches in neocortical circuits. *J. Neurosci.* **23**:11167–11177. [12]

Bell, C., V. Han, Y. Sugawara, and K. Grant. 1997. Synaptic plasticity in a cerebellum-like structure depends on temporal order. *Nat. Neurosci.* **387**:278–281. [3]

Bellman, R. E. 1961. Adaptive Control Processes. Princeton, NJ: Princeton Univ. Press. [1]

Bellugi, U., L. Lichtenberger, W. Jones, Z. Lai, and M. St George. 2000. I. The neurocognitive profile of Williams syndrome: A complex pattern of strengths and weaknesses. *J. Cogn. Neurosci.* **12(1)**:7–29. [18]

Benton, R. 2006. On the origin of smell: Odorant receptors in insects. *Cell. Mol. Life Sci.* **63**:1579–1585. [5]

Benucci, A., R. A. Frazor, and M. Carandini. 2007. Standing waves and traveling waves distinguish two circuits in visual cortex. *Neuron* **55**:103–117. [12]

Berry, M. J., D. K. Warland, and M. Meister. 1997. The structure and precision of retinal spike trains. *PNAS* **94**:5411–5416. [3]

Bi, G. Q., and M. Poo. 1998. Synaptic modifications in cultured hippocampal neurons: Dependence on spike timing, synaptic strength, and postsynaptic cell type. *J. Neurosci.* **18**:10464–10472. [3]

Bi, G. Q., and J. Rubin. 2005. Timing in synaptic plasticity: From detection to integration. *Trends Neurosci.* **28(5)**:222–228. [16]

Bialek, W., and F. Rieke. 1992. Reliability and information transmission in spiking neurons. *Trends Neurosci.* **15(11)**:428–433. [3]

Biederlack, J., M. Castelo-Branco, S. Neuenschwander, et al. 2006. Brightness induction: Rate enhancement and neuronal synchronization as complementary codes. *Neuron* **52**:1073–1083. [1, 11]

Bienenstock, E. 1996. Composition. In: Brain Theory: Biological Basis and Computational Theory of Vision, ed. A. Aertsen and V. Braitenberg, pp. 269–300. Amsterdam: Elsevier. [12]

Binzegger, T., R. J. Douglas, and K. A. C. Martin. 2007. Stereotypical bouton clustering of individual neurons in cat primary visual cortex. *J. Neurosci.* **27**:12242–12254. [6]

Bird, C. D., and N. J. Emery. 2009. Insightful problem solving and creative tool modification by captive nontool-using rooks. *PNAS* **106(25)**:10370–10375. [4]

Bishop, C. M. 2006. Pattern Recognition and Machine Learning. Heidelberg: Springer. [10]

Bjursten, L. M., K. Norrsell, and U. Norrsell. 1976. Behavioural repertory of cats without cerebral cortex from infancy. *Exp. Brain Res.* **25**:115–130. [5]

Blakemore, S.-J., and J. Decety. 2001. From the perception of action to the understanding of intention. *Nat. Rev. Neurosci.* **2**:561–567. [17]

Bleuler, E. 1911. Dementia Praecox oder Gruppe der Schizophrenien. Leipzig: F. Deuticke. [18]

Bliss, T. V. P., and G. L. Collingridge. 1993. A synaptic model of memory: Long-term potentiation in the hippocampus. *Nature* **361**:31–39. [3]

Blum, H. J. 1967. A new model of global brain function. *Persp. Biol. Med.* **10**:381–407. [14]

Bod, R., J. Hay, and S. Jannedy. 2003. Probabilistic Linguistics. Cambridge, MA: MIT Press. [5]

Börgers, C., S. Epstein, and N. J. Kopell. 2008. Gamma oscillations mediate stimulus competition and attentional selection in a cortical network model. *PNAS* **105**:18023–18028. [13]

Börgers, C., and N. J. Kopell. 2008. Gamma oscillations and stimulus coupling. *Neural Comput.* **20**:383–414. [11, 13]

Bowers, J. S. 2009. On the biological plausibility of grandmother cells: Implications for neural network theories in psychology and neuroscience. *Psychol. Rev.* **116**:220–251. [6]

Bradley, M. M., S. Hamby, A. Löw, and P. J. Lang. 2007. Brain potentials in perception: Picture complexity and emotional arousal. *Psychophysiol.* **44**:364–373. [18]

Bragin, A., G. Jando, Z. Nadasdy, et al. 1995. Gamma (40–100 Hz) oscillation in the hippocampus of the behaving rat. *J. Neurosci.* **15**:47–60. [7]

Braun, M. 1995. Picturing Time: The Work of Etienne-Jules Marey (1830–1904). Chicago: Univ. of Chicago Press. [14]

Breakspear, M. J., A. Daffertshofer, and P. Ritter. 2009. BrainModes: A principled approach to modeling and measuring large-scale neuronal activity. *J. Neurosci. Meth.* **183(1)**:1–4. [13]

Brecht, M., W. Singer, and A. K. Engel. 1998. Correlation analysis of corticotectal interactions in the cat visual system. *J. Neurophysiol.* **79**:2394–2407. [13]

Bressler, S. L., and J. A. Kelso. 2001. Cortical coordination dynamics and cognition. *Trends Cogn. Sci.* **5**:26–36. [18]

Bressler, S. L., W. Tang, C. M. Sylvester, G. L. Shulman, and M. Corbetta. 2008. Top-down control of human visual cortex by frontal and parietal cortex in anticipatory visual spatial attention. *J. Neurosci.* **28(40)**:10056–10061. [16]

Breuer, J., and S. Freud. 1895. Studien über Hysterie Vienna: Deuticke. [17]

Briggman, K. L., and W. B. Kristan. 2008. Multifunctional pattern-generating circuits. *Ann. Rev. Neurosci.* **31**:271–294. [18]

Bringuier, V., F. Chavane, L. Glaeser, and Y. Frégnac. 1999. Horizontal propagation of visual activity in the synaptic integration field of area 17 neurons. *Science* **283**:695–699. [12]

Bringuier, V., Y. Frégnac, A. Baranyi, D. Debanne, and D. E. Shulz. 1997. Synaptic origin and stimulus dependency of neuronal oscillatory activity in the primary visual cortex of the cat. *J. Physiol.* **500(3)**:751–774. [12]

Brooks, R. A. 1991. Intelligence without representation. *Art. Intell.* **47**:139–160. [5]

Brosch, M., E. Budinger, and H. Scheich. 2002. Stimulus-related gamma oscillations in primate auditory cortex. *J. Neurophysiol.* **87**:2715–2725. [18]

Brovelli, A., M. Ding, A. Ledberg, et al. 2004. Beta oscillations in a large-scale sensorimotor cortical network: Directional influences revealed by Granger causality. *PNAS* **101**:9849–9854. [18]

Brown, N. R., and D. Schopflocher. 1998. Event clusters: An organization of personal events in autobiographical memory. *Psychol. Sci.* **6**:470–475. [17]

Brown, P. 2003. Oscillatory nature of human basal ganglia activity: Relationship to the pathophysiology of Parkinson's disease. *Mov. Disord.* **8**:357–363. [18]

Brown, P., and C. D. Marsden. 1998. What do the basal ganglia do? *Lancet* **351**:1801–1804. [18]

Brown, P., A. Oliviero, P. Mazzone, et al. 2001. Dopamine dependence of oscillations between subthalamic nucleus and pallidum in Parkinson's disease. *J. Neurosci.* **21**:1033–1038. [18]

Brunel, N. 2000. Dynamics of sparsely connected networks of excitatory and inhibitory spiking neurons. *J. Comput. Neurosci.* **8**:183–208. [3]

Bibliography

Brunel, N., and X. Wang. 2003. What determines the frequency of fast network oscillations with irregular discharges? I. Synaptic dynamics and excitation-inhibition balance. *J. Neurophysiol.* **90**:415–430. [6]

Bruns, A., and R. Eckhorn. 2004. Task-related coupling from high- to low-frequency signals among visual cortical areas in human subdural recordings. *Intl. J. Psychophysiol.* **51(2)**:97–116. [16]

Brzustowicz, L. M. 2008. NOS1AP in schizophrenia. *Curr. Psychiatry Rep.* **10**:158–163. [17]

Bugnyar, T., and B. Heinrich. 2005. Ravens, *Corvus corax*, differentiate between knowledgeable and ignorant competitors. *Proc. R. Soc. B.* **272**:1641–1646. [5]

Buhl, E., and W. Singer. 1989. The callosal projection in cat visual cortex as revealed by a combination of retrograde tracing and intracellular injection. *Exp. Brain Res.* **75**:470–476. [13]

Bullier, J. 2001. Feedback connections and conscious vision. *Trends Cogn. Sci.* **5(9)**:369–370. [16]

Bullock, T. H., J. Z. Achimowicz, R. B. Duckrow, S. S. Spencer, and V. J. Iragui-Madoz. 1997. Bicoherence of intracranial EEG in sleep, wakefulness and seizures. *Electroenceph. Clin. Neurophysiol.* **103(6)**:661–678. [16]

Busch, N. A., J. Dubois, and R. VanRullen. 2009. The phase of ongoing EEG oscillations predicts visual perception. *J. Neurosci.* **29(24)**:7869–7876. [13]

Buschman, T. J., and E. K. Miller. 2007. Top-down versus bottom-up control of attention in the prefrontal and posterior parietal cortices. *Science* **315(5820)**:1860–1862. [13, 16, 18]

———. 2009. Serial, covert, shifts of attention during visual search are reflected by the frontal eye fields and correlated with population oscillations. *Neuron* **63**:386–396. [13, 18]

Butko, N., and J. Triesch. 2007. Learning sensory representations with intrinsic plasticity. *Neurocomputing* **70**:1130–1138. [10]

Buzsáki, G. 1989. Two-stage model of memory trace formation: A role for "noisy" brain states. *Neuroscience* **31**:551–570. [7]

———. 2002. Theta oscillations in the hippocampus. *Neuron* **33**:325–340. [7]

———. 2005. Theta rhythm of navigation: Link between path integration and landmark navigation, episodic and semantic memory. *Hippocampus* **15**:827–840. [7]

———. 2006. Rhythms of the Brain. Oxford: Oxford Univ. Press. [1, 18]

Buzsáki, G., and A. Draguhn. 2004. Neuronal oscillations in cortical networks. *Science* **304(5679)**:1926–1929. [8, 16, 18]

Buzsáki, G., Z. Horvath, R. Urioste, J. Hetke, and K. Wise. 1992. High frequency network oscillation in the hippocampus. *Science* **256**:1025–1027. [7, 8]

Buzsáki, G., L. W. Leung, and C. H. Vanderwolf. 1983. Cellular bases of hippocampal EEG in the behaving rat. *Brain Res.* **287**:139–171. [7]

Camperi, M., and X.-J. Wang. 1998. A model of visuospatial working memory in prefrontal cortex: Recurrent network and cellular bistability. *J. Comp. Neurosci.* **5**:383–405. [6]

Canolty, R. T., E. Edwards, S. S. Dalal, et al. 2006. High gamma power is phase-locked to theta oscillations in human neocortex. *Science* **313(5793)**:1626–1628. [16]

Capaday, C., C. Ethier, L. Brizzi, et al. 2009. On the nature of the intrinsic connectivity of the cat motor cortex: Evidence for a recurrent neural network topology. *J. Neurophysiol.* **102**:2131–2141. [2]

Capotosto, P., C. Babiloni, G. L. Romani, and M. Corbetta. 2009. Frontoparietal cortex controls spatial attention through modulation of anticipatory alpha rhythms. *J. Neurosci.* **29**:5863–5872. [13]

Cardin, J. A., M. Carlén, K. Meletis, et al. 2009. Driving fast-spiking cells induces gamma rhythm and controls sensory responses. *Nature* **459**:663–667. [11]

Carr, C. E. 1993. Processing of temporal information in the brain. *Ann. Rev. Neurosci.* **16**:223–243. [3]

Carr, V., and J. Wale. 1986. Schizophrenia: An information processing model. *Aust. NZ J. Psychiatry* **20**:136–155. [17]

Cassenaer, S., and G. Laurent. 2007. Hebbian STDP in mushroom bodies facilitates the synchronous flow of olfactory information in locusts. *Nature* **448**:709–713. [3, 5]

Cassidy, M., P. Mazzone, A. Oliviero, et al. 2002. Movement-related changes in synchronization in the human basal ganglia. *Brain* **125**:1235–1246. [18]

Castelo-Branco, M., R. Goebel, S. Neuenschwander, and W. Singer. 2000. Neural synchrony correlates with surface segregation rules. *Nature* **405(6787)**:685–689. [11, 18]

Castelo-Branco, M., S. Neuenschwander, and W. Singer. 1998. Synchronization of visual responses between the cortex, lateral geniculate nucleus, and retina in the anesthetized cat. *J. Neurosci.* **18(16)**:6395–6410. [11]

Catmur, C., V. Walsh, and C. Heyes. 2009. Associative sequence learning: The role of experience in the development of imitation and the mirror system. *Phil. Trans. R. Soc. Lond. B* **364**:2369–2380. [5]

Cenier, T., F. David, P. Litaudon, et al. 2009. Respiration-gated formation of gamma and beta neural assemblies in the mammalian olfactory bulb. *Eur. J. Neurosci.* **29(5)**:921–930. [16]

Chandrasekaran, C., and A. A. Ghazanfar. 2009. Different neural frequency bands integrate faces and voices differently in the superior temporal sulcus. *J. Neurophysiol.* **101(2)**:773–788. [16]

Chaumon, M., V. Drouet, and C. Tallon-Baudry. 2008. Unconscious associative memory affects visual processing before 100 ms. *J. Vision* **8(3)**:1–10. [16, 18]

Chaumon, M., D. Hasboun, M. Baulac, C. Adam, and C. Tallon-Baudry. 2009. Unconscious contextual memory affects early responses in the anterior temporal lobe. *Brain Res.* **1285**:77–87. [18]

Chaumon, M., D. Schwartz, and C. Tallon-Baudry. 2009. Unconscious learning versus visual perception: Dissociable roles for gamma oscillations revealed in MEG. *J. Cogn. Neurosci.* **21**:2287–2299. [18]

Chavane, F., C. Monier, V. Bringuier, et al. 2000. The visual cortical association field: A Gestalt concept or a psychophysiological entity? *J. Physiol.* **94(5–6)**:333–342. [12]

Chawla, D., K. J. Friston, and E. D. Lumer. 2001. Zero-lag synchronous dynamics in triplets of interconnected cortical areas. *Neural Netw.* **14**:727–735. [13]

Chen, Y., M. Ding, and J. A. S. Kelso. 2003. Task-related power and coherence changes in neuromagnetic activity during visuomotor coordination. *Exp. Brain Res.* **148**:105–116. [18]

Chikkerur, S., C. Tan, T. Serre, and T. Poggio. 2009. An integrated model of visual attention using shape-based features. Massachusetts Institute of Technology. CBCL paper #278/CSAIL Technical Report #2009-029. Cambridge, MA. [13]

Chrobak, J. J., and G. Buzsáki. 1994. Selective activation of deep layer (V–VI) retrohippocampal cortical neurons during hippocampal sharp waves in the behaving rat. *J. Neurosci.* **14**:6160–6170. [7]

———. 1998a. Gamma oscillations in the entorhinal cortex of the freely behaving rat. *J. Neurosci.* **18(1)**:388–398. [16]

———. 1998b. Operational dynamics in the hippocampal-entorhinal axis. *Neurosci. Biobehav. Rev.* **22**:303–310. [7]

Ciaramelli, E., F. Leo, M. M. Del Viva, D. C. Burr, and E. Ladavas. 2007. The contribution of prefrontal cortex to global perception. *Exp. Brain Res.* **181**:427–434. [17, 18]

Clayton, N. S., and A. Dickinson. 1998. Episodic-like memory during cache recovery by scrub jays. *Nature* **395**:272–274. [4]

———. 1999. Memory for the content of caches by scrub jays (*Aphelocoma coerulescens*). *J. Exp. Psychol. Anim. Behav. Proc.* **25**:82–91. [4]

Clayton, N. S., K. S. Yu, and A. Dickinson. 2001. Scrub jays (*Aphelocoma coerulescens*) form integrated memories of the multiple features of caching episodes. *J. Exp. Psychol. Anim. Behav. Proc.* **27**:17–29. [4]

———. 2003. Interacting cache memories: Evidence for flexible memory use by Western scrub-jays (*Aphelocoma californica*). *J. Exp. Psychol. Anim. Behav. Proc.* **29**:14–22. [4]

Clowes, M. B. 1971. On seeing things. *Art. Intell.* **2**:79–116. [1]

Cobb, S. R., E. H. Buhl, K. Halasy, O. Paulsen, and P. Somogyi. 1995. Synchronization of neuronal activity in hippocampus by individual GABAergic interneurons. *Nature* **378**:75–78. [18]

Collins, C. E., A. Hendrickson, and J. H. Kaas. 2005. Overview of the visual system of tarsius. *Anat. Rec.* **287A**:1013–1025. [2]

Colzato, L. S., N. C. v. Wouwe, and B. Hommel. 2007. Feature binding and affect: Emotional modulation of visuo-motor integration. *Neuropsychologia* **45**:440–446. [18]

Connors, B. W., and M. J. Gutnick. 1990. Intrinsic firing patterns of diverse cortical neurons. *Trends Neurosci.* **13**:99–104. [3]

Cook, E. P., and J. H. R. Maunsell. 2002. Attentional modulation of behavioral performance and neuronal responses in middle temporal and ventral intraparietal areas of macaque monkey. *J. Neurosci.* **22**:1994–2004. [11]

Corbetta, M., and G. L. Shulman. 2002. Control of goal-directed and stimulus-driven attention in the brain. *Nat. Rev. Neurosci.* **3**:201–215. [18]

Crick, F. 1994. The Astonishing Hypothesis. London: Simon and Schuster. [10]

Crick, F., and C. Koch. 1990. Towards a neurobiological theory of consciousness. *Semin. Neurosci.* **2**:263–275. [16]

———. 1995. Cortical areas in visual awareness. *Nature* **377**:294–295. [14]

Croft, R. J., J. D. Williams, C. Haenschel, and J. H. Gruzelier. 2002. Pain perception, hypnosis and 40 Hz oscillations. *Intl. J. Psychophysiol.* **46**:101–108. [17]

Crone, N. E., D. L. Miglioretti, B. Gordon, and R. P. Lesser. 1998. Functional mapping of human sensorimotor cortex with electrocorticographic spectral analysis. II. Event-related synchronization in the gamma band. *Brain* **121**:2301–2315. [13]

Croner, L. J., K. Purpura, and E. Kaplan. 1993. Response variability in retinal ganglion cells of primates. *PNAS* **90**:8128–8130. [12]

Crystal, J. D. 2009. Elements of episodic-like memory in animal models. *J. Exp. Psychol. Anim. Behav. Proc.* **80**:269–277. [4]

Csibra, G., G. Davis, M. W. Spratling, and M. H. Johnson. 2000. Gamma oscillations and object processing in the infant brain. *Science* **290**:1582–1585. [18]

Csicsvari, J., H. Hirase, A. Czurko, A. Mamiya, and G. Buzsáki. 1999a. Fast network oscillations in the hippocampal CA1 region of the behaving rat. *J. Neurosci.* **19**:RC20. [7]

———. 1999b. Oscillatory coupling of hippocampal pyramidal cells and interneurons in the behaving rat. *J. Neurosci.* **19(274–287)**. [7]

Csicsvari, J., B. Jamieson, K. D. Wise, and G. Buzsáki. 2003. Mechanisms of gamma oscillations in the hippocampus of the behaving rat. *Neuron* **37**:311–322. [18]

Csicsvari, J., J. O'Neill, K. Allen, and T. Senior. 2007. Place-selective firing contributes to the reverse-order reactivation of CA1 pyramidal cells during sharp waves in open-field exploration. *Eur. J. Neurosci.* **26(704–716)**. [7]

Cunningham, M. O., D. D. Pervouchine, C. Racca, et al. 2006. Neuronal metabolism governs cortical network response state. *PNAS* **103**:5597–5601. [8]

Cunningham, M. O., M. A. Whittington, A. Bibbig, et al. 2004. A role for fast rhythmic bursting neurons in cortical gamma oscillations *in vitro*. *PNAS* **101(18)**:7152–7157. [16]

Curry, R. L., A. Towsend Peterson, and T. A. Langen. 2002. Western scrub-jay. In: The Birds of North America, ed. A. Poole and F. Gill, pp. 1–36. Philadelphia: Acad. Nat. Sci., Am. Ornithologists Union. [4]

Cusick, C., H. J. Gould, and J. H. Kaas. 1984. Interhemispheric connections of visual cortex of owl monkeys (*Aotus trivirgatus*), marmosets (*Calithrix jacchus*), and galagos (*Galago crassicaudatus*). *J. Comp. Neurol.* **230**:311–336. [2]

Cusick, C., and J. H. Kaas. 1988. Surface view patterns of intrinsic and extrinsic cortical connections of area 17 in a prosimian primate. *Brain Res.* **458**:383–388. [2]

Dale, R. A., and M. J. Spivey. 2005. From apples and oranges to symbolic dynamics: A framework for conciliating notions of cognitive representation. *J. Exp. Theor. Art. Intell.* **17**:317–342. [5]

Damasio, A. R. 1989. The brain binds entities and events by multiregional activation from convergence zones. *Neural Comput.* **1**:123–132. [16]

Dambacher, M., M. Rolfs, K. Gollner, R. Kliegl, and A. M. Jacobs. 2009. Event-related potentials reveal rapid verification of predicted visual input. *PLoS ONE* **4(3)**:e5047. [16]

Danion, J. M., L. Rizzo, and A. Bruant. 1999. Functional mechanisms underlying impaired recognition memory and conscious awareness in patients with schizophrenia. *Arch. Gen. Psychiatry* **56**:639–644. [17]

Debener, S., C. S. Herrmann, C. Kranczioch, D. Gembris, and A. K. Engel. 2003. Top-down attentional processing enhances auditory evoked gamma band activity. *NeuroReport* **14**:683–686. [18]

deCharms, R. C., and M. M. Merzenich. 1996. Primary cortical representation of sounds by the coordination of action-potential timing. *Nature* **381**:610–613. [12, 13]

Dechter, R. 1992. From local to global consistency. *Art. Intell.* **55**:87–107. [1]

Deco, G., and M. Corbetta. 2010. The dynamical balance of the brain at rest. *Neuroscientist*, in press. [13]

Deco, G., and A. Thiele. 2009. Attention: Oscillations and neuropharmacology. *Eur. J. Neurosci.* **30**:347–354. [1]

DeCoteau, W. E., C. Thorn, D. J. Gibson, et al. 2007. Learning-related coordination of striatal and hippocampal theta rhythms during acquisition of a procedural maze task. *PNAS* **104**:5644–5649. [13]

DeFelipe, J. 2005. Reflections on the structure of the cortical minicolumn. In: Neocortical Modularity and the Cell Minicolumn, ed. M. F. Casanova, pp. 57–91. Hauppauge, NY: Nova Science Publishers. [2]

———. 2006. Double-bouquet cells in the monkey and human cerebral cortex with special reference to areas 17 and 18. *Prog. Brain Res.* **154**:15–31. [6]

DeFelipe, J., M. del Rio, M. Gonzilez-Albo, and G. M. Elston. 1999. Distribution and patterns of connectivity of interneurons containing calbindin, calretinin and parvalbumin in visual areas of the occipital and temporal lobes of the macaque monkey. *J. Comp. Neurol.* **412**:515–526. [2]

Dehaene, S., M. Kerszberg, and J. P. Changeux. 1998. A neuronal model of a global workspace in effortful cognitive tasks. *PNAS* **95(24)**:14529–14534. [16]

Delage, Y. 1919. Le Rêve: Etude Psychologique, Philosophique et Litteraire. Paris: Presses Universitaires de France. [12]

Del Viva, M. M., R. Igliozzi, R. Tancredi, and D. Brizzolara. 2006. Spatial and motion integration in children with autism. *Vision Res.* **46**:1242–1252. [17]

Deneve, S. 2005. Bayesian inferences in spiking neurons. *Adv. Neural Info. Process. Sys.* **17**:353–360. [15]

Dennett, D. C. 1991. Consciousness Explained. Boston: Little, Brown & Co. [16]

———. 1995. Darwin's Dangerous Idea: Evolution and the Meanings of Life. New York: Simon & Schuster. [5]

Dere, E., E. Kart-Teke, J. P. Huston, and M. A. D. Silva. 2006. The case for episodic memory in animals. *Neurosci. Biobehav. Rev.* **30**:1206–1224. [4]

Desbordes, G., J. Jin, C. Weng, et al. 2008. Timing precision in population coding of natural scenes in the early visual system. *PLoS Biol.* **6(12)**:e324. [16]

Desimone, R., and J. Duncan. 1995. Neural mechanisms of selective visual attention. *Ann. Rev. Neurosci.* **18**:193–222. [13]

Destexhe, A., D. Contreras, and M. Steriade. 1999. Cortically-induced coherence of a thalamic-generated oscillation. *Neuroscience* **92**:427–443. [7]

Diba, K., and G. Buzsáki. 2007. Forward and reverse hippocampal place-cell sequences during ripples. *Nat. Neurosci.* **10**:1241–1242. [7]

———. 2008. Hippocampal network dynamics constrain the time lag between pyramidal cells across modified environments. *J. Neurosci.* **28**:13448–13456. [7]

Dickinson, A. 1980. Contemporary Animal Learning Theory. Cambridge: Cambridge Univ. Press. [4]

Dickinson, P. S., C. Mecsas, and E. Marder. 1990. Neuropeptide fusion of two motor-pattern generator circuits. *Nature* **344**:155–158. [12]

Dickinson, P. S., and F. Nagy. 1983. Control of a central pattern generator by an identified modulatory interneurone in crustacea. II. Induction and modification of plateau properties in pyloric neurones. *J. Exp. Biol.* **105**:59–82. [12]

Diesmann, M., M.-O. Gewaltig, and A. Aertsen. 1999. Stable propagation of synchronous spiking in cortical neural networks. *Nature* **402**:529–533. [3]

Di Russo, F., A. Martinez, M. I. Sereno, S. Pitzalis, and S. A. Hillyard. 2001. Cortical sources of the early components of the visual evoked potential. *Hum. Brain Mapp.* **15**:95–111. [17]

Djurfeldt, M., M. Lundqvist, C. Johansson, et al. 2008. Brain-scale simulation of the neocortex on the IBM Blue Gene/L supercomputer. *IBM J. Res. Dev.* **52**:31–41. [6]

Doesburg, S. M., A. B. Roggeveen, K. Kitajo, and L. M. Ward. 2008. Large-scale gamma-band phase synchronization and selective attention. *Cereb. Cortex* **18(2)**:386–396. [16]

Doherty, M. J., N. M. Campbell, H. Tsuji, and W. A. Phillips. 2010. Size illusions deceive adults but not young children. *Devel. Sci.* doi:10.1111/j.1467-7687.2009.00931.x. [1]

Doig, N. M., J. Moss, and J. P. Bolam. 2009. Medium-sized spiny neurons of the direct and indirect pathways in the striatum receive excitatory input from both cortical and thalamic afferents. *Soc. Neurosci. Abst.* **845**:19/Q17. [5]

Doischer, D., J. A. Hosp, Y. Yanagawa, et al. 2008. Postnatal differentiation of basket cells from slow to fast signaling devices. *J. Neurosci.* **28**:12956–12968. [18]

Douglas, R. J., and K. A. C. Martin. 2007. Mapping the matrix: The ways of neocortex. *Neuron* **56**:226–238. [1, 2]

Doupe, A., and P. Kuhl. 1999. Birdsong and human speech: Common themes and mechanisms. *Ann. Rev. Neurosci.* **22**:567–631. [5]

Doursat, R. 1991. Contribution à l'étude des représentations dans le système nerveux et dans les réseaux de neurones formels, Thèse de Doctorat, Paris VI. [12]

Dräger, U. C., and D. H. Hubel. 1975. Responses to visual stimulation and relationship between visual, auditory, and somatosensory inputs in mouse superior colliculus. *J. Neurophysiol.* **38**:690–713. [10]

Dragoi, G., and G. Buzsáki. 2006. Temporal encoding of place sequences by hippocampal cell assemblies. *Neuron* **50**:145–157. [7]

Duncan, J. 2006. Brain mechanisms of attention. *Q. J. Exp. Psychol.* **59**:2–27. [1]

Duncan, S. 1997. Early parent-child interaction grammar prior to language acquisition. *Lang. Comm.* **17**:149–164. [5]

Duvdevani-Bar, S., and S. Edelman. 1999. Visual recognition and categorization on the basis of similarities to multiple class prototypes. *Intl. J. Comp. Vision* **33**.201–228. [13]

Eckhorn, R., R. Bauer, W. Jordan, et al. 1988. Coherent oscillations: A mechanism of feature linking in the visual cortex? *Biol. Cyber.* **60**:121–130. [3]

Edelman, S. 1999. Representation and Recognition in Vision. Cambridge, MA: MIT Press. [5]

———. 2003. But will it scale up? Not without representations. A commentary on "The dynamics of active categorical perception in an evolved model agent" by R. Beer. *Adap. Behav.* **11**:273–275. [5]

———. 2008a. Computing the Mind. New York: Oxford Univ. Press. [1, 5, 18]

———. 2008b. On the nature of minds, or truth and consequences. *J. Exp. Theor. Art. Intell.* **20**:181–196. [5]

Edelman, S., and S. Duvdevani-Bar. 1997. A model of visual recognition and categorization. *Phil. Trans. R. Soc. Lond. B* **1358**:1191–1202. [13]

Edelman, S., and N. Intrator. 2003. Towards structural systematicity in distributed, statically bound visual representations. *Cogn. Sci.* **27**:1, 73–109. [5, 13]

Edinger, L. 1908. Vorlesungen über den Bau der Nervösen Zentralorgane des Menschen und der Tiere für Ärtze und Studierende. Leipzig: Vogel Verlag. [5]

Ekeberg, O., P. Wallen, A. Lansner, et al. 1991. A computer based model for realistic simulations of neural networks. *Biol. Cyber.* **65**:81–90. [3]

Ekstrom, L. B., P. R. Roelfsema, J. T. Arsenault, G. Bonmassar, and W. Vanduffel. 2008. Bottom-up dependent gating of frontal signals in early visual cortex. *Science* **321**:414–417. [13]

El Boustani, S., O. Marre, S. Béhuret, et al. 2009. Network-state modulation of power-law frequency-scaling in visual cortical neurons. *PLoS Comput. Biol.* **5(9)**:e1000519. [12]

Elston, G. N. 2003. Cortex, cognition and the cell: New insights into the pyramidal neuron and prefrontal function. *Cereb. Cortex* **13**:1124–1138. [2]

Elston, G. N., A. Elston, V. A. Casagrande, and J. H. Kaas. 2005. Areal specialization of pyramidal cell structure in the visual cortex of the tree shrew: A new twist revealed in the evolution of cortical circuits. *Exp. Brain Res.* **163**:13–20. [2]

Elston, G. N., R. Tweedale, and M. G. P. Rosa. 1999. Cortical integration in the visual system of the macaque monkey: Large-scale morphological differences of pyramidal neurons in the occipital, parietal and temporal lobes. *Proc. R. Soc. B.* **266**:1367–1374. [2]

Emery, N. J., and N. S. Clayton. 2001. Effects of experience and social context on prospective caching strategies by scrub jays. *Nature* **414**:443–446. [4, 5]

———. 2004. The mentality of crows: Convergent evolution of intelligence in corvids and apes. *Science* **306**:1903–1907. [4]

Emes, R. D., A. J. Pocklington, C. N. G. Anderson, et al. 2008. Evolutionary expansion and anatomical specialization of synapse proteome complexity. *Nat. Neurosci.* **11**:799–806. [5]

Engel, A. K., P. Fries, and W. Singer. 2001. Dynamic predictions: Oscillations and synchrony in top-down processing. *Nat. Rev. Neurosci.* **2(10)**:704–716. [6, 11, 18]

Engel, A. K., P. König, A. K. Kreiter, T. B. Schillen, and W. Singer. 1992. Temporal coding in the visual cortex: New vistas on integration in the nervous system. *Trends Neurosci.* **15**:218–226. [14, 18]

Engel, A. K., P. König, A. K. Kreiter, and W. Singer. 1991. Interhemispheric synchronization of oscillatory neuronal responses in cat visual cortex. *Science* **252**:1177–1179. [11, 13, 18]

Engel, A. K., P. König, and W. Singer. 1991. Direct physiological evidence for scene segmentation by temporal coding. *PNAS* **88**:9136–9140. [3, 11, 18]

Engel, A. K., C. K. Moll, I. Fried, and G. A. Ojemann. 2005. Invasive recordings from the human brain: Clinical insights and beyond. *Nat. Rev. Neurosci.* **6**:35–47. [18]

Engel, A. K., and W. Singer. 2001. Temporal binding and the neural correlates of sensory awareness. *Trends Cogn. Sci.* **5(1)**:16–25. [16]

Epstein, R., C. E. Kirshnit, R. P. Lazna, and L. C. Rubin. 1984. "Insight" in the pigeon: Antecedents and determinants of an intelligent performance. *Nature* **308**:61–62. [4]

Ermentrout, G. B., and N. J. Kopell. 1998. Fine structure of neural spiking and synchronization in the presence of conduction delays. *PNAS* **95**:1259–1264. [13]

Ernst, M. O., M. S. Banks, and H. H. Bülthoff. 2000. Touch can change visual slant perception. *Nat. Neurosci.* **3**:69–73. [15]

Fan, J., J. Byrne, M. S. Worden, et al. 2007. The relation of brain oscillations to attentional networks. *J. Neurosci.* **27(23)**:6197–6206. [16]

Fang, P.-C., N. Jain, and J. H. Kaas. 2002. Few intrinsic connections cross the hand-face border of area 3b of New World Monkeys. *J. Comp. Neurol.* **454**:310–319. [2]

Farmer, S. F. 1998. Rhythmicity, synchronization and binding in human and primate motor systems. *J. Physiol.* **509**:3–14. [18]

Farran, E. K. 2005. Perceptual grouping ability in Williams syndrome: Evidence for deviant patterns of performance. *Neuropsychologia* **43**:815–822. [17]

Favorov, O. V., and D. G. Kelly. 1994. Minicolumnar organization within somatosensory cortical segregates. I. Development of afferent connections. *Cereb. Cortex* **4**:408–427. [6]

Feenders, G., M. Liedvogel, M. Rivas, et al. 2008. Molecular mapping of movement-associated areas in the avian brain: A motor theory for vocal learning origin. *PLoS ONE* **3**:e1768. [5]

Fell, J., G. Fernandez, P. Klaver, C. E. Elger, and P. Fries. 2003. Is synchronized neuronal gamma activity relevant for selective attention? *Brain Res. Rev.* **42(3)**:265–272. [10]

Fell, J., P. Klaver, K. Lehnertz, et al. 2001. Human memory formation is accompanied by rhinal-hippocampal coupling and decoupling. *Nat. Neurosci.* **4**:1259–1264. [13, 18]

Felleman, D. J., and D. C. Van Essen. 1991. Distributed hierarchical processing in the primate cerebral cortex. *Cereb. Cortex* **1**:1–47. [16]

Ferrarelli, F., M. Massimini, M. J. Peterson, et al. 2008. Reduced evoked gamma oscillations in the frontal cortex in schizophrenia patients: A TMS/EEG study. *Am. J. Psychiatry* **165**:996–1005. [18]

Ferrari, P. F., L. Bonini, and L. Fogassi. 2009. From monkey mirror neurons to primate behaviours: Possible "direct" and "indirect" pathways. *Phil. Trans. R. Soc. Lond. B* **364**:2311–2323. [5]

Fiala, A. 2007. Olfaction and olfactory learning in Drosophila: Recent progress. *Curr. Opin. Neurobiol.* **17**:720–726. [5]

Field, D. J., A. Hayes, and R. F. Hess. 1993. Contour integration by the human visual system: Evidence for a local "association field." *Vision Res.* **33(2)**:173–193. [18]

Fingelkurts, A. A., S. Kallio, and A. Revonsuo. 2007. Cortex functional connectivity as a neurophysiological correlate of hypnosis: An EEG case study. *Neuropsychologia* **45**:1452–1462. [17]

Finger, S. 1994. Origins of Neuroscience. New York: Oxford Univ. Press. [1]

Fink, M. M. 1996. Time reversal in acoustics. *Cont. Physics* **37**:95–109. [12]

Fiorito, G., and P. Scotto. 1992. Observational learning in *Octopus vulgaris*. *Science* **256**:545–547. [5]

Fischer, I., R. Vicente, J. M. Buldú, et al. 2006. Zero-lag long-range synchronization via dynamical relaying. *Phys. Rev. Lett.* **97**:123901–123904. [13]

FitzHugh, R. 1961. Impulses and physiological states in models of nerve membrane. *Biophys. J.* **1**:445–466. [3]

Fitzpatrick, D. 1996. The functional organization of local circuits in visual cortex: Insights from the study of tree shrew striate cortex. *Cereb. Cortex* **6**:329–341. [6]

Flynn, G., D. Alexander, A. Harris, et al. 2008. Increased absolute magnitude of gamma synchrony in first-episode psychosis. *Schizophr. Res.* **105**:262–271. [17]

Fodor, J. A., and Z. W. Pylyshyn. 1988. Connectionism and cognitive architecture: A critical analysis. *Cognition* **28**:3–71. [12]

Foster, D. J., and M. A. Wilson. 2006. Reverse replay of behavioural sequences in hippocampal place cells during the awake state. *Nature* **440**:680–683. [7]

Foucher, J. R., P. Vidailhet, S. Chanraud, et al. 2005. Functional integration in schizophrenia: Too little or too much? Preliminary results on fMRI data. *NeuroImage* **26**:374–388. [17]

Fox, M. D., and M. E. Raichle. 2007. Spontaneous fluctuations in brain activity observed with functional magnetic resonance imaging. *Nat. Rev. Neurosci.* **8**:700–711. [13]

Foxe, J. J., G. M. Doniger, and D. C. Javitt. 2001. Early visual processing deficits in schizophrenia: Impaired P1 generation revealed by high-density electrical mapping. *NeuroReport* **12**:3815–3820. [17]

Fransén, E., and A. Lansner. 1998. A model of cortical associative memory based on a horizontal network of connected columns. *Netw. Comp. Neural Sys.* **9**:235–264. [6]

Frégnac, Y. 1991. How many cycles make an oscillation? In: Representations of Vision: Trends and Tacit Assumptions in Vision Research, ed. A. Gorea, Y. Frégnac, Z. Kapoula and J. Finlay, pp. 97–109. Cambridge: Cambridge Univ. Press. [12]

———. 2002. Hebbian synaptic plasticity. In: The Handbook of Brain Theory and Neural Networks, ed. M. A. Arbib, pp. 515–522. Cambridge, MA: MIT Press. [12]

Frégnac, Y., P. Baudot, F. Chavane, et al. 2010. Multiscale functional imaging in V1 and cortical correlates of apparent motion. In: Dynamics of Visual Motion Processing, ed. G. Masson and U. Ilg, pp. 73–94. Berlin: Springer-Verlag. [12, 13]

Frégnac, Y., P. Baudot, M. Levy, and O. Marre. 2005. An intracellular view of time coding and sparseness in V1 during virtual oculomotor exploration of natural scenes. 2nd Intl. Cosyne Conf. in Computational and Systems Neuroscience. Salt Lake City. [12, 13]

Frégnac, Y., M. Blatow, J. P. Changeux, et al. 2006. Ups and downs in the genesis of cortical computation. In: Microcircuits: The Interface between Neurons and Global Brain Function Microcircuits, ed. S. Grillner and A. Graybiel, pp. 397–437. Dahlem Workshop Reports. Cambridge, MA: MIT Press. [12]

Frégnac, Y., and V. Bringuier. 1996. Spatio-temporal dynamics of synaptic integration in cat visual cortical receptive fields. In: Brain Theory: Biological Basis and Computational Theory of Vision, ed. A. Aertsen and V. Braitenberg, pp. 143–199. Amsterdam: Springer-Verlag. [12]

Freunberger, R., W. Klimesch, B. Griesmayr, P. Sauseng, and W. Gruber. 2008. Alpha phase coupling reflects object recognition. *NeuroImage* **42(2)**:928–935. [16]

Freund, T. F., and M. Antal. 1988. GABA-containing neurons in the septum control inhibitory interneurons in the hippocampus. *Nature* **336**:170–173. [13]

Freund, T. F., and G. Buzsáki. 1996. Interneurons of the hippocampus. *Hippocampus* **6**:347–470. [7]

Friedman, J. I., C. Tang, D. Carpenter, et al. 2008. Diffusion tensor imaging findings in first-episode and chronic schizophrenia patients. *Am. J. Psychiatry* **165**:1024–1032. [17]

Fries, P. 2005. A mechanism for cognitive dynamics: Neuronal communication through neuronal coherence. *Trends Cogn. Sci.* **9**:474–480. [1, 7, 8, 13, 18]

———. 2009. Neuronal gamma-band synchronization as a fundamental process in cortical computation. *Ann. Rev. Neurosci.* **32**:209–224. [12, 13]

Fries, P., S. Neuenschwander, A. K. Engel, R. Goebel, and W. Singer. 2001. Rapid feature selective neuronal synchronization through correlated latency shifting. *Nat. Neurosci.* **4(2)**:194–200. [11, 16]

Fries, P., D. Nikolić, and W. Singer. 2007. The gamma cycle. *Trends Neurosci.* **30(7)**:309–316. [6, 8, 11]

Fries, P., J. H. Reynolds, A. E. Rorie, and R. Desimone. 2001. Modulation of oscillatory neuronal synchronization by selective visual attention. *Science* **291**:1560–1563. [1, 6, 11, 13, 18]

Fries, P., P. R. Roelfsema, A. K. Engel, P. König, and W. Singer. 1997. Synchronization of oscillatory responses in visual cortex correlates with perception in interocular rivalry. *PNAS* **94**:12699–12704. [11, 13, 18]

Fries, P., J. H. Schroder, P. R. Roelfsema, W. Singer, and A. K. Engel. 2002. Oscillatory neuronal synchronization in primary visual cortex as a correlate of stimulus selection. *J. Neurosci.* **22(9)**:3739–3754. [11]

Fries, P., T. Womelsdorf, R. Oostenveld, and R. Desimone. 2008. The effects of visual stimulation and selective visual attention on rhythmic neuronal synchronization in macaque area V4. *J. Neurosci.* **28**:4823–4835. [13]

Friston, K. J. 1999. The disconnection hypothesis. *Schizophr. Res.* **30**:115–125. [1, 18]

———. 2009. The free-energy principle: A rough guide to the brain? *Trends Cogn. Sci.* **13(7)**:293–301. [1]

Friston, K. J., C. Buechel, G. R. Fink, et al. 1997. Psychophysiological and modulatory interactions in neuroimaging. *NeuroImage* **6**:218–229. [13]

Friston, K. J., L. Harrison, and W. Penny. 2003. Dynamic causal modelling. *NeuroImage* **19**:1273–1302. [13]

Friston, K. J., and S. Keibel. 2009. Predictive coding under the free-energy principle. *Phil. Trans. R. Soc. Lond. B* **364**:1211–1221. [1]

Friston, K. J., and K. E. Stephan. 2007. Free-energy and the brain. *Synthese* **159**:417–458. [18]

Frith, C., and R. J. Dolan. 1997. Brain mechanisms associated with top–down processes in perception. *Phil. Trans. R. Soc. Lond. B* **352**:1221–1230. [18]

FRVT. 2002. Face recognition vendor test results. www.frvt.org/FRVT2002/documents. htm (accessed Jan. 2, 2010). [13]

Fuchs, E. C., H. Doheny, H. Faulkner, et al. 2001. Genetically altered AMPA-type glutamate receptor kinetics in interneurons disrupts long-range synchrony of gamma oscillations. *PNAS* **98**:3571–3576. [8]

Fuster, J. M. 1989. The Prefrontal Cortex. New York: Raven. [18]

———. 2003. Cortex and Mind: Unifying Cognition. New York: Oxford Univ. Press. [17]

Gabbiani, F., I. Cohen, and G. Laurent. 2005. Time-dependent activation of feed-forward inhibition in a looming-sensitive neuron. *J. Neurophysiol.* **94**:2150–2161. [5]

Gaillard, R., S. Dehaene, C. Adam, et al. 2009. Converging intracranial markers of conscious access. *PLoS Biol.* **7(3)**:e61. [16]

Galaburda, A. M., and U. V. Bellugi. 2000. V. Multi-level analysis of cortical neuroanatomy in Williams syndrome. *J. Cogn. Neurosci.* **12(1)**:74–88. [18]

Gallese, V., L. Fadiga, L. Fogassi, and G. Rizzolatti. 1996. Action recognition in the premotor cortex. *Brain Res. Rev.* **119**:593–609. [5]

Gallinat, J., G. Winterer, C. S. Herrmann, and D. Senkowski. 2004. Reduced oscillatory gamma-band responses in unmedicated schizophrenic patients indicate impaired frontal network processing. *Clin. Neurophys.* **115**:1863–1874. [18]

Gelperin, A., and D. W. Tank. 1990. Odor-modulated collective network oscillations of olfactory interneurons in a terrestrial mollusc. *Nature* **345**:437–440. [5]

Geman, S., E. Bienenstock, and R. Doursat. 1992. Neural networks and the bias/variance dilemma. *Neural Comput.* **4**:1–58. [10]

Georges, S., P. Sèries, Y. Frégnac, and J. Lorenceau. 2002. Orientation dependent modulation of apparent speed: Psychophysical evidence. *Vision Res.* **42**:2757–2772. [12]

Georgopoulos, A. P. 1995. Current issues in directional motor control. *Trends Neurosci.* **18**:506–510. [18]

Georgopoulos, A. P., A. B. Schwartz, and R. E. Kettner. 1986. Neuronal population coding of movement direction. *Science* **233**:1416–1419. [3, 13]

Gerhardtstein, P., I. Kovács, J. Ditre, and A. Fehér. 2004. Detection of contour continuity and closure in 3-month-old human infants. *Vision Res.* **44**:2881–2988. [18]

Gerstner, W. 1995. Time structure of the activity in neural network models. *Phys. Rev. E* **51**:738–758. [3]

Gerstner, W., J. L. van Hemmen, and J. D. Cowan. 1996. What matters in neuronal locking. *Neural Comput.* **8**:1653–1676. [3]

Gervan, P., and I. Kovács. 2010. Sleep-dependent learning in visual contour integration. *J. Vision*, in press. [18]

Giersch, A., G. Humphreys, M. Boucart, and I. Kovács. 2000. The computation of occluded contours in visual agnosia: Evidence for early computation prior to shape binding and figure-ground coding. *Cogn. Neuropsych.* **17**:731–759. [18]

Gilbert, C. D. 1983. Microcircuitry of visual cortex. *Ann. Rev. Neurosci.* **6**:217–247. [9]

———. 1992. Horizontal integration and cortical dynamics. *Neuron* **9(1)**:1–13. [1, 2, 14]

Gilbert, C. D., and T. N. Wiesel. 1983. Functional organization of the visual cortex. *Prog. Brain Res.* **58**:209–218. [9]

Glezer, I. I., M. S. Jacobs, and P. J. Morgane. 1988. Implications of the "initial brain" concept for brain evolution in Cetacea. *Behav. Brain Sci.* **11**:75–116. [2]

Goldberg, I. I., M. Harel, and R. Malach. 2006. When the brain loses its self: Prefrontal inactivation during sensorimotor processing. *Neuron* **50(2)**:329–339. [16]

Gonchar, Y. A., P. B. Johnson, and R. J. Weinberg. 1995. GABA-immunopositive neurons in rat neocortex with contralateral projections to S-I. *Brain Res.* **697**:27–34. [13]

Gonzalez-Burgos, G., and D. A. Lewis. 2008. GABA neurons and the mechanisms of network oscillations: Implications for understanding cortical dysfunction in schizophrenia. *Schizophr. Bull.* **34**:944–961. [17]

Goodhill, G. J. 2007. Contributions of theoretical modeling to the understanding of neural map development. *Neuron* **56**:301–311. [10]

Goodman, C. S. 1978. Isogenic grasshoppers: Genetic variability in the morphology of identified neurons. *J. Comp. Neurol.* **182**:681–705. [5]

Gori, M., M. M. Del Viva, G. Sandini, and D. C. Burr. 2008. Young children do not integrate visual and haptic form information. *Curr. Biol.* **18(9)**:694–698. [15]

Gray, C. M., P. König, A. K. Engel, and W. Singer. 1989. Oscillatory responses in cat visual cortex exhibit inter-columnar synchronization which reflects global stimulus properties. *Nature* **338**:334–337. [3, 11, 12, 18]

Gray, C. M., and W. Singer. 1989. Stimulus-specific neuronal oscillations in orientation columns of cat visual cortex. *PNAS* **86**:1698–1702. [8, 11]

Green, J. J., and J. J. McDonald. 2008. Electrical neuroimaging reveals timing of attentional control activity in human brain. *PLoS Biol.* **6(4)**:730–738. [16]

Gregoriou, G. G., S. J. Gotts, H. Zhou, and R. Desimone. 2009. High-frequency, long-range coupling between prefrontal and visual cortex during attention. *Science* **324**:1207–1210. [13, 18]

Grenier, F. O., I. Timofeev, and M. Steriade. 2001. Focal synchronization of ripples (80–200 Hz) in neocortex and their neuronal correlates. *J. Neurophysiol.* **86(4)**:1884–1898. [16]

Grent-'t-Jong, T., and M. G. Woldorff. 2007. Timing and sequence of brain activity in top-down control of visual-spatial attention. *PLoS Biol.* **5(1)**:e12. [16]

Grice, S. J., M. W. Spratling, A. Karmiloff-Smith, et al. 2001. Disordered visual processing and oscillatory activity in autism and Williams syndrome. *NeuroReport* **12**:2697–2700. [17]

Grillner, S. 2003. The motor infrastructure: From ion channels to neuronal networks. *Nat. Rev. Neurosci.* **4**:573–586. [5]

———. 2006. Biological pattern generation: The cellular and computational logic of networks in motion. *Neuron* **52**:751–766. [5]

Grimault, S., N. Robitaille, C. Grova, et al. 2009. Oscillatory activity in parietal and dorsolateral prefrontal cortex during retention in visual short-term memory: Additive effects of spatial attention and memory load. *Hum. Brain Mapp.* **30(10)**:3378–3392. [16]

Grinvald, A., A. Areli, M. Tsodyks, and T. Kenet. 2003. Neuronal assemblies: Single cortical neurons are obedient members of a huge orchestra. *Biopolymers* **68**:422–436. [6]

Grinvald, A., E. Lieke, R. D. Frostig, and R. Hildesheim. 1994. Cortical point-spread function and long-range lateral interactions revealed by real-time optical imaging of macaque monkey primary visual cortex. *J. Neurosci.* **14**:2545–2568. [12]

Gross, J., F. Schmitz, I. Schnitzler, et al. 2004. Modulation of long-range neural synchrony reflects temporal limitations of visual attention in humans. *PNAS* **101(35)**:13050–13055. [16]

Grossberg, S. 1980. How does the brain build a cognitive code? *Psychol. Rev.* **87**:1–51. [18]

Gruber, T., and M. M. Müller. 2005. Oscillatory brain activity dissociates between associative stimulus content in a repetition priming task in the human EEG. *Cereb. Cortex* **16**:109–116. [18]

Gulyás, A. I., N. Hájos, I. Katona, and T. F. Freund. 2003. Interneurons are the local targets of hippocampal inhibitory cells which project to the medial septum. *Eur. J. Neurosci.* **17**:1861–1872. [13]

Guterman, Y. 2007. A neural plasticity perspective on the schizophrenic condition. *Conscious Cogn.* **16**:400–420. [17]

Hadjikhani, N., A. K. Liu, A. M. Dale, P. Cavanagh, and R. B. H. Tootell. 1998. Retinotopy and color sensitivity in human visual cortical area V8. *Nat. Neurosci.* **1(3)**:235–241. [16]

Hadjipapas, A., P. Adjamian, J. B. Swettenham, I. E. Holliday, and G. R. Barnes. 2007. Stimuli of varying spatial scale induce gamma activity with distinct temporal characteristics in human visual cortex. *NeuroImage* **35(2)**:518–530. [16]

Hagmann, P., L. Cammoun, X. Gigandet, et al. 2008. Mapping the structural core of human cerebral cortex. *PLoS Biol.* **6**:e159. [13]

Hajós, M., W. E. Hoffman, and B. Kocsis. 2008. Activation of cannabinoid-1 receptors disrupts sensory gating and neuronal oscillation: Relevance to schizophrenia. *Biol. Psychiatry* **63**:1075–1083. [17]

Haken, H., J. A. S. Kelso, A. Fuchs, and A. Pandya. 1990. Dynamic pattern recognition of coordinated biological motion. *Neural Netw.* **3**:395–401. [14]

Hall, S. D., I. E. Holliday, A. Hillebrand, et al. 2005. The missing link: Analogous human and primate cortical gamma oscillations. *NeuroImage* **26(1)**:13–17. [16]

Hamker, F. H. 2005. The reentry hypothesis: The putative interaction of the frontal eye field, ventrolateral prefrontal cortex, and areas V4, IT for attention and eye movement. *Cereb. Cortex* **15**:431–447. [13]

Hampton, R. R. 2001. Rhesus monkeys know when they remember. *PNAS* **98**:5359–5362. [4]

Hamzei-Sichani, F., N. Kamasawa, W. G. Jannsen, et al. 2007. Gap junctions on mossy fiber axons demonstrated by thin section electron microscopy and freeze-fracture replica immunogold labelling. *PNAS* **104**:12548–12553. [8]

Hancock, P. J. B., L. Walton, G. Mitchell, Y. Plenderleith, and W. A. Phillips. 2008. Segregation by onset asynchrony. *J. Vision* **8(7)**:1–21. [1]

Happe, F., and U. Frith. 2006. The weak coherence account: Detail-focused cognitive style in autism spectrum disorders. *J. Autism Devel. Disord.* **36**:5–25. [17]

Hare, B., J. Call, B. Agnetta, and M. Tomasello. 2000. Chimpanzees know what conspecifics do and do not see. *Anim. Behav.* **59**:771–785. [4]

Hare, B., J. Call, and M. Tomasello. 2001. Do chimpanzees know what conspecifics know? *Anim. Behav.* **61**:139–151. [5]

Harris, K. D., J. Csicsvari, H. Hirase, G. Dragoi, and G. Buzsáki. 2003. Organization of cell assemblies in the hippocampus. *Nature* **424**:552–556. [7]

Hartline, D. K., and D. R. Colman. 2007. Rapid conduction and the evolution of giant axons and myelinated fibers. *Curr. Biol.* **17**:R29–35. [5]

Hashimoto, T., Q. L. Nguyen, D. Rotaru, et al. 2009. Protracted developmental trajectories of GABA(A) receptor alpha1 and alpha2 subunit expression in primate prefrontal cortex. *Biol. Psychiatry* **65**:1015–1023. [18]

Hässler, R. 1967. Comparative anatomy in the central visual systems in day- and night-active primates. In: Evolution of the Forebrain, ed. R. Hässler and H. Stephan, pp. 419–434. Stuttgart: Thieme. [2]

Haxby, J. V., M. I. Gobbini, M. L. Furey, et al. 2001. Distributed and overlapping representations of faces and objects in ventral temporal cortex. *Science* **293(5539)**:2425–2430. [16]

Haykin, S. 1994. Neural Networks: A Comprehensive Foundation. London: Macmillan. [10]

Haynes, J. D., J. Tregellas, and G. Rees. 2005. Attentional integration between anatomically distinct stimulus representations in early visual cortex. *PNAS* **102**:14925–14930. [13]

He, B. J., A. Z. Snyder, J. L. Vincent, et al. 2007. Breakdown of functional connectivity in frontoparietal networks underlies behavioral deficits in spatial neglect. *Neuron* **53**:905–918. [13]

Hebb, D. O. 1949. The Organization of Behavior. New York: Wiley. [3, 6, 12, 17]

Hemsley, D. R. 2005. The development of a cognitive model of schizophrenia: Placing it in context. *Neurosci. Biobehav. Rev.* **29**:977–988. [17]

Herculano-Houzel, S., C. E. Collins, P. Wong, and J. H. Kaas. 2007. Cellular scaling rules for primate brains. *PNAS* **104**:3562–3567. [2]

Herculano-Houzel, S., C. E. Collins, P. Wong, J. H. Kaas, and R. Lent. 2008. The basic nonuniformity of the cerebral cortex. *PNAS* **105**:12593–12598. [2]

Herculano-Houzel, S., M. H. J. Munk, S. Neuenschwander, and W. Singer. 1999. Precisely synchronized oscillatory firing patterns require electroencephalographic activation. *J. Neurosci.* **19(10)**:3992–4010. [11]

Herrmann, C. S., D. Lenz, S. Junge, N. A. Busch, and B. Maess. 2004. Memory-matches evoke human gamma-responses. *BMC Neurosci.* **5**:13. [18]

Herrmann, C. S., M. H. Munk, and A. K. Engel. 2004. Cognitive functions of gamma-band activity: Memory match and utilization. *Trends Cogn. Sci.* **8**:347–355. [18]

Herz, A. V. M., B. Sulzer, R. Kühn, and J. L. van Hemmen. 1989. Hebbian learning reconsidered: Representation of static and dynamic objects in associative neural nets. *Biol. Cyber.* **60**:457–467. [3]

Hess, R. F., A. Hayes, and D. J. Field. 2003. Contour integration and cortical processing. *J. Physiol.* **97**:105–119. [18]

Heyes, C. 2001. Causes and consequences of imitation. *Trends Cogn. Sci.* **5**:253–261. [5]

———. 2009. Evolution, development and intentional control of imitation. *Phil. Trans. R. Soc. Lond. B* **364**:2293–2298. [5]

Hikosàka, O., S. Miyauchi, and S. Shimojo. 1993. Focal visual attention produces illusory temporal order and motion sensation. *Vision Res.* **33**:1219–1240. [12]

Hilgetag, C. C., M. A. Oneill, and M. P. Young. 1996. Indeterminate organization of the visual system. *Science* **271**:776–777. [16]

Hillebrand, A., K. D. Singh, I. E. Holliday, P. L. Furlong, and G. R. Barnes. 2005. A new approach to neuroimaging with magnetoencephalography. *Hum. Brain Mapp.* **25(2)**:199–211. [16]

Hochstein, S., and M. Ahissar. 2002. View from the top: Hierarchies and reverse hierarchies in the visual system. *Neuron* **36(5)**:791–804. [17]

Hof, P. R., I. I. Glezer, F. Condé, et al. 1999. Cellular distribution of the calcium-binding proteins parvalbumin, calbindin, and calretinin in the neocortex of mammals: Phylogenetic and developmental patterns. *J. Chem. Neuroanat.* **16**:77–116. [2]

Hof, P. R., and E. Van der Gucht. 2007. Structure of the cerebral cortex of the humpback whale, *Megaptera novaeangliae* (Cetacea, Mysticeti, Balaenopteridae). *Anat. Rec.* **290**:1–31. [2]

Hoffman, D. A., R. Sprengel, and B. Sakmann. 2002. Molecular dissection of hippocampal theta-burst pairing potentiation. *PNAS* **99**:7740–7745. [6]

Hoffman, K. P., and J. Stone. 1971. Conduction velocity of afferents to cat visual cortex: A correlation with cortical receptive field properties. *Brain Res.* **32**:460–466. [12]

Hoffman, R. E., and T. H. McGlashan. 2001. Neural network models of schizophrenia. *Neuroscientist* **7**:441–454. [17]

Holscher, C., R. Anwyl, and M. J. Rowan. 1997. Stimulation on the positive phase of hippocampal theta rhythm induces long-term potentiation that can be depotentiated by stimulation on the negative phase in area CA1 *in vivo*. *J. Neurosci.* **17**:6470–6477. [7]

Honey, C., R. Kötter, M. Breakspear, and O. Sporns. 2007. Network structure of cerebral cortex shapes functional connectivity on multiple time scales. *PNAS* **104**:10240–10245. [6]

Hoogenboom, N., J. M. Schoffelen, R. Oostenveld, L. M. Parkes, and P. Fries. 2006. Localizing human visual gamma-band activity in frequency, time and space. *NeuroImage* **29(3)**:764–773. [16]

Hooper, S., and M. Moulins. 1989. Switching of a neuron from one network to another by sensory-induced changes in membrane properties. *Science* **244**:1587–1589. [12]

Hopcroft, J. E., and J. D. Ullman. 1979. Introduction to Automata Theory, Languages, and Computation. Reading, MA: Addison-Wesley. [5]

Hopfield, J. J. 1982. Neural networks and physical systems with emergent collective computational abilities. *PNAS* **79**:2554–2558. [6, 10]

Howard, M. W., D. S. Rizzuto, J. B. Caplan, et al. 2003. Gamma oscillations correlate with working memory load in humans. *Cereb. Cortex* **13(12)**:1369–1374. [14, 16]

Hubel, D. H., and T. N. Wiesel. 1959. Receptive fields of single neurons in the cat's visual cortex. *J. Physiol.* **148**:574–591. [14]

———. 1977. The functional architecture of the macaque visual cortex. The Ferrier lecture. *Proc. R. Soc. Sci. B* **198**:1–59. [6]

Huber, L., F. Range, B. Voelkl, et al. 2009. The evolution of imitation: What do the capacities of non-human animals tell us about the mechanisms of imitation? *Phil. Trans. R. Soc. Lond. B* **364**:2299–2309. [5]

Hubl, D., T. Koenig, W. Strik, et al. 2004. Pathways that make voices: White matter changes in auditory hallucinations. *Arch. Gen. Psychiatry* **61(7)**:658–668. [17]

Huerta, P. T., and J. E. Lisman. 1993. Heightened synaptic plasticity of hippocampal CA1 neurons during a cholinergically induced rhythmic state. *Nature* **364**:723–725. [18]

———. 1996. Low-frequency stimulation at the troughs of theta-oscillation induces long-term depression of previously potentiated CA1 synapses. *J. Neurophysiol.* **75**:877–884. [7]

Hummel, J. E., and I. Biederman. 1992. Dynamic binding in a neural network for shape recognition. *Psychol. Rev.* **99**:480–517. [13]

Hummel, J. E., and K. J. Holyoak. 1993. Distributing structure over time. *Behav. Brain Sci.* **16**:464. [1]

———. 2005. Relational reasoning in a neurally plausible cognitive architecture. *Curr. Dir. Psychol. Sci.* **14**:153–157. [1]

Hunt, G. R. 1996. Manufacture and use of hook-tools by New Caledonian crows. *Nature* **379**:249–251. [4]

Hupé, J.-M., L.-M. Joffo, and D. Pressnitzer. 2008. Bistability for audiovisual stimuli: Perceptual decision is modality specific. *J. Vision* **8(7)**:1–15. [1]

Hutcheon, B., and Y. Yarom. 2000. Resonance, oscillation and the intrinsic frequency preference of neurons. *Trends Neurosci.* **23**:216–222. [8]

Ikegaya, Y., G. Aaron, R. Cossart, et al. 2004. Synfire chains and cortical songs: Temporal modules of cortical activity. *Science* **304**:559–564. [6]

Imaizumi, K., and G. S. Pollack. 1999. Neural coding of sound frequency by cricket auditory receptors. *J. Neurosci.* **19**:1508–1516. [5]

Ishikane, H., M. Gangi, S. Honda, and M. Tachibana. 2005. Synchronized retinal oscillations encode essential information for escape behavior in frogs. *Nat. Neurosci.* **8**:1087–1095. [11, 13]

Isomura, Y., A. Sirota, S. Ozen, et al. 2006. Integration and segregation of activity in entorhinal-hippocampal subregions by neocortical slow oscillations. *Neuron* **52**:871–882. [7]

Izhikevich, E. 2001. Resonate-and-fire neurons. *Neural Netw.* **14**:883–894. [3]

Jacob, V., D. J. Brasier, I. Erchova, D. Feldman, and D. E. Shulz. 2007. Spike timing-dependent synaptic depression in the *in vivo* barrel cortex of the rat. *J. Neurosci.* **27**:1271–1284. [13]

Jacobs, G. A., and F. E. Theunissen. 1996. Functional organization of a neural map in the cricket cercal sensory system. *J. Neurosci.* **16**:769–784. [5]

Jacobs, R. A., M. I. Jordan, S. J. Nowlan, and G. E. Hinton. 1991. Adaptive mixtures of local experts. *Neural Comput.* **3**:79–87. [10]

Jain, N., T. M. Preuss, and J. H. Kaas. 1994. Subdivisions of the visual system labeled with the Cat-301 antibody in tree shrews. *Vis. Neurosci.* **11**:731–741. [2]

Jancke, D., F. Chavane, S. Naaman, and A. Grinvald. 2004. Imaging cortical correlates of illusion in early visual cortex. *Nature* **428**:423–426. [12]

Janet, P. 1889. L'Automatisme Psychologique. Paris: Felix Akan. [17]

Janik, V. M., and P. J. B. Slater. 2000. The different roles of social learning in vocal communication. *Anim. Behav.* **60**:1–11. [5]

Jansen, B. H., and M. E. Brandt. 1991. The effect of the phase of prestimulus alpha activity on the averaged visual evoked response. *Electroenceph. Clin. Neurophysiol.* **80**:241–250. [16]

Jarvis, E. D. 2004. Learned birdsong and the neurobiology of human language. *Ann. NY Acad. Sci.* **1016**:749–777. [5]

Jarvis, E. D., O. Güntürkün, L. Bruce, et al. 2005. Avian brains and a new understanding of vertebrate brain evolution. *Nat. Rev. Neurosci.* **6**:151–159. [5]

Jensen, O., J. Gelfand, J. Kounios, and J. E. Lisman. 2002. Oscillations in the alpha band (9–12 Hz) increase with memory load during retention in a short-term memory task. *Cereb. Cortex* **12(8)**:877–882. [16]

Jensen, O., J. Kaiser, and J. P. Lachaux. 2007. Human gamma-frequency oscillations associated with attention and memory. *Trends Neurosci.* **30(7)**:317–324. [16]

Jensen, O., and J. E. Lisman. 2005. Hippocampal sequence-encoding driven by a cortical multi-item working memory buffer. *Trends Neurosci.* **28**:67–72. [18]

Jiménez, L., ed. 2003. Attention and Implicit Learning. Amsterdam: Benjamins. [10]

Jinno, S., T. Klausberger, L. F. Marton, et al. 2007. Neuronal diversity in GABAergic long-range projections from the hippocampus. *J. Neurosci.* **27**:8790–8804. [13]

Johansson, G. 1973. Visual perception of biological motion and a model for its analysis. *Percept. Psychophys.* **14**:201–211. [14, 18]

Johnson, S. C., N. Lowery, C. Kohler, and B. I. Turetsky. 2005. Global-local visual processing in schizophrenia: Evidence for an early visual processing deficit. *Biol. Psychiatry* **58**:937–946. [17]

Jokisch, D., and O. Jensen. 2007. Modulation of gamma and alpha activity during a working memory task engaging the dorsal or ventral stream. *J. Neurosci.* **27(12)**:3244–3251. [16]

Jones, M. W., and M. A. Wilson. 2005. Theta rhythms coordinate hippocampal-prefrontal interactions in a spatial memory task. *PLoS Biol.* **3**:e402. [13]

Jortner, R. A., S. S. Farivar, and G. Laurent. 2007. A simple connectivity scheme for sparse coding in an olfactory system. *J. Neurosci.* **27**:1659–1669. [5]

Jung, C. 1907. Über die Psychologie der Dementia Praecox: Ein Versuch. Halle: Marhold. [17]

Kaas, J. H. 2007. Reconstructing the organization of the forebrain of the first mammals. In: Evolution of Nervous Systems, ed. J. H. Kaas and L. A. Krubitzer, pp. 27–48. London: Elsevier. [2]

Kaiser, J., B. Rahm, and W. Lutzenberger. 2009. Temporal dynamics of stimulus-specific gamma-band activity components during auditory short-term memory. *NeuroImage* **44(1)**:257–264. [16]

Kalatsky, V. A., and M. P. Stryker. 2003. New paradigm for optical imaging: Temporally encoded maps of intrinsic signal. *Neuron* **38**:529–545. [12]

Kanter, I. 1988. Potts-glass models of neural networks. *Phys. Rev. A* **37(7)**:2739–2742. [6]

Kapadia, M. K., M. Ito, C. D. Gilbert, and G. Westheimer. 1995. Improvement of visual sensitivity by changes in local context: Parallel studies in human observers and in V1 of alert monkeys. *Neuron* **15**:843–856. [18]

Karni, A., G. Meyer, C. Rey-Hipolito, et al. 1998. The acquisition of skilled motor performance: Fast and slow experience-driven changes in primary motor cortex. *PNAS* **95**:861–868. [18]

Karni, A., and D. Sagi. 1993. The time course of learning a visual skill. *Nature* **365**:250–252. [18]

Katchalsky, A., L. E. Scriven, and R. Blumenthal. 1974. Dynamic patterns of brain cell assemblies. II. Concept of dynamic patterns. Terminology and essential features. Multiple modes of communication. *Neurosc. Res. Prog. Bull.* **12**:58–59. [18]

Kay, J., D. Floreano, and W. A. Phillips. 1998. Contextually guided unsupervised learning using local multivariate binary processors. *Neural Netw.* **11**:117–140. [18]

Kayser, C., and P. König. 2004. Stimulus locking and feature selectivity prevail in complementary frequency ranges of V1 local field potentials. *Eur. J. Neurosci.* **19**:485–489. [18]

Kayser, C., and N. K. Logothetis. 2009. Directed interactions between auditory and superior temporal cortices and their role in sensory integration. *Front. Integr. Neurosci.* **3**:7. [18]

Kayser, C., R. F. Salazar, and P. König. 2003. Responses to natural scenes in cat V1. *J. Neurophysiol.* **90**:1910–1920. [18]

Kelly, S. P., M. Gomez-Ramirez, and J. J. Foxe. 2008. Spatial attention modulates initial afferent activity in human primary visual cortex. *Cereb. Cortex* **18(11)**:2629–2636. [16]

Kelso, J. A. S. 1995. Dynamic Patterns: The Self-Organization of Brain and Behavior. Cambridge, MA: MIT Press. [1, 14, 18]

Kent, J. P. 1987. Experiments on the relationship between the hen and chick (*Gallus gallus*): The role of the auditory mode in recognition and the effects of maternal separation. *Behaviour* **102**:1–14. [5]

———. 1989. On the acoustic basis of recognition of the mother hen by the chick in the domestic fowl (*Gallus gallus*). *Behaviour* **108**:1–9. [5]

Kerns, J., J. Cohen, V. Stenger, and C. Carter. 2004. Prefrontal cortex guides context-appropriate responding during language production. *Neuron* **43**:283–291. [1]

Kersten, D., P. Mamassian, and A. Yuille. 2004. Object perception as Bayesian inference. *Ann. Rev. Psychol.* **55**:271–304. [15]

Kersten, D., and A. Yuille. 2003. Bayesian models of object perception. *Curr. Opin. Neurobiol.* **13(2)**:150–158. [15]

Kilgard, M. P., and M. M. Merzenich. 1988. Cortical map reorganization enabled by nucleus basalis activity. *Science* **279**:1714–1718. [12]

Kim, U., and F. F. Ebner. 1999. Barrels and septa: Separate circuits in rat barrel field cortex. *J. Comp. Neurol.* **408**:489–505. [2]

Kimura, F., and R. W. Baughman. 1997. GABAergic transcallosal neurons in developing rat neocortex. *Eur. J. Neurosci.* **9**:1137–1143. [13]

King, R. 2000. Brunelleschi's Dome: How a Renaissance Genius Reinvented Architecture. New York: Penguin Books. [14]

Kiorpes, L. 2006. Visual processing in amblyopia: Animal studies. *Strabismus* **14**:3–10. [17]

Kjelstrup, K. B., T. Solstad, V. H. Brun, et al. 2008. Finite scale of spatial representation in the hippocampus. *Science* **321**:140–143. [7]

Klausberger, T., and P. Somogyi. 2008. Neuronal diversity and temporal dynamics: The unity of hippocampal circuit operations. *Science* **321**:53–57. [7]

Klimesch, W., M. Doppelmayr, and S. Hanslmayr. 2006. Upper alpha ERD and absolute power: Their meaning for memory performance. *Prog. Brain Res.* **159**:151–165. [18]

Klimesch, W., M. Doppelmayr, T. Pachinger, and B. Ripper. 1997. Brain oscillations and human memory: EEG correlates in the upper alpha and theta band. *Neurosci. Lett.* **238(1–2)**:9–12. [16]

Klimesch, W., M. Doppelmayr, H. Schimke, and B. Ripper. 1997. Theta synchronization and alpha desynchronization in a memory task. *Psychophysiol.* **34**:169–176. [16]

Klimesch, W., M. Doppelmayr, J. Schwaiger, P. Auinger, and T. Winkler. 1999. "Paradoxical" alpha synchronization in a memory task. *Cogn. Brain Res.* **7(4)**:493–501. [16]

Klimesch, W., P. Sauseng, and S. Hanslmayr. 2007. EEG alpha oscillations: The inhibition-timing hypothesis. *Brain Res. Rev.* **53(1)**:63–88. [16]

Knight, R. A., and S. M. Silverstein. 1998. The role of cognitive psychology in guiding research on cognitive deficits in schizophrenia. In: Origins and Development of Schizophrenia, ed. M. Lenzenweger and R. H. Dworkin, pp. 247–295. Advances in Experimental Psychopathology. Washington, D.C.: APA Press. [17]

Knyazev, G. G. 2007. Motivation, emotion, and their inhibitory control mirrored in brain oscillations. *Neurosci. Biobehav. Rev.* **31**:377–395. [18]

Koch, C., M. Rapp, and I. Segev. 1996. A brief history of time (constants). *Cereb. Cortex* **6**:93–101. [16]

Koch, S. P., S. Koendgen, R. Bourayou, J. Steinbrink, and H. Obrig. 2008. Individual alpha-frequency correlates with amplitude of visual evoked potential and hemodynamic response. *NeuroImage* **41(2)**:233–242. [16]

Koenig, T., L. Prichep, T. Dierks, et al. 2005. Decreased EEG synchronization in Alzheimer's disease and mild cognitive impairment. *Neurobiol. Aging* **26**:165–171. [11]

Koerding, K. P., U. Beierholm, W. J. Ma, et al. 2007. Causal inference in multisensory perception. *PLoS ONE* **2(9)**:e943. [15]

Koffka, K. 1935. Principles of Gestalt Psychology. London: Routledge & Kegan Paul. [10, 12, 14]

Köhler, W. 1925. The Mentality of Apes. New York: Harcourt Brace. [10]

———. 1947. Gestalt Psychology: An Introduction to New Concepts in Modern Psychology. New York: Liveright Publishing Corp. [12]

Komatsu, H. 2006. The neural mechanism of perceptual filling. *Nat. Rev. Neurosci.* **6**:220–231. [14]

König, P., A. K. Engel, S. Löwel, and W. Singer. 1993. Squint affects synchronization of oscillatory responses in cat visual cortex. *Eur. J. Neurosci.* **5**:501–508. [18]

Kopell, N. J., G. B. Ermentrout, M. A. Whittington, and R. D. Traub. 2000. Gamma rhythms and beta rhythms have different synchronization properties. *PNAS* **97(4)**:1867–1872. [8, 11, 16]

Körding, K. P., and P. König. 2000. Learning with two sites of synaptic integration. *Netw. Comp. Neural Sys.* **11**:1–15. [1]

Körding, K. P., and D. M. Wolpert. 2004. Bayesian integration in sensorimotor learning. *Nature* **427**:244–247. [1]

Kotter, R. 2004. Online retrieval, processing, and visualization of primate connectivity data from the CoCoMac database. *Neuroinformatics* **2**:127–144. [13]

Kouider, S., S. Dehaene, A. Jobert, and D. Le Bihan. 2007. Cerebral bases of subliminal and supraliminal priming during reading. *Cereb. Cortex* **17(9)**:2019–2029. [16]

Kourtzi, Z., A. S. Tolias, C. F. Altmann, M. Augath, and N. K. Logothetis. 2003. Integration of local features into global shapes: Monkey and human fMRI studies *Neuron* **37**:333–346. [17]

Kovács, I. 1996. Gestalten of today: Early processing of visual contours and surfaces. *Behav. Brain Sci.* **82(1)**:1–11. [1]

———. 2000. Human development of perceptual organization. *Vision Res.* **40**:1301–1310. [1, 18]

———. 2004. Visual Integration: Development and Impairments. Budapest: Akademiai. [18]

Kovács, I., A. Fehér, and B. Julesz. 1998. Medial-point description of shape: A representation for action coding and its psychophysical correlates. *Vision Res.* **38(15–16)**:2429–2454. [14]

Kovács, I., and B. Julesz. 1993. A closed curve is much more than an incomplete one: Effect of closure in figure-ground segmentation. *PNAS* **90(16)**:7495–7497. [14, 18]

———. 1994. Perceptual sensitivity maps within globally defined visual shapes. *Nature* **70**:644–646. [14]

Kovács, I., P. Kozma, Á. Fehér, and G. Benedek. 1999. Late maturation of visual spatial integration in humans. *PNAS* **96**:12204–12209. [18]

Kovács, I., T. V. Papathomas, M. Yang, and Á. Fehér. 1996. When the brain changes its mind: Interocular grouping during binocular rivalry. *PNAS* **93**:15508–15511. [14]

Kovács, I., U. Polat, P. M. Pennefather, A. Chandna, and A. M. Norcia. 2000. A new test of contour integration deficits in patients with a history of disrupted binocular experience during visual development. *Vision Res.* **40**:1775–1783. [17, 18]

Kozlov, A., M. Huss, A. Lansner, J. Hellgren Kotaleski, and S. Grillner. 2009. Simple cellular and network control principles govern complex patterns of motor behavior. *PNAS* **106(47)**:20027–20032. [6]

Kramer, M. A., A. K. Roopun, L. M. Carracedo, et al. 2008. Rhythm generation through period concatenation in rat somatosensory cortex. *PLoS Comput. Biol.* **4(9)**:e1000169. [8]

Kreiter, A. K., and W. Singer. 1996. Stimulus-dependent synchronization of neuronal responses in the visual cortex of the awake macaque monkey. *J. Neurosci.* **16**:2381–2396. [11, 12, 18]

Krishnan, G. P., J. L. Vohs, W. P. Hetrick, et al. 2005. Steady state visual evoked potential abnormalities in schizophrenia. *Clin. Neurophys.* **116**:614–624. [18]

Krubitzer, L. A., and J. H. Kaas. 1990. Cortical connections of MT in four species of primates: Areal, modular, and retinotopic patterns. *Vis. Neurosci.* **5**:165–204. [2]

Kruger, N., and F. Wörgötter. 2002. Multi-modal estimation of collinearity and parallelism in natural image sequences. *Netw. Comp. Neural Sys.* **13**:553–576. [1]

Krupa, M., N. Popovic, N. J. Kopell, and R. H. G. 2008. Mixed-mode oscillations in a three timescale model for the dopaminergic neuron. *Chaos* **18**:015106. [8]

Kubicki, M., R. McCarley, C. F. Westin, et al. 2007. A review of diffusion tensor imaging studies in schizophrenia. *J. Psychiatry Res.* **41**:15–30. [18]

Kuffler, S. W. 1953. Discharge patterns and functional organization of mammalian retina. *J. Neurophysiol.* **16**:37–68. [13]

Kwon, J. S., B. F. O'Donnell, G. V. Wallenstein, et al. 1999. Gamma frequency-range abnormalities to auditory stimulation in schizophrenia. *Arch. Gen. Psychiatry* **56**:1001–1005. [18]

Lachaux, J. P., N. George, C. Tallon-Baudry, et al. 2005. The many faces of the gamma band response to complex visual stimuli. *NeuroImage* **25(2)**:491–501. [16]

Lachaux, J. P., E. Rodriguez, J. Martinerie, et al. 2000. A quantitative study of gamma-band activity in human intracranial recordings triggered by visual stimuli. *Eur. J. Neurosci.* **12(7)**:2608–2622. [16]

Lachaux, J. P., E. Rodriguez, J. Martinerie, and F. Varela. 1999. Measuring phase synchrony in brain signals. *Hum. Brain Mapp.* **8(4)**:194–208. [16]

Lakatos, P., A. S. Shah, K. H. Knuth, I. Ulbert, and G. Karmos. 2005. An oscillatory hierarchy controlling neuronal excitability and stimulus processing in the auditory cortex. *J. Neurophysiol.* **94**:1904–1911. [8]

Lamme, V. A. F. 2004. Beyond the classical receptive field: Contextual modulation of V1 response. In: The Visual Neurosciences, ed. L. M. Chalupa and J. S. Werner, pp. 720–732. Cambidge, MA: MIT Press. [1]

Lamme, V. A. F., and P. R. Roelfsema. 2000. The distinct modes of vision offered by feedforward and recurrent processing. *Trends Neurosci.* **23(11)**:571–579. [1, 16]

Lamme, V. A. F., and H. Spekreijse. 2000. Contextual modulation in primary visual cortex and scene perception. In: The New Cognitive Neurosciences, ed. M. S. Gazzaniga. Cambridge, MA: MIT Press. [1]

Landy, M. S., L. T. Maloney, E. B. Johnston, and M. P. Young. 1995. Measurement and modeling of depth cue combination: In defense of weak fusion. *Vision Res.* **35(3)**:389–412. [15]

Lansner, A., and E. Fransén. 1992. Modeling Hebbian cell assemblies comprised of cortical neurons. *Netw. Comp. Neural Sys.* **3**:105–119. [6]

Lashley, K. S. 1931. Mass action in cerebral function. *Science* **73(1888)**:245–254. [16]

———. 1951. The problem of serial order in behavior. In: Cerebral Mechanisms in Behavior, ed. L. A. Jeffress, pp. 112–131. New York: Wiley. [5]

Laudet, V., C. Hänni, D. Stéhelin, and M. Duterque-Coquillaud. 1999. Molecular phylogeny of the ETS gene family. *Oncogene* **18**:1351–1359. [5]

Laurent, G. 1996. Dynamical representation of odors by oscillating and evolving neural assemblies. *Trends Neurosci.* **19**:489–496. [3]

———. 1999. Dendritic processing in invertebrates: A link to function. In: Dendrites, ed. G. Stuart, N. Spruston and M. Hausser, pp. 290–309. Oxford: Oxford Univ. Press. [5]

———. 2002. Olfactory network dynamics and the coding of multidimensional signals. *Nat. Rev. Neurosci.* **3**:884–895. [3, 5]

Laurent, G., and H. Davidowitz. 1994. Encoding of olfactory information with oscillating neural assemblies. *Science* **265**:1872–1875. [5]

Lee, S. H., J. K. Wynn, M. F. Green, et al. 2006. Quantitative EEG and low resolution electromagnetic tomography (LORETA) imaging of patients with persistent auditory hallucinations. *Schizophr. Res.* **83**:111–119. [18]

Lee, T. S. 2003. Computations in the early visual cortex. *J. Physiol.* **97(2–3)**:121–139. [14]

Lee, T. S., D. Mumford, R. Romero, and V. A. F. Lamme. 1998. The role of primary visual cortex in object representation. *Vision Res.* **38(15–16)**:2429–2454. [14]

Lefèbvre, A. A., C. Cellard, S. Tremblay, et al. 2009. The process of binding in schizophrenia: Examination of associations with temporal and spatial context in working memory. *Schizophr. Bull.* **35(1)**:282. [17]

Lehmann, D., W. K. Strik, B. Henggeler, T. Koenig, and M. Koukkou. 1998. Brain electric microstates and momentary conscious mind states as building blocks of spontaneous thinking. 1. Visual imagery and abstract thoughts. *Intl. J. Psychophysiol.* **29**:1–11. [6]

Lennie, P. 2003. The cost of cortical computation. *Curr. Biol.* **13**:493–497. [6]

Leopold, D. A., and N. K. Logothetis. 1996. Activity changes in early visual cortex reflect monkeys' percepts during binocular rivalry. *Nature* **379**:549–553. [14]

Levitt, J. B., D. A. Lewis, T. Yoshioka, and J. S. Lund. 1993. Topography of pyramidal neuron intrinsic connections in macaque monkey prefrontal cortex (areas 9 and 46). *J. Comp. Neurol.* **338**:360–376. [2]

Lewis, D. A., T. Hashimoto, and D. W. Volk. 2005. Cortical inhibitory neurons and schizophrenia. *Nat. Rev. Neurosci.* **6**:312–324. [1, 18]

Li, W., V. Piech, and C. D. Gilbert. 2004. Perceptual learning and top-down influences in primary visual cortex. *Nat. Neurosci.* **7(6)**:651–657. [16]

Li, Z. 2005. The primary visual cortex creates a bottom-up saliency map. In: Neurobiology of Attention, ed. L. Itti, G. Rees and J. K. Tsotsos, pp. 570–575. Amsterdam: Elsevier. [14]

Limousin, P., P. Pollak, A. Benazzouz, et al. 1995. Effect on parkinsonian signs and symptoms of bilateral subthalamic nucleus stimulation. *Lancet* **345**:91–95. [18]

Linkenkaer-Hansen, K., V. V. Nikulin, S. Palva, R. J. Ilmoniemi, and J. M. Palva. 2004. Prestimulus oscillations enhance psychophysical performance in humans. *J. Neurosci.* **10**:10186–10190. [18]

Lisman, J. E., and G. Buzsáki. 2008. A neural coding scheme formed by the combined function of gamma and theta oscillations. *Schizophr. Bull.* **34**:974–980. [6]

Lisman, J. E., J. T. Coyle, R. W. Green, et al. 2008. Circuit-based framework for understanding neurotransmitter and risk gene interactions in schizophrenia. *Trends Neurosci.* **31**:234–242. [17]

Lisman, J. E., and M. A. P. Idiart. 1995. Storage of 7 ± 2 short-term memories in oscillatory subcycles. *Science* **267**:1512–1515. [7, 8, 16]

Liu, D., J. Diorio, J. C. Day, D. D. Francis, and M. J. Meaney. 2000. Maternal care, hippocampal synaptogenesis and cognitive development in rats. *Nat. Neurosci.* **3**:799–806. [17]

Llinás, R. 1991. The noncontinuous nature of movement execution. In: Motor Control: Concepts and Issues, ed. D. R. Humphrey and H.-J. Freund, pp. 223–242. New York: Wiley. [3]

Loffler, G. 2008. Perception of contours and shapes: Low and intermediate stage mechanisms. *Vision Res.* **48**:2106–2127. [14]

Logothetis, N. K., J. Pauls, H. H. Bulthoff, and T. Poggio. 1994. View-dependent object recognition by monkeys. *Curr. Biol.* **4**:401–414. [11]

Long, K., G. Kennedy, and E. Balaban. 2001. Transferring an inborn auditory perceptual preference with interspecies brain transplants. *PNAS* **98**:5862–5867. [5]

Löwel, S., and W. Singer. 1992. Selection of intrinsic horizontal connections in the visual cortex by correlated neuronal activity. *Science* **255**:209–212. [11, 18]

Lubenov, E. V., and A. G. Siapas. 2009. Hippocampal theta oscillations are travelling waves. *Nature* **459**:534–539. [3, 13]

Luck, S. J., and E. K. Vogel. 1999. The capacity of visual working memory for features and conjunctions. *Nature* **390**:279–281. [1]

Lücke, J., C. Keck, and C. von der Malsburg. 2008. Rapid convergence to feature layer correspondence. *Neural Comput.* **20**:2441–2463. [1]

Lund, J. S., T. Yoshioka, and J. B. Levitt. 1993. Comparison of intrinsic connectivity in different areas of macaque monkey cerebral cortex. *Cereb. Cortex* **3**:148–162. [2]

Lundqvist, M., M. Rehn, M. Djurfeldt, and A. Lansner. 2006. Attractor dynamics in a modular network model of the neocortex. *Netw. Comp. Neural Sys.* **17**:253–276. [6]

Ma, W. J., J. M. Beck, P. E. Latham, and A. Pouget. 2006. Bayesian inference with probabilistic population codes. *Nat. Neurosci.* **9(11)**:1432–1438. [15]

Maass, W., T. Natschlaeger, and H. Markram. 2002. Real-time computing without stable states: A new framework for neural computation based on perturbations. *Neural Comput.* **14**:2531–2560. [3, 6]

Macagno, E. R., V. LoPresti, and C. Levinthal. 1973. Structure and development of neuronal connections in isogenic organisms: Variations and similarities in the optic system of Daphnia magna. *PNAS* **70**:57. [5]

Mainy, N., P. Kahane, L. Minotti, et al. 2007. Neural correlates of consolidation in working memory. *Hum. Brain Mapp.* **28(3)**:183–193. [16]

Makeig, S., M. Westerfield, T. P. Jung, et al. 2002. Dynamic brain sources of visual evoked responses. *Science* **295(5555)**:690–694. [16]

Maldonado, P., C. Babul, W. Singer, et al. 2008. Synchronization of neuronal responses in primary visual cortex of monkeys viewing natural images. *J. Neurophysiol.* **100**:1523–1532. [11]

Mantini, D., M. G. Perrucci, C. Del Gratta, G. L. Romani, and M. Corbetta. 2007. Electrophysiological signatures of resting state networks in the human brain. *PNAS* **104(32)**:13170–13175. [16]

Marder, E., and D. Bucher. 2007. Understanding circuit dynamics using the stomatogastric ganglion of lobsters and crabs. *Ann. Rev. Physiol.* **69**:291–316. [5]

Marino, L. 2007. Cetacean brain evolution. In: Evolution of Nervous Systems, ed. J. H. Kaas and L. A. Krubitzer, pp. 261–266. London: Elsevier. [2]

Markram, H. 2006. The blue brain project. *Nature Rev. Neurosci.* **7**:153–160. [12]

Markram, H., J. Lubke, M. Frotscher, and B. Sakmann. 1997. Regulation of synaptic efficacy by coincidence of postsynaptic APs and EPSPs. *Science* **275(5297)**:213–215. [3, 11, 16, 18]

Markram, H., and M. Tsodyks. 1996. Redistribution of synaptic efficacy between neocortical pyramidal neurons. *Nature* **382**:807–810. [11]

Marre, O., P. Baudot, M. Levy, and Y. Frégnac. 2005. High timing precision and reliability, low redundancy and low entropy code in V1 neurons during visual processing of natural scenes. *Soc. Neurosci. Abstr.* **285**:5. [12]

Marsden, J. F., P. Limousin-Dowsey, P. Ashby, P. Pollak, and P. Brown. 2001. Subthalamic nucleus, sensorimotor cortex and muscle interrelationships in Parkinson's disease. *Brain* **124**:378–388. [18]

Martens, M. A., S. J. Wilson, and D. C. Reutens. 2008. Williams syndrome: A critical review of the cognitive, behavioral, and neuroanatomical phenotype. *J. Child Psychol. Psychiatry* **49**:576–608. [17]

Martin-Ordas, G., J. Call, and F. Colmenares. 2008. Tubes, tables and traps: Great apes solve two functionally equivalent trap tasks but show no evidence of transfer across tasks. *Anim. Cogn.* **11**:423–430. [4, 5]

Massimini, M., R. Huber, F. Ferrarelli, S. Hill, and G. Tononi. 2004. The sleep slow oscillation as a traveling wave. *J. Neurosci.* **24**:6862–6870. [7]

Mathes, B., and M. Fahle. 2007. Closure facilitates contour integration. *Vision Res.* **47**:818–827. [14]

Mathewson, K. E., G. Gratton, M. Fabiani, D. M. Beck, and T. Ro. 2009. To see or not to see: Prestimulus alpha phase predicts visual awareness. *J. Neurosci.* **29**:2725–2732. [13]

Matsumoto, A., Y. Ichikawa, N. Kanayama, H. Ohira, and T. Iidaka. 2006. Gamma band activity and its synchronization reflect the dysfunctional emotional processing in alexithymic persons. *Psychophysiol.* **43**:533–540. [17]

Maunsell, J. H. R. 2004. The role of attention in visual cerebral cortex. In: The Visual Neurosciences, ed. L. M. Chalupa and J. S. Werner, pp. 1538–1545. Cambridge MA: MIT Press. [1]

Maunsell, J. H. R., and S. Treue. 2006. Feature-based attention in visual cortex. *Trends Neurosci.* **29**:317–322. [13]

Maurer, A. P., S. R. Vanrhoads, G. R. Sutherland, P. Lipa, and B. L. McNaughton. 2005. Self-motion and the origin of differential spatial scaling along the septo-temporal axis of the hippocampus. *Hippocampus* **15**:841–852. [7]

Mazaheri, A., and O. Jensen. 2006. Posterior alpha activity is not phase-reset by visual stimuli. *PNAS* **103(8)**:2948–2952. [17]

———. 2008. Asymmetric amplitude modulations of brain oscillations generate slow evoked responses. *J. Neurosci.* **28(31)**:7781–7787. [16]

Mazor, O., and G. Laurent. 2005. Transient dynamics versus fixed points in odor representations by locust antennal lobe projection neurons. *Neuron* **48**:661–673. [3]

McCormick, D. A. 1992. Cellular mechanisms underlying cholinergic and noradrenergic modulation of neuronal firing mode in the cat and guinea pig dorsal lateral geniculate nucleus. *J. Neurosci.* **12**:278–289. [12]

McGowan, P. O., M. J. Meaney, and M. Szyf. 2008. Diet and the epigenetic (re) programming of phenotypic differences in behavior. *Brain Res.* **27**:12–24. [17]

McGowan, P. O., A. Sasaki, A. C. D'Alessio, et al. 2009. Epigenetic regulation of the glucocorticoid receptor in human brain associates with childhood abuse. *Nat. Neurosci.* **12**:342–348. [17]

Medendorp, W. P., G. F. Kramer, O. Jensen, et al. 2007. Oscillatory activity in human parietal and occipital cortex shows hemispheric lateralization and memory effects in a delayed double-step saccade task. *Cereb. Cortex* **17(10)**:2364–2374. [16]

Mehta, M. R., M. Quirk, and M. Wilson. 2000. Experience-dependent asymmetric shape of hippocampal receptive fields. *Neuron* **25**:707–715. [3]

Melloni, L., C. Molina, M. Pena, et al. 2007. Synchronization of neural activity across cortical areas correlates with conscious perception. *J. Neurosci.* **27(11)**:2858–2865. [1, 11, 16]

Meltzer, J. A., H. P. Zaveri, I. I. Goncharova, et al. 2008. Effects of working memory load on oscillatory power in human intracranial EEG. *Cereb. Cortex* **18(8)**:1843–1855. [16]

Mendola, J. D., I. P. Conner, A. Roy, et al. 2005. Voxel-based analysis of MRI detects abnormal visual cortex in children and adults with amblyopia. *Hum. Brain Mapp.* **25**:222–236. [17]

Merriam-Webster. 2009. http://www.merriam-webster.com/dictionary/coordination (accessed Feb. 23, 2010). [15]

Messer, K., J. Kittler, M. Sadeghi, et al. 2004. Face authentication test on the BANCA database. Proc. 17th Intl. Conf. on Pattern Recognition. pp. 523–532. IEEE Xplore. [13]

Meyrand, P., J. Simmers, and M. Moulins. 1991. Construction of a pattern-generating circuit with neurons of different networks. *Nature* **351**:60–62. [12, 13]

Miller, E. K. 2000. The prefrontal cortex and cognitive control. *Nat. Rev. Neurosci.* **1**:59–65. [18]

Miller, E. K., and J. D. Cohen. 2001. An integrative theory of prefrontal cortex function. *Ann. Rev. Neurosci.* **24**:167–202. [18]

Milner, P. M. 1974. A model for visual shape recognition. *Psychol. Rev.* **81**:521–535. [12]

Mima, T., T. Oluwatimilehin, T. Hiraoka, and M. Hallett. 2001. Transient interhemispheric neuronal synchrony correlates with object recognition. *J. Neurosci.* **21(11)**:3942–3948. [16, 18]

Minsky, M. 1988. The Society of Mind. London: Simon and Schuster. [10]

Mirnics, K., F. A. Middleton, D. A. Lewis, and P. Levitt. 2001. Analysis of complex brain disorders with gene expression microarrays: Schizophrenia as a disease of the synapse. *Trends Neurosci.* **24**:479–486. [17]

Mitchell, L. G., J. A. Mutchmor, and W. D. Dolphin. 1988. Zoology. Redwood City, CA: Addison-Wesley. [5]

Mitchison, G., and F. Crick. 1982. Long axons within the striate cortex: Their distribution, orientation, and patterns of connection. *PNAS* **79**:3661–3665. [12]

Mizuseki, K., A. Sirota, E. Pastalkova, and G. Buzsáki. 2009. Theta oscillations provide temporal windows for local circuit computation in the entorhinal-hippocampal loop. *Neuron* **64**:267–280. [7]

Mobbs, D., M. A. Eckert, V. Menon, et al. 2007. Reduced parietal and visual cortical activation during global processing in Williams syndrome. *Devel. Med. Child Neurol.* **49**:433–438. [17]

Molle, M., L. Marshall, S. Gais, and J. Born. 2002. Grouping of spindle activity during slow oscillations in human non-rapid eye movement sleep. *J. Neurosci.* **22**:10941–10947. [7]

Molnar, G., S. Olah, G. Komlosi, et al. 2008. Complex events intiated by individual spikes in the human cerebral cortex. *PLoS Biol.* **6**:e222. [8]

Monier, C., F. Chavane, P. Baudot, L. J. Graham, and Y. Frégnac. 2003. Orientation and direction selectivity of excitatory and inhibitory inputs in visual cortical neurons: A diversity of combinations produces spike tuning. *Neuron* **37**:663–680. [12]

Montgomery, S. M., M. I. Betancur, and G. Buzsáki. 2009. Behavior-dependent coordination of multiple theta dipoles in the hippocampus. *J. Neurosci.* **29**:1381–1394. [7]

Monto, S., S. Palva, J. Voipio, and J. M. Palva. 2008. Very slow EEG fluctuations predict the dynamics of stimulus detection and oscillation amplitudes in humans. *J. Neurosci.* **28(33)**:8268–8272. [14, 16]

Moore, C. I., and S. B. Nelson. 1998. Spatio-temporal subthreshold receptive fields in the vibrissa representation of rat primary somatosensory cortex. *J. Neurophysiol.* **80**:2882–2892. [12]

Moore, T., and K. M. Armstrong. 2003. Selective gating of visual signals by microstimulation of frontal cortex. *Nature* **421**:370–373. [18]

Morrot, G., F. Brochet, and D. Dubourdieu. 2001. The color of odors. *Brain Lang.* **79**:309–320. [12]

Mountcastle, V. B. 1978. An organizing principle for cerebral function: The unit module and distributed function. In: The Mindful Brain, ed. G. Edelman and V. B. Mountcastle, pp. 7–50. Cambridge, MA: MIT Press. [6, 8]

———. 1997. The columnar organization of the neocortex. *Brain* **120**:701–722. [2]

Moyaux, T., B. L. Smith, S. Paurobally, V. Tamma, and M. Wooldridge. 2006. Towards service-oriented ontology-based coordination. Intl. Conf. on Web Services: IEEE. [1]

Mukamel, R., H. Gelbard, A. Arieli, et al. 2005. Coupling between neuronal firing, field potentials, and FMRI in human auditory cortex. *Science* **309(5736)**:951–954. [16]

Müller, M. M., T. Gruber, and A. Keil. 2000. Modulation of induced gamma band activity in the human EEG by attention and visual information processing. *Int. J. Psychophysiol.* **38**:283–299. [18]

Müller, M. M., M. Junghöfer, T. Elbert, and B. Rockstroh. 1997. Visually induced gamma-band responses to coherent and incoherent motion: A replication study. *NeuroReport* **8**:2575–2579. [18]

Mumford, D. 1992. On the computational architecture of the neocortex. *Biol. Cyber.* **66**:241–251. [18]

Mundhenk, T. N., and L. Itti. 2005. Computational modeling and exploration of contour integration for visual saliency. *Biol. Cyber.* **93**:188–212. [14]

Murphy, R. R. 1996. Biological and cognitive foundations of intelligent sensor fusion. *IEE Trans. Systems, Man & Cybernetics* **26(1)**:42–51. [15]

Murthy, M., I. Fiete, and G. Laurent. 2008. Testing odor response stereotypy in the Drosophila mushroom body. *Neuron* **59**:1009–1023. [5]

Nakagawa, T., and L. B. Vosshall. 2009. Controversy and consensus: Noncanonical signaling mechanisms in the insect olfactory system. *Curr. Opin. Neurobiol.* **19**:284–292. [5]

Nardini, M., P. Jones, R. Bedford, and O. Braddick. 2008. Development of cue integration in human navigation. *Curr. Biol.* **18(9)**:689–693. [15]

Nauhaus, I., L. Busse, M. Carandini, and D. L. Ringach. 2009. Stimulus contrast modulates functional connectivity in visual cortex. *Nat. Neurosci.* **12(1)**:70–76. [12]

Naya, Y., K. Sakai, and Y. Miyashita. 1996. Activity of primate inferotemporal neurons related to a sought target in pair-association task. *PNAS* **93**:2664–2669. [18]

Neil, A. P., C. Chee-Ruiter, C. Scheier, J. D. Lewkowicz, and S. Shimojo. 2006. Development of multisensory spatial integration and perception in humans. *Devel. Sci.* **9(5)**:454–464. [15]

Neuenschwander, S., and W. Singer. 1996. Long-range synchronization of oscillatory light responses in the cat retina and lateral geniculate nucleus. *Nature* **379**:728–733. [11]

Niedermeyer, E., and F. L. D. Silva, eds. 2005. Electroencephalography: Basic Principles, Clinical Applications, and Related Fields. New York: Lippincott Williams & Wilkins. [18]

Niessing, J., B. Ebisch, K. E. Schmidt, et al. 2005. Hemodynamic signals correlate tightly with synchronized gamma oscillations. *Science* **309(5736)**:948–951. [16]

Nikulin, V. V., K. Linkenkaer-Hansen, G. Nolte, et al. 2007. A novel mechanism for evoked responses in the human brain. *Eur. J. Neurosci.* **25(10)**:3146–3154. [16]

Nimchinsky, E. A., E. P. E. Gilissen, J. M. Allman, et al. 1999. A neuronal morphologic type unique to humans and great apes. *PNAS* **96**:5268–5273. [2]

Nir, Y., R. Mukamel, I. Dinstein, et al. 2008. Interhemispheric correlations of slow spontaneous neuronal fluctuations revealed in human sensory cortex. *Nat. Neurosci.* **11(9)**:1100–1108. [16]

Nowak, L. G., and J. Bullier. 1997. The timing of information transfer in the visual system. In: Extrastriate Visual Cortex in Primates, ed. J. H. Kaas, K. L. Rockland and A. L. Perters, pp. 205–241. New York: Plenum. [12]

O'Keefe, J. 1993. Hippocampus, theta, and spatial memory. *Curr. Opin. Neurobiol.* **3**:917–924. [3]

O'Keefe, J., and L. Nadel. 1978. The Hippocampus as a Cognitive Map. Oxford: Clarendon Press. [7]

O'Keefe, J., and M. L. Recce. 1993. Phase relationship between hippocampal place units and the EEG theta rhythm. *Hippocampus* **3**:317–330. [7]

Olshausen, B., C. Anderson, and D. C. Van Essen. 1993. A neurobiological model of visual attention and invariant recognition based on dynamic routing of information. *J. Neurosci.* **13**:4700–4719. [13]

Osipova, D., A. Takashima, R. Oostenveld, et al. 2006. Theta and gamma oscillations predict encoding and retrieval of declarative memory. *J. Neurosci.* **26(28)**:7523–7531. [16]

Palva, J. M., S. Palva, and K. Kaila. 2005. Phase synchrony among neuronal oscillations in the human cortex *J. Neurosci.* **25(15)**:3962–3972. [8, 16]

Palva, S., and J. M. Palva. 2007. New vistas for alpha-frequency band oscillations. *Trends Neurosci.* **30(4)**:150–158. [16, 18]

Park, T., and E. Balaban. 1991. Relative salience of species maternal calls in neonatal Gallinaceous birds: A direct comparison of Japanese quail (*Coturnix coturnix japonica*) and domestic chickens (*Gallus gallus domesticus*). *J. Comp. Psychol.* **105**:45–54. [5]

Parnas, J., P. Vianin, D. Saebye, et al. 2001. Visual binding abilities in the initial and advanced stages of schizophrenia. *Acta Psychiatr. Scand.* **103**:171–180. [17]

Pastalkova, E., V. Itskov, A. Amarasingham, and G. Buzsáki. 2008. Internally generated cell assembly sequences in the rat hippocampus. *Science* **321**:1322–1327. [7]

Pearl, J. 1988. Probabilistic Reasoning in Intelligent Systems: Networks of Plausible Inference. San Mateo, CA: Morgan Kaufmann. [15]

Pearson, J. C., D. Lemons, and W. McGinnis. 2005. Modulating Hox gene functions during animal body patterning. *Nat. Rev. Gen.* **6**:893–904. [5]

Pediaditakis, N. 2006. Considering the major mental disorders as clinical expressions of periodic pathological oscillations of the overall operating mode of brain function. *Med. Hypoth.* **67**:395–400. [17]

Penn, D. C., K. J. Holyoak, and D. J. Povinelli. 2008. Darwin's mistake: Explaining the discontinuity between human and nonhuman minds. *Behav. Brain Sci.* **31**:109–178. [1]

Penn, D. C., and D. J. Povinelli. 2007. On the lack of evidence that non-human animals possess anything remotely resembling a theory of mind. *Phil. Trans. R. Soc. Lond. B* **362**:731–744. [4]

Penny, G. R., M. Conley, D. E. Schmechel, and I. T. Diamond. 1984. The distribution of glutamic acid decarboxylase immunoreactivity in the diencephalon of the opossum and rabbit. *J. Comp. Neurol.* **228**:38–56. [2]

Pepperberg, I. M. 2006. Grey parrot numerical competence: A review. *Anim. Cogn.* **9**:377–391. [5]

Perrin, J. S., G. Leonard, M. Perron, et al. 2009. Sex differences in the growth of white matter during adolescence. *NeuroImage* **45**:1055–1066. [18]

Pesaran, B., J. S. Pezaris, M. Sahani, P. P. Mitra, and R. A. Andersen. 2002. Temporal structure in neuronal activity during working memory in macaque parietal cortex. *Nat. Neurosci.* **5**:805–811. [13]

Peters, A., and E. G. Jones. 1984. Cellular components of the cerebral cortex. In: Cerebral Cortex, ed. E. G. Jones and A. Peters. New York: Plenum Press. [2]

Petsche, H., S. Kaplan, A. von Stein, and O. Filz. 1997. The possible meaning of the upper and lower alpha frequency ranges for cognitive and creative tasks. *Intl. J. Psychophysiol.* **26(1–3)**:77–97. [16]

Phillips, P. J., P. J. Flynn, W. T. Scruggs, et al. 2005. Overview of the face recognition grand challenge. IEEE Computer Society Conf. on Computer Vision and Pattern Recognition, at San Diego. pp. 947–954. San Diego: IEEE. [13]

Phillips, W. A., and B. J. Craven. 2000. Interactions between coincident and orthogonal cues to texture boundaries. *Percept. Psychophys.* **62**:1019–1038. [1]

Phillips, W. A., and S. M. Silverstein. 2003. Convergence of biological and psychological perspectives on cognitive coordination in schizophrenia. *Behav. Brain Sci.* **26**:65–82; discussion 82–137. [1, 8, 17, 18]

Phillips, W. A., and W. Singer. 1997. In search of common foundations for cortical computation. *Behav. Brain Sci.* **20**:657–683; discussion 683–722. [1, 14, 17, 18]

Pipa, G., A. Riehle, and S. Grün. 2007. Validation of task-related excess of spike coincidences based on NeuroXidence. *Neurocomputing* **70**:2064–2068. [11]

Pipa, G., D. W. Wheeler, W. Singer, and D. Nikolić. 2008. NeuroXidence: Reliable and efficient analysis of an excess or deficiency of joint-spike events. *J. Comp. Neurosci.* **25**:64–88. [11]

Plenz, D., and T. C. Thiagarajan. 2007. The organizing principles of neuronal avalanches: Cell assemblies in the cortex? *Trends Neurosci.* **30**:101–110. [12]

Poggio, T., and S. Edelman. 1990. A network that learns to recognize three-dimensional objects. *Nature* **343**:263–266. [13]

Poghosyan, V., and A. A. Ioannides. 2008. Attention modulates earliest responses in the primary auditory and visual cortices. *Neuron* **58(5)**:802–813. [16]

Polat, U., and D. Sagi. 1994. The architecture of perceptual spatial interactions. *Vision Res.* **34**:73–78. [18]

Porter, M. A., and M. Coltheart. 2006. Global and local processing in Williams syndrome, autism and Down syndrome: Perception, attention, and construction. *Devel. Neuropsychol.* **30**:771–789. [17]

Poulet, J. F. A., and B. Hedwig. 2002. A corollary discharge maintains auditory sensitivity during sound production. *Nature* **418**:872–876. [5]

———. 2006. The cellular basis of a corollary discharge. *Science* **311**:518–522. [5]

Praamstra, P., and P. Pope. 2007. Slow brain potential and oscillatory EEG manifestations of impaired temporal preparation in Parkinson's disease. *J. Neurophysiol.* **98(5)**:2848–2857. [16]

Prather, J. F., S. Peters, S. Nowicki, and R. Mooney. 2008. Precise auditory-vocal mirroring in neurons for learned vocal communication. *Nature* **451**:305–310. [5]

Preuss, T. M., and G. Coleman. 2002. Human-specific organization of primary visual cortex: Alternating compartments of dense Cat-301 and calbindin immunoreactivity in layer 4A. *Cereb. Cortex* **12**:671–691. [2]

Preuss, T. M., H.-X. Qi, and J. H. Kaas. 1999. Distinctive compartmental organization of human primary visual cortex. *PNAS* **96**:11601–11606. [2]

Qi, H.-X., N. Jain, T. M. Preuss, and J. H. Kaas. 1999. Inverted pyramidal neurons in chimpanzee sensorimotor cortex are revealed by immunostaining with monoclonal antibody SMI-32. *Somato Motor Res.* **16**:49–59. [2]

Quiroga, R. Q., L. Reddy, G. Kreiman, C. Koch, and I. Fried. 2005. Invariant visual representation by single neurons in the human brain. *Nature* **435(7045)**:1102–1107. [11]

Rabiner, L., and B. Juang. 1986. An introduction to hidden Markov models. *IEEE ASSP Mag.* **3**:4–16. [10]

Rabinovich, M., R. Huerta, and G. Laurent. 2008. Transient dynamics for neural processing. *Science* **321**:48–50. [3]

Rakic, P. 2008. Confusing cortical columns. *PNAS* **105**:12099–12100. [1, 2]

Ramon y Cajal, S. 1911. Histologie du Systeme Nerveux de l'Homme et des Vertebres, vol. 2. Paris: Maloine. [5]

Rapoport, J., A. Chavez, D. Greenstein, A. Addington, and N. Gogtay. 2009. Autism spectrum disorders and childhood-onset schizophrenia: Clinical and biological contributions to a relation revisited. *J. Am. Acad. Child Adolesc. Psychiatry* **48**:10–18. [17]

Read, J., J. v. Os, A. P. Morrison, and C. A. Ross. 2005. Childhood trauma, psychosis and schizophrenia: A literature review with theoretical and clinical implications. *Acta Psychiatr. Scand.* **112**:330–350. [17]

Reiner, A., D. J. Perkel, L. L. Bruce, et al. 2004. Revised nomenclature for avian telencephalon and some related brainstem nuclei. *J. Comp. Neurol.* **473**:377–414. [4]

Reiss, A. L., S. Eliez, J. E. Schmitt, et al. 2000. IV. Neuroanatomy of Williams syndrome: A high-resolution MRI study. *J. Cogn. Neurosci.* **12(1)**:65–73. [18]

Ress, D., and D. J. Heeger. 2003. Neuronal correlates of perception in early visual cortex. *Nat. Neurosci.* **6(4)**:414–420. [16]

Reynolds, C. W. 1987. Folcks, Herds, and Schools: A distributed behavioral model. *Comp. Graph.* **21(4)**:25–34. [16]

Reynolds, J. H., and R. Desimone. 1999. The role of neural mechanisms of attention in solving the binding problem. *Neuron* **24**:19–29. [18]

Reynolds, J. H., and D. J. Heeger. 2009. The normalization model of attention. *Neuron* **61**:1689–1685. [1, 13]

Rind, F. C. 1984. A chemical synapse between two motion detecting neurones in the locust brain. *J. Exp. Biol.* **110**:143–167. [5]

Risner, M. L., C. J. Aura, J. E. Black, and T. J. Gawne. 2009. The visual evoked potential is independent of surface alpha rhythm phase. *NeuroImage* **45(2)**:463–469. [16]

Ritz, R., W. Gerstner, U. Fuentes, and J. v. Hemmen. 1994. A biologically motivated and analytically soluble model of collective oscillations in the cortex. II. Application to binding and pattern segmentation. *Biol. Cyber.* **71**:349–358. [6]

Rizzolatti, G., L. Riggio, I. Dascola, and C. Umiltá. 1987. Reorienting attention across the horizontal and vertical meridians: Evidence in favor of a premotor theory of attention. *Neuropsychologia* **25**:31–40. [18]

Robbe, D., S. M. Montgomery, A. Thome, et al. 2006. Cannabinoids reveal importance of spike timing coordination in hippocampal function. *Nat. Neurosci.* **9**:1526–1533. [13]

Roberts, A. C., and D. L. Glanzmann. 2003. Learning in *Aplysia*: Looking at synaptic plasticity from both sides. *Trends Neurosci.* **26**:662–670. [5]

Rockel, A. J., R. W. Hiorns, and T. P. S. Powell. 1980. The basic uniformity in structure of the neocortex. *Brain* **103**:221–244. [2]

Rockland, K., and N. Ichinohe. 2004. Some thoughs on cortical minicolumns. *Exp. Brain Res.* **158**:265–277. [6]

Rockland, K., and J. S. Lund. 1982. Widespread periodic intrinsic connections in tree shrew visual cortex (area 17). *Science* **215**:1532–1534. [2]

Roebroeck, A., E. Formisano, and R. Goebel. 2005. Mapping directed influence over the brain using Granger causality and fMRI. *NeuroImage* **25**:230–242. [13]

Roelfsema, P. R., A. K. Engel, P. König, and W. Singer. 1997. Visuomotor integration is associated with zero time-lag synchronization among cortical areas. *Nature* **385**:157–161. [11, 13, 18]

Roelfsema, P. R., P. König, A. K. Engel, R. Sireteanu, and W. Singer. 1994. Reduced synchronization in the visual cortex of cats with strabismic amblyopia. *Eur. J. Neurosci.* **6**:1645–1655. [17, 18]

Roland, P. E. 2002. Dynamic depolarization fields in the cerebral cortex. *Trends Neurosci.* **25(4)**:183–190. [12]

Rolls, E. T. 2006. Consciousness absent and present: A neurophysiological exploration of masking. In: The First Half Second: The Microgenesis and Temporal Dynamics of Unconscious and Conscious Visual Processes, ed. H. Ögmen and B. G. Breitmeyer, pp. 89–108. Cambridge, MA: MIT Press. [17]

Roopun, A. K., M. O. Cunningham, C. Racca, et al. 2008. Region-specific changes in gamma and beta2 rhythms in NMDA receptor dysfunction models of schizophrenia. *Schizophr. Bull.* **34**:962–973. [1, 8, 11, 17]

Roopun, A. K., M. A. Kramer, L. M. Carracedo, et al. 2008. Temporal interactions between cortical rhythms. *Front. Neurosci.* **2(2)**:145–154. [8, 11]

Roopun, A. K., S. J. Middleton, M. O. Cunningham, et al. 2006. A beta2-frequency (20–30 Hz) oscillation in nonsynaptic networks of somatosensory cortex. *PNAS* **103(42)**:15646–15650. [8, 16]

Rothkopf, C., T. Weisswange, and J. Triesch. 2010. Computational modeling of multisensory object perception. In: Multisensory Object Perception in the Primate Brain, ed. J. Kaiser and M. J. Naumer. New York: Springer. [15]

Rotstein, H. G., D. D. Pervouchine, C. D. Acker, et al. 2005. Slow and fast inhibition and an H-current interact to create a theta rhythm in a model of CA1 interneuron network. *J. Neurophysiol.* **94**:1509–1518. [8]

Ruff, C. C., F. Blankenburg, O. Bjoertomt, et al. 2006. Concurrent TMS-fMRI and psychophysics reveal frontal influences on human retinotopic visual cortex. *Curr. Biol.* **16**:1479–1488. [13]

Ryan, T. J., R. D. Emes, S. G. N. Grant, and N. H. Komiyama. 2008. Evolution of NMDA receptor cytoplasmic interaction domains: Implications for organisation of synaptic signalling complexes. *BMC Neurosci.* **9**:6. [5]

Ryan, T. J., and S. G. N. Grant. 2009. The origin and evolution of synapses. *Nat. Rev. Neurosci.* **10**:701–712. [5]

Sachs, B. D. 1988. The development of grooming and its expression in adult animals. *Ann. NY Acad. Sci.* **525**:1–17. [5]

Salin, P.-A., and J. Bullier. 1995. Corticocortical connections in the visual system: Structure and function. *Physiol. Rev.* **75**:107–154. [2]

Salinas, E., and T. J. Sejnowski. 2001a. Correlated neuronal activity and the flow of neural information. *Nat. Rev. Neurosci.* **2**:539–550. [18]

———. 2001b. Gain modulation in the central nervous system: Where behavior, neurophysiology, and computation meet. *Neuroscientist* **7**:430–440. [1]

Salzman, C. D., C. M. Murasugi, K. H. Britten, and W. T. Newsome. 1992. Microstimulation in visual area MT: Effects on direction discrimination performance. *J. Neurosci.* **12**:2331–2355. [11]

Samonds, J. M., Z. Zhou, M. R. Bernard, and A. B. Bonds. 2006. Synchronous activity in cat visual cortex encodes collinear and cocircular contours. *J. Neurophysiol.* **95**:2602–2616. [11]

Sandberg, A., A. Lansner, and J. Tegnér. 2003. A working memory model based on fast Hebbian learning. *Netw. Comp. Neural Sys.* **14**:789–802. [6]

Sanes, J. N., and J. P. Donoghue. 1993. Oscillations in local field potentials of the primate motor cortex during voluntary movement. *PNAS* **90**:4470–4474. [18]

Sarter, M., and J. P. Bruno. 2002. The neglected constituent of the basal forebrain corticopetal projection system: GABAergic projections. *Eur. J. Neurosci.* **15**:1867–1873. [13]

Sass, L., and J. Parnas. 2003. Schizophrenia, consciousness, and the self. *Schizophr. Bull.* **29**:427–444. [17]

Sauseng, P., W. Klimesch, W. Gruber, and N. Birbaumer. 2008. Cross-frequency phase synchronization: A brain mechanism of memory matching and attention. *NeuroImage* **40(1)**:308–317. [16]

Sauseng, P., W. Klimesch, W. Gruber, et al. 2002. The interplay between theta and alpha oscillations in the human electroencephalogram reflects the transfer of information between memory systems. *Neurosci. Lett.* **324(2)**:121–124. [16]

Schall, S., C. Quigley, S. Onat, and P. König. 2009. Visual stimulus locking of EEG is modulated by temporal congruency of auditory stimuli. *Exp. Brain Res.* **198**:137–151. [18]

Schenkel, L. S., W. D. Spaulding, D. DiLillo, and S. M. Silverstein. 2005. Histories of childhood maltreatment in schizophrenia: Relationships with premorbid functioning, symptomatology, and cognitive deficits. *Schizophr. Res.* **76**:273–286. [17]

Schillen, T. B., and P. König. 1991. Stimulus-dependent assembly formation of oscillatory responses. II. Desynchronization. *Neural Comput.* **3**:167–178. [13]

Schmidt, K. E., W. Singer, and R. A. W. Galuske. 2004. Processing deficits in primary visual cortex of amblyopic cats. *J. Neurophysiol.* **91**:1661–1671. [18]

Schmolesky, M. T., Y. C. Wang, D. P. Hanes, et al. 1998. Signal timing across the macaque visual system. *J. Neurophysiol.* **79(6)**:3272–3278. [16]

Schneider, T. R., S. Debener, R. Oostenveld, and A. K. Engel. 2008. Enhanced EEG gamma-band activity reflects multisensory semantic matching in visual-to-auditory object priming. *NeuroImage* **42**:1244–1254. [18]

Schoffelen, J. M., R. Oostenveld, and P. Fries. 2005. Neuronal coherence as a mechanism of effective corticospinal interaction. *Science* **308(5718)**:111–113. [11]

Schroeder, C. E., and P. Lakatos. 2009. Low-frequency neuronal oscillations as instruments of sensory selection. *Trends Neurosci.* **32(1)**:9–18. [16]

Schroeder, C. E., P. Lakatos, Y. Kajikawa, S. Partan, and A. Puce. 2008. Neuronal oscillations and visual amplification of speech. *Trends Cogn. Sci.* **12**:106–113. [18]

Schupp, H. T., M. Junghöfer, A. I. Weike, and A. O. Hamm. 2003. Emotional facilitation of sensory processing in the visual cortex. *Psychol. Sci.* **14**:7–13. [18]

Schwartz, B. L., M. L. Hoffman, and S. Evans. 2005. Episodic-like memory in a gorilla: A review and new findings. *Learn. Motiv.* **36**:226–244. [4]

Schwartz, O., A. Hsu., and P. Dayan. 2007. Space and time in visual context. *Nat. Rev. Neurosci.* **8**:522–535. [1]

Sederberg, P. B., M. J. Kahana, M. W. Howard, E. J. Donner, and J. R. Madsen. 2003. Theta and gamma oscillations during encoding predict subsequent recall. *J. Neurosci.* **23(34)**:10809–10814. [16]

Sederberg, P. B., A. Schulze-Bonhage, J. R. Madsen, et al. 2007. Hippocampal and neocortical gamma oscillations predict memory formation in humans. *Cereb. Cortex* **17(5)**:1190–1196. [16]

Seed, A. M., J. Call, N. J. Emery, and N. S. Clayton. 2009. Chimpanzees solve the trap problem when the confound of tool-use is removed. *J. Exp. Psychol. Anim. Behav. Proc.* **35**:23–34. [4, 5]

Seed, A. M., N. J. Emery, and N. S. Clayton. 2009. Intelligence in corvids and apes: A case of convergent evolution? *Ethology* **115**:401–420. [4]

Seed, A. M., S. Tebbich, N. J. Emery, and N. S. Clayton. 2006. Investigating physical cognition in rooks (*Corvus frugilegus*). *Curr. Biol.* **16**:697–701. [4, 5]

Seeley, W. W., D. A. Carlin, J. M. Allman, et al. 2006. Early frontotemporal dementia targets neurons unique to apes and humans. *Ann. Neurol.* **60**:660–667. [2]

Selemon, L. D. 2001. Regionally diverse cortical pathology in schizophrenia: Clues to the etiology of the disease. *Schizophr. Bull.* **27**:349–377. [17]

Seriés, P., J. Lorenceau, and Y. Frégnac. 2003. The "silent" surround of V1 receptive fields: Theory and experiments. *J. Physiol.* **97**:453–474. [1]

Serre, T., L. Wolf, S. Bileschi, M. Riesenhuber, and T. Poggio. 2007. Robust object recognition with cortex-like mechanisms. *IEE Trans. Systems, Man & Cybernetics* **29**:411–426. [13]

Sesma, M. A., V. A. Casagrande, and J. H. Kaas. 1984. Cortical connections of area 17 in tree shrews. *J. Comp. Neurol.* **230**:337–351. [2]

Shah, A. S., S. L. Bressler, K. H. Knuth, et al. 2004. Neural dynamics and the fundamental mechanisms of event-related brain potentials. *Cereb. Cortex* **14(5)**:476–483. [16]

Shams, L., and U. Beierholm. 2009. Brain's strategy for perceptual estimates: Model averaging, model selection, or prob matching? Conference Abstract: Computational and Systems Neuroscience 2009, doi: 10.3389/conf.neuro.06.2009.03.220. Salt Lake City: Frontiers in Systems Neuroscience. [15]

Shannon, C. E. 1948. A mathematical theory of communication. *Bell Sys. Tech. J.* **27**:379–423. [3]

Sherman, S. M., and R. W. Guillery. 1998. On the actions that one nerve cell can have on another: Distinguishing "drivers" from "modulators." *PNAS* **95**:7121–7126. [1]

Sherwood, C. C., and P. R. Hof. 2007. The evolution of neuron types and cortical histology in apes and humans. In: Evolution of Nervous Systems, ed. J. H. Kaas and T. M. Preuss, pp. 355–378. London: Elsevier. [2]

Sherwood, C. C., P. W. H. Lee, C.-B. Rivara, et al. 2003. Evolution of specialized pyramidal neurons in primate visual and motor cortex. *Brain Behav. Evol.* **61**:28–44. [2]

Siapas, A. G., E. V. Lubenov, and M. A. Wilson. 2005. Prefrontal phase locking to hippocampal theta oscillations. *Neuron* **46**:141–151. [7]

Siapas, A. G., and M. A. Wilson. 1998. Coordinated interactions between hippocampal ripples and cortical spindles during slow-wave sleep. *Neuron* **21**:1123–1128. [7]

Siddiqi, K., and S. M. Pizer. 2007. Medial Representations: Mathematics, Algorithms and Applications. New York: Kluwer Academic. [14]

Siegel, M., T. H. Donner, R. Oostenveld, P. Fries, and A. K. Engel. 2007. High-frequency activity in human visual cortex is modulated by visual motion strength. *Cereb. Cortex* **17**:732–741. [18]

———. 2008. Neuronal synchronization along the dorsal visual pathway reflects the focus of spatial attention. *Neuron* **60(4)**:709–719. [16, 18]

Siegel, M., and P. König. 2003. A functional gamma-band defined by stimulus-dependent synchronization in area 18 of awake behaving cats. *J. Neurosci.* **23**:4251–4260. [18]

Siegel, M., M. R. Warden, and E. K. Miller. 2009. Phase-dependent neuronal coding of objects in short-term memory. *PNAS* **106**:21341–21346. [18]

Sik, A., A. Ylinen, M. Penttonen, and G. Buzsáki. 1994. Inhibitory CA1-CA3-hilar region feedback in the hippocampus. *Science* **265**:1722–1724. [13]

Silva, F. J., D. M. Page, and K. M. Silva. 2005. Methodological-conceptual problems in the study of chimpanzees' folk physics: How studies with adult humans can help. *Learn. Behav.* **33**:47–58. [4]
Silverstein, S. M. 1993. Methodological and empirical considerations in assessing the validity of psychoanalytic theories of hypnosis. *Genet. Soc. Gen. Psychol. Monogr.* **19**:5–54. [17]
———. 2008. Measuring specific, rather than generalized, cognitive deficits, and maximizing between-group effect size in studies of cognition and cognitive change. *Schizophr. Bull.* **34**:645–655. [17]
Silverstein, S. M., S. Berten, B. Essex, et al. 2009. An fMRI examination of visual integration in schizophrenia. *J. Integr. Neurosci.* **8**:175–202. [17, 18]
Silverstein, S. M., I. Kovács, R. Corry, and C. Valone. 2000. Perceptual organization, the disorganization syndrome, and context processing in chronic schizophrenia. *Schizophr. Res.* **43**:11–20. [17]
Silverstein, S. M., and D. R. Palumbo. 1995. Nonverbal perceptual organization output disability and schizophrenia spectrum symptomatology. *Psychiatry* **66**:66–81. [18]
Silverstein, S. M., and L. Schenkel. 1997. Schizophrenia as a model of context-deficient cortical computation. *Behav. Brain Sci.* **20**:696–697. [18]
Silverstein, S. M., P. J. Uhlhaas, B. Essex, et al. 2006. Perceptual organization in first episode schizophrenia and ultra-high-risk states. *Schizophr. Res.* **83**:41–52. [17]
Simon, H. A. 1973. The organization of complex systems. In: Hierarchy Theory: The Challenge of Complex Systems, ed. H. H. Pattee, pp. 1–28. New York: George Braziller. [5]
Singer, W. 1999. Neuronal synchrony: A versatile code for the definition of relations? *Neuron* **24(1)**:49–65. [11, 18]
———. 2004. Synchrony, oscillations, and relational codes. In: The Visual Neurosciences, ed. L. M. Chalupa and J. S. Werner, pp. 1665–1681. Cambridge, MA: MIT Press. [1]
Singer, W., and C. M. Gray. 1995. Visual feature integration and the temporal correlation hypothesis. *Ann. Rev. Neurosci.* **18**:555–586. [2, 3, 13, 16, 18]
Singh, M., and M. N. Huhns. 2005. Service-Oriented Computing: Semantics, Processes, Agents. Chicester: Wiley. [1]
Sirota, A., and G. Buzsáki. 2005. Interaction between neocortical and hippocampal networks via slow oscillations. *Thalamus Relat. Syst.* **3**:245–259. [7]
Sirota, A., J. Csicsvari, D. Buhl, and G. Buzsáki. 2003. Communication between neocortex and hippocampus during sleep in rodents. *PNAS* **100**:2065–2069. [7]
Sirota, A., S. Montgomery, S. Fujisawa, et al. 2008. Entrainment of neocortical neurons and gamma oscillations by the hippocampal theta rhythm. *Neuron* **60**:683–697. [6, 7, 16]
Skaggs, W. E., B. L. McNaughton, M. A. Wilson, and C. A. Barnes. 1996. Theta phase precession in hippocampal neuronal populations and the compression of temporal sequences. *Hippocampus* **6**:149–172. [7]
Slewa-Younan, S., E. Gordon, A. W. Harris, et al. 2004. Sex differences in functional connectivity in first-episode and chronic schizophrenia patients. *Am. J. Psychiatry* **161**:1595–1602. [18]
Sloman, A., and R. Chrisley. 2005. More things than are dreamt of in your biology: Information processing in biologically-inspired robots. *Cogn. Sys. Res.* **6**:145–174. [5]
Smith, M. L., F. Gosselin, and P. G. Schyns. 2006. Perceptual moments of conscious visual experience inferred from oscillatory brain activity. *PNAS* **103**:5626–5631. [8]

Smyth, D., W. A. Phillips, and J. Kay. 1996. Measures for investigating the contextual modulation of information transmission. *Netw. Comp. Neural Sys.* **7**:307–316. [1]

Sohal, V. S., F. Zhang, O. Yizhar, and K. Deisseroth. 2009. Parvalbumin neurons and gamma rhythms synergistically enhance cortical circuit performance. *Nature* **459**:698–702. [11]

Sommer, F. T., and T. Wennekers. 2001. Associative memory in networks of spiking neurons. *Neural Netw.* **14**:825–834. [6]

Spelke, E. S. 1998. Nativism, empiricism, and the origins of knowledge. *Infant Beh. Devel.* **21**:181–200. [10]

Spencer, K. M., P. G. Nestor, M. A. Niznikiewicz, et al. 2003. Abnormal neural synchrony in schizophrenia. *J. Neurosci.* **23**:7407–7411. [18]

Spencer, K. M., P. G. Nestor, R. Perlmutter, et al. 2004. Neural synchrony indexes disordered perception and cognition in schizophrenia. *PNAS* **101**:17288–17293. [17, 18]

Spencer, K. M., D. F. Salisbury, M. E. Shenton, and R. W. McCarley. 2008. Gamma-band auditory steady-state responses are impaired in first episode psychosis. *Biol. Psychiatry* **64**:369–375. [18]

Sporns, O., D. R. Chialvo, M. Kaiser, and C. C. Hilgetag. 2004. Organisation, development and function of complex brain networks. *Trends Cogn. Sci.* **8**:418–415. [8]

Spratling, M. W. 2008. Predictive coding as a model of biased competition in visual attention. *Vision Res.* **48**:1391–1408. [1]

Spratling, M. W., and M. H. Johnson. 2006. A feedback model of perceptual learning and categorisation. *Vis. Cogn.* **13**:129–165. [1]

Sprong, M., H. E. Becker, P. F. Schothorst, et al. 2008. Pathways to psychosis: A comparison of the pervasive developmental disorder subtype "multiple complex developmental disorder" and the "at risk mental state." *Schizophr. Res.* **99**:38–47. [17]

Spruston, N., D. B. Jaffe, S. H. Williams, and D. Johnston. 1993. Voltage- and space-clamp errors associated with the measurement of electronically remote synaptic events. *J. Neurophysiol.* **70**:781–802. [12]

Stam, C. J., B. F. Jones, G. Nolte, M. Breakspear, and P. H. Scheltens. 2007. Small-world networks and functional connectivity in Alzheimer's disease. *Cereb. Cortex* **17**:92–99. [11]

Stein, B. E., and M. A. Meredith. 1993. The Merging of the Senses. Cambridge, MA: MIT Press. [15]

Stephan, K. E., K. J. Friston, and C. D. Frith. 2009. Dysconnection in schizophrenia: From abnormal synaptic plasticity to failures of self-monitoring. *Schizophr. Bull.* **35**:509–527. [17]

Steriade, M. 1996. Arousal: Revisiting the reticular activating system. *Science* **272**:225–226. [12]

Steriade, M., A. Nunez, and F. Amzica. 1993a. Intracellular analysis of relations between the slow (< 1 Hz) neocortical oscillation and other sleep rhythms of the electroencephalogram. *J. Neurosci.* **13**:3266–3283. [7]

———. 1993b. A novel slow (< 1 Hz) oscillation of neocortical neurons *in vivo*: Depolarizing and hyperpolarizing components. *J. Neurosci.* **13**:3252–3265. [7]

Steriade, M., and I. Timofeev. 2003. Neuronal plasticity in thalamocortical networks during sleep and waking oscillations. *Neuron* **37**:563–576. [7]

Stickgold, R., D. Whidbee, B. Schirmer, V. Patel, and J. A. Hobson. 2000. Visual discrimination task improvement: A multi-step process occurring during sleep. *J. Cogn. Neurosci.* **12**:246–254. [18]

Stopfer, M., S. Bhagavan, B. H. Smith, and G. Laurent. 1997. Impaired odour discrimination on desycnhronization of odour-encoding neural assemblies. *Nature* **390**:70–74. [11]

Stopfer, M., V. Jayaraman, and G. Laurent. 2003. Intensity versus identity coding in an olfactory system. *Neuron* **39**:991–1004. [3]

Sutton, R. S., and A. G. Barto. 1998. Reinforcement Learning: An Introduction. A Bradford Book. Cambridge, MA: MIT Press. [15]

Swinnen, S. P. 2002. Intermanual coordination: From behavioural principles to neural network interactions. *Nat. Rev. Neurosci.* **3**:350–361. [18]

Symond, M. P., A. Harris, E. Gordon, and L. M. Williams. 2005. "Gamma synchrony" in first-episode schizophrenia: A disorder of temporal connectivity? *Am. J. Psychiatry* **162**:459–465, 1042. [17]

Tallon-Baudry, C. 2004. Attention and awareness in synchrony. *Trends Cogn. Sci.* **8(12)**:523–525. [16]

———. 2009. The roles of gamma-band oscillatory synchrony in human visual cognition. *Front. Biosci.* **14**:321–332. [16]

Tallon-Baudry, C., and O. Bertrand. 1999. Oscillatory gamma activity in humans and its role in object representation. *Trends Cogn. Sci.* **3(4)**:151–162. [12, 18]

Tallon-Baudry, C., O. Bertrand, C. Delpuech, and J. Pernier. 1996. Stimulus specificity of phase-locked and non-phase-locked 40 Hz visual responses in human. *J. Neurosci.* **16**:4240–4249. [18]

Tallon-Baudry, C., O. Bertrand, and C. Fischer. 2001. Oscillatory synchrony between human extrastriate areas during visual short-term memory maintenance. *J. Neurosci.* **21**:171–175. [16]

Tallon-Baudry, C., O. Bertrand, M. A. Henaff, J. Isnard, and C. Fischer. 2005. Attention modulates gamma-band oscillations differently in the human lateral occipital cortex and fusiform gyrus. *Cereb. Cortex* **15**:654–662. [16]

Tallon-Baudry, C., O. Bertrand, F. Peronnet, and J. Pernier. 1998. Induced gamma-band activity during the delay of a visual short-term memory task in humans. *J. Neurosci.* **18(11)**:4244–4254. [16]

Tallon-Baudry, C., S. Mandon, W. A. Freiwald, and A. K. Kreiter. 2004. Oscillatory synchrony in the monkey temporal lobe correlates with performance in a visual short-term memory task. *Cereb. Cortex* **14(7)**:713–720. [16]

Tanji, K., K. Suzuki, A. Delorme, H. Shamoto, and N. Nakasato. 2005. High-frequency gamma-band activity in the basal temporal cortex during picture-naming and lexical-decision tasks. *J. Neurosci.* **25(13)**:3287–3293. [16]

Taylor, A. H., G. R. Hunt, F. S. Medina, and R. D. Gray. 2009. Do New Caledonian crows solve physical problems through causal reasoning? *Proc. R. Soc. Sci. B* **276**:247–254. [4, 5]

Tessmar-Raible, K., and D. Arendt. 2003. Emerging systems: Between vertebrates and arthropods, the Lophotrochozoa. *Curr. Opin. Gen. Devel.* **13**:331–340. [5]

Theunissen, F., and J. P. Miller. 1995. Temporal encoding in nervous systems: A rigorous definition. *J. Comput. Neurosci.* **2**:149–162. [3]

Thompson, E., and F. Varela. 2001. Radical embodiment: Neural dynamics and consciousness. *Trends Cogn. Sci.* **5**:418–425. [5]

Thomson, A. M., and C. Lamy. 2007. Functional maps of neocortical local circuitry. *Front. Neurosci.* **1**:19–42. [2]

Thomson, A. M., D. C. West, Y. Wang, and A. P. Bannister. 2002. Synaptic connections and small circuits involving excitatory and inhibitory neurons in layer 2–5 of adult rat and cat neocortex: Triple intracellular recordings and biocytin labelling *in vitro*. *Cereb. Cortex* **12**:936–953. [6]

Thorpe, S., D. Fize, and C. Marlot. 1996. Speed of visual processing in the human visual system. *Nature* **381**:520–522. [11, 16]

Thorpe, W. H. 1964. Learning and Instinct in Animals. London: Methuen. [4]

Thut, G., A. Nietzel, S. A. Brandt, and A. Pascual-Leone. 2006. Alpha-band electroencephalographic activity over occipital cortex indexes visuospatial attention bias and predicts visual target detection. *J. Neurosci.* **26(37)**:9494–9502. [16, 18]

Tiesinga, P., J.-M. Fellous, E. Salinas, J. Jose, and T. Sejnowski. 2005. Inhibitory synchrony as a mechanism for attentional gain modulation. *J. Physiol.* **98**:296–314. [1]

Tiesinga, P., J.-M. Fellous, and T. J. Sejnowski. 2008. Regulation of spike timing in visual cortical circuits. *Nat. Rev. Neurosci.* **9**:97–109. [1, 18]

Tiesinga, P., and T. J. Sejnowski. 2004. Rapid temporal modulation of synchrony by competition in cortical interneuron networks. *Neural Comput.* **16**:251–275. [1]

Titone, D., T. Ditman, P. S. Holzman, H. Eichenbaum, and D. Levy. 2004. Transitive inference in schizophrenia: Impairments in relational memory organization. *Schizophr. Res.* **68**:235–247. [17]

Toates, F. 1980. Animal Behaviour: A Systems Approach. Chichester: Wiley. [10]

Tognoli, E., J. Lagarde, G. C. DeGuzman, and J. A. S. Kelso. 2007. The phi complex as a neuromarker of human social coordination. *PNAS* **104**:8190–8195. [18]

Toledo-Rodriguez, M., B. Blumenfeld, C. Wu, et al. 2004. Correlation maps allow neuronal electrical properties to be predicted from single-cell gene expression profiles in rat neocortex. *Cereb. Cortex* **14**:1310–1327. [12]

Tomioka, R., K. Okamoto, T. Furuta, et al. 2005. Demonstration of long-range GABAergic connections distributed throughout the mouse neocortex. *Eur. J. Neurosci.* **21**:1587–1600. [13]

Tootell, R. B., M. S. Silverman, E. Switkes, and R. L. De Valois. 1982. Deoxyglucose analysis of retinotopic organization in primate striate cortex. *Science* **218**:902–904. [18]

Tort, A. B., M. A. Kramer, C. Thorn, et al. 2008. Dynamic cross-frequency couplings of local field potential oscillations in rat striatum and hippocampus during performance of a T-maze task. *PNAS* **105**:20517–20522. [13]

Traub, R. D., M. O. Cunningham, T. Gloveli, F. E. LeBeau, and A. Bibbig. 2003. GABA-enhanced collective behavior in neuronal axons underlies persistent gamma-frequency oscillations. *PNAS* **100**:11047–11052. [8]

Traub, R. D., W. D. Knowles, R. Miles, and R. K. S. Wong. 1987. Models of cellular mechanisms underlying propagations of epileptiform activity in the CA2-CA3 region of the hippocampal slice. *Neuroscience* **21**:457–470. [8]

Traub, R. D., M. A. Whittington, I. M. Stanford, and J. G. Jefferys. 1996. A mechanism for generation of long-range synchronous fast oscillations in the cortex. *Nature* **383**:621–624. [8, 13]

Treisman, A. 1999. Solutions to the binding problem: Progress through controversy and convergence. *Neuron* **24(1)**:105–125. [1, 11]

Triesch, J., D. H. Ballard, and R. A. Jacobs. 2002. Fast temporal dynamics of visual cue integration. *Perception* **31(4)**:421–434. [15]

Triesch, J., and C. von der Malsburg. 2001. Democratic integration: Self-organized integration of adaptive cues. *Neural Comput.* **13(9)**:2049–2074. [15]

Troje, N. F. 2002. Decomposing biological motion: A framework for analysis and synthesis of human gait patterns. *J. Vision* **2(371–387)**. [14, 18]

Tschacher, W., P. Dubouloz, R. Meier, and U. Junghan. 2008. Altered perception of apparent motion in schizophrenia spectrum disorder. *Psychiatry Res.* **159**:290–299. [17]

Tsuda, I. 2001. Toward an interpretation of dynamic neural activity in terms of chaotic dynamical systems. *Behav. Brain Sci.* **24**:793–810; discussion 810–848. [13]

Tsunoda, K., Y. Yamane, M. Nishizaki, and M. Tanifuji. 2001. Complex objects are represented in macaque inferotemporal cortex by the combination of feature columns. *Nat. Neurosci.* **4(8)**:832–838. [11]

Tulving, E. 1972. Episodic and semantic memory. In: Organisation of Memory, ed. E. Tulving and W. Donaldson, pp. 381–403. New York: Academic Press. [4]

———. 2005. Episodic memory and autonoesis: Uniquely human. In: The Missing Link in Cognition: Evolution of Self-knowing Consciousness, ed. H. Terrace and J. Metcalfe, pp. 3–56. Oxford: Oxford Univ. Press. [4]

Turner, G. C., M. Bazhenov, and G. Laurent. 2008. Olfactory representations by Drosophila mushroom body neurons. *J. Neurophysiol.* **99**:734–746. [5]

Uchida, N., A. Kepecs, and Z. F. Mainen. 2006. Seeing at a glance, smelling in a whiff: Rapid forms of perceptual decision making. *Nat. Rev. Neurosci.* **7(6)**:485–491. [16]

Uhlhaas, P. J., C. Haenschel, D. Nikolić, and W. Singer. 2008. The role of oscillations and synchrony in cortical networks and their putative relevance for the pathophysiology of schizophrenia. *Schizophr. Bull.* **34**:927–943. [17]

Uhlhaas, P. J., D. Linden, W. Singer, et al. 2006. Dysfunctional long-range coordination of neural activity during Gestalt perception in schizophrenia. *J. Neurosci.* **26(31)**:8168–8175. [11, 16, 18]

Uhlhaas, P. J., J. Pantel, H. Lanfermann, et al. 2008. Visual perceptual organization deficits in Alzheimer's dementia. *Dement. Geriatr. Cogn. Disord.* **25**:465–485. [17]

Uhlhaas, P. J., W. A. Phillips, G. Mitchell, and S. M. Silverstein. 2006. Perceptual grouping in disorganized schizophrenia. *Schizophr. Res.* **145**:105–117. [17]

Uhlhaas, P. J., F. Roux, W. Singer, et al. 2009. The development of neural synchrony reflects late maturation and restructuring of functional networks in humans. *PNAS* **106**:9866–9871. [18]

Uhlhaas, P. J., and S. M. Silverstein. 2005. Perceptual organization in schizophrenia spectrum disorders: Empirical research and theoretical implications. *Psychol. Bull.* **131(4)**:618–632. [17, 18]

Uhlhaas, P. J., S. M. Silverstein, W. A. Phillips, and P. G. Lovell. 2004. Evidence for impaired visual context processing in schizotypy with thought disorder. *Schizophr. Res.* **68**:249–260. [17]

Uhlhaas, P. J., and W. Singer. 2006. Neural synchrony in brain disorders: Relevance for cognitive dysfunctions and pathophysiology. *Neuron* **52**:155–168. [11, 17]

———. 2010. Abnormal neural oscillations and synchrony in schizophrenia. *Nat. Rev. Neurosci.* **11**:100–113. [11]

Ullman, S. 1995. Sequence seeking and counter streams: A computational model for bidirectional information flow in the visual cortex. *Cereb. Cortex* **5**:1–11. [18]

———. 2007. Object recognition and segmentation by a fragment-based hierarchy. *Trends Cogn. Sci.* **11(2)**:58–64. [14]

Vaadia, E., I. Haalman, M. Abeles, et al. 1995. Dynamics of neuronal interactions in monkey cortex in relation to behavioural events. *Nature* **373**:515–518. [12]

van Assche, M., and A. Giersch. 2009. Visual organization processes in schizophrenia. *Schizophr. Bull.* doi:10.1093/schbul/sbp084 [17] van der Hart, O., and B. Friedman. 1989. A reader's guide to Pierre Janet on dissociation: A neglected intellectual heritage. *Dissociation* **2**:1–16. [17]

Van Essen, D. C. 1979. Visual areas of the mammalian cerebral cortex. *Ann. Rev. Neurosci.* **2**:227–263. [16]

van Hemmen, J. L., W. Gerstner, A. V. M. Herz, et al. 1990. Encoding and decoding of patterns which are correlated in space and time. In: Konnektionismus in Artificial Intelligence und Kognitionsforschung, ed. G. Dorffner, pp. 153–162. Heidelberg Springer. [3]

Van Hooser, S. D. 2007. Similarity and diversity in visual cortex: Is there a unifying theory of cortical computation? *Neuroscientist* **13**:630–656. [2]

Van Hooser, S. D., J. A. Heimel, S. Chung, and S. B. Nelson. 2006. Lack of patchy horizontal connectivity in primary visual cortex of a mammal without orientation maps. *J. Neurosci.* **26**:7680–7692. [2]

VanRullen, R., T. Carlson, and P. Cavanagh. 2007. The blinking spotlight of attention. *PNAS* **104**:19204–19209. [18]

VanRullen, R., L. Reddy, and C. Koch. 2006. The continuous wagon wheel illusion is associated with changes in electroencephalogram power at approximately 13 Hz. *J. Neurosci.* **26(2)**:502–507. [16]

VanRullen, R., and S. Thorpe. 2001. Rate coding versus temporal order coding: What the retinal ganglion cells tell the visual cortex. *Neural Comput.* **13**:1255–1283. [11, 13]

Varela, F., J. P. Lachaux, E. Rodriguez, and J. Martinerie. 2001. The brainweb: Phase synchronization and large-scale integration. *Nat. Rev. Neurosci.* **2(4)**:229–239. [12, 13, 16, 18]

Vasconcelos, M. 2008. Transitive inference in non-human animals: An empirical and theoretical analysis. *Behav. Proc.* **78**:313–334. [4, 5]

Vicente, R., L. L. Gollo, C. R. Mirasso, I. Fischer, and G. Pipa. 2008. Dynamical relaying can yield zero time lag neuronal synchrony despite long conduction delays. *PNAS* **105**:17157–17162. [11, 13]

Vidal, J. R., M. Chaumon, J. K. O'Regan, and C. Tallon-Baudry. 2006. Visual grouping and selective attention induce gamma-band oscillations at different frequencies in human MEG signals. *J. Cogn. Neurosci.* **18**:1850–1862. [16]

Vierling-Claassen, D., P. Siekmeier, S. Stufflebeam, and N. J. Kopell. 2008. Modeling GABA alterations in schizophrenia: A link between impaired inhibition and altered gamma and beta range auditory entrainment. *J. Neurophysiol.* **99**:2656–2671. [11]

Vinje, W. E., and J. L. Gallant. 2000. Sparse coding and decorrelation in primary visual cortex during natural vision. *Science* **287**:1273–1276. [12, 13]

Visalberghi, E., and L. Limongelli. 1994. Lack of comprehension of cause-effect relations in tool-using capuchin monkeys (*Cebus apella*). *J. Comp. Psychol.* **108**:15–22. [4]

Volgushev, M., M. Chistiakova, and W. Singer. 1998. Modification of discharge patterns of neocortical neurons by induced oscillations of the membrane potential. *Neuroscience* **83(1)**:15–25. [11]

von der Malsburg, C. 1981/1994. The correlation theory of brain function (reprint of Internal Report 81-2, Max-Planck-Institute for Biophysical Chemistry, 1981). In: Models of Neural Networks II, ed. E. Domany, J. L. v. Hemmen and K. Schulten, pp. 95–119 Heidelberg: Springer. [1, 3, 12, 13, 14]

———. 1986. Am I thinking assemblies? In: Brain Theory, ed. G. Palm and A. Aertsen, pp. 161–176. Berlin: Springer-Verlag. [12]

von der Malsburg, C., and W. Schneider. 1986. A neural cocktail-party processor. *Biol. Cyber.* **54**:29–40. [1, 18]

von Stein, A., C. Chiang, and P. König. 2000. Top-down processing mediated by interareal synchronization. *PNAS* **97**:14748–14753. [18]

von Stein, A., P. Rappelsberger, J. Sarnthein, and H. Petsche. 1999. Synchronization between temporal and parietal cortex during multimodal object processing in man. *Cereb. Cortex* **9**:137–150. [17]

von Stein, A., and J. Sarnthein. 2000. EEG frequency and the size of cognitive neuronal assemblies. *Behav. Brain Sci.* **23(3)**:413–414, 432–437. [16]

Walker, M. P., T. Brakefield, A. Morgan, J. A. Hobson, and R. Stickgold. 2002. Practice with sleep makes perfect: Sleep-dependent motor skill learning. *Neuron* **35**:205–211. [18]

Walker, M. P., T. Brakefield, J. Seidman, et al. 2003. Sleep and the time course of motor skill learning. *Learn. Mem.* **10**:275–284. [18]

Wang, D. L. 2005. The time dimension for scene analysis. *IEEE Trans. Neural Netw.* **16**:1401–1426. [11]

Wang, X. J., and G. Buzsáki. 1996. Gamma oscillation by synaptic inhibition in a hippocampal interneuronal network model. *J. Neurosci.* **16**:6402–6413. [18]

Ward, P. B. 2009. Neural plasticity and structural brain change in schizophrenia: Are we asking the right questions? *Schizophr. Bull.* **35(1)**:218. [17]

Warnking, J., M. Dojat, A. Guerin-Dugue, et al. 2002. fMRI retinotopic mapping-step by step. *NeuroImage* **17**:1665–1683. [12]

Waters, F. A. V., M. T. Maybery, J. C. Badcock, and P. T. Michie. 2004. Context memory and binding in schizophrenia. *Schizophr. Res.* **68**:119–125. [17]

Watt, R. J., and W. A. Phillips. 2000. The function of dynamic grouping in vision. *Trends Cogn. Sci.* **4(12)**:447–454. [1]

Wehr, M., and G. Laurent. 1996. Odour encoding by temporal sequences of firing in oscillating neural assemblies. *Nature* **384**:162–166. [18]

Weiller, C., and F. Chollet. 1994. Imaging recovery of function following brain injury. *Curr. Opin. Neurobiol.* **4**:226–230. [10]

Weir, A. A. S., J. Chappell, and A. Kacelnik. 2002. Shaping of hooks in new Caledonian crows. *Science* **297**:981. [4, 5]

Weiss, Y., E. P. Simoncelli, and E. H. Adelson. 2002. Motion illusions as optimal percepts. *Nat. Neurosci.* **5(6)**:598–604. [15]

Weisswange, T., C. Rothkopf, T. Rodemann, and J. Triesch. 2009. Can reinforcement learning explain the development of causal inference in multisensory integration? 8th Intl. Conference on Development and Learning. Aachen: IEEE. [15]

Wertheimer, M. 1912. Experimentelle studien über das sehen von beuegung. *Zt. Psychol. Physiol. Sinnesorgane* **61**:161–265. [12]

Wespatat, V., F. Tennigkeit, and W. Singer. 2004. Phase sensitivity of synaptic modifications in oscillating cells of rat visual cortex. *J. Neurosci.* **24**:9067–9075. [1, 11, 18]

Whitfield-Gabrieli, S., H. W. Thermenos, S. Milanovic, et al. 2009. Hyperactivity and hyperconnectivity of the default network in schizophrenia and in first-degree relatives of persons with schizophrenia. *PNAS* **106**:1279–1284, 4572. [17]

Whittington, M. A., H. C. Doheny, R. D. Traub, F. E. N. LeBeau, and E. H. Buhl. 2001. Differential expression of synaptic and nonsynaptic mechanisms underlying stimulus-induced gamma oscillations *in vitro*. *J. Neurosci.* **21(5)**:1727–1738. [11]

Whittington, M. A., and R. D. Traub. 2003. Interneuron diversity series: Inhibitory interneurons and network oscillations *in vitro*. *Trends Neurosci.* **26**:676–682. [1]

Wicher, D., R. Schafer, R. Bauernfeind, et al. 2008. Drosophila odorant receptors are both ligand-gated and cyclic-nucleotide-activated cation channels. *Nature* **452**:1007–1011. [5]

Williams, L. M., A. Sidis, E. Gordon, and R. A. Meares. 2006. "Missing links" in borderline personality disorder: Loss of neural synchrony relates to lack of emotion regulation and impulse control. *J. Psychiatry Neurosci.* **31**:181–188. [17]

Wilson, M. A., and B. L. McNaughton. 1994. Reactivation of hippocampal ensemble memories during sleep. *Science* **265**:676–679. [7]

Wilson, T. W., D. C. Rojas, M. L. Reite, P. D. Teale, and S. J. Rogers. 2007. Children and adolescents with autism exhibit reduced MEG steady-state gamma responses. *Biol. Psychiatry* **62**:192–197. [11]

Wiskott, L. 2006. How does our visual system achieve shift and size invariance? In: 23 Problems in Systems Neuroscience, ed. J. L. van Hemmen and T. J. Sejnowski. New York: Oxford Univ. Press. [10]

Wiskott, L., and T. J. Sejnowski. 2002. Slow feature analysis: Unsupervised learning of invariances. *Neural Comput.* **14**:715–770. [8]

Wiskott, L., and C. von der Malsburg. 1996. Face recognition by dynamic link matching. In: Lateral Interactions in the Cortex: Structure and Function, ed J. Sirosh, R. Miikkulainen and Y. Choe. http://www.cs.utexas.edu/users/nn/web-pubs/htmlbook96/. [13]

Wolfrum, P., C. Wolff, J. Lücke, and C. von der Malsburg. 2008. A recurrent dynamic model for correspondence-based face recognition. *J. Vision* **8(7)**:1–18. [1, 10, 13]

Wolpert, D. H., and W. G. Macready. 1997. No free lunch theorems for optimization. *IEEE Trans. Evol. Comp.* **1**:67–82. [10]

Womelsdorf, T., P. Fries, P. P. Mitra, and R. Desimone. 2006. Gamma-band synchronization in visual cortex predicts speed of change detection. *Nature* **439**:733–736. [18]

Womelsdorf, T., J. M. Schoffelen, R. Oostenveld, et al. 2007. Modulation of neuronal interactions through neuronal synchronization. *Science* **316(5831)**:1609–1612. [1, 7, 8, 11, 18]

Wong, P., and J. H. Kaas. 2008. Architectonic subdivisions of neocortex in the gray squirrel (*Sciurus carolinensis*). *Anat. Rec.* **10**:1301–1333. [2]

———. 2009a. Architectonic study of neocortex in the short-tailed opossums (*Monodelphis domestica*). *Brain Behav. Evol.* **73**:206–228. [2]

———. 2009b. Architectonic subdivisions of neocortex in tree shrew (*Tupaia belangeri*). *Anat. Rec.* **292**:994–1027. [2]

Wyart, V., and C. Tallon-Baudry. 2008. A neural dissociation between visual awareness and spatial attention. *J. Neurosci.* **28**:2667–2679. [18]

———. 2009. How ongoing fluctuations in human visual cortex predict perceptual awareness: Baseline shift versus decision bias. *J. Neurosci.* **29(27)**:8715–8725. [16, 18]

Wynn, J. K., G. A. Light, B. Breitmeyer, K. H. Nuechterlein, and M. F. Green. 2005. Event-related gamma activity in schizophrenia patients during a visual backward-masking task. *Am. J. Psychiatry* **162**:2330–2336. [18]

Xu, W., X. Huang, K. Takagaki, and J. Y. Wu. 2007. Compression and reflection of visually evoked cortical waves. *Neuron* **55**:119–129. [12]

Yabuta, N. H., and E. M. Callaway. 1998. Cytochrome-oxidase blobs and intrinsic horizontal connections of layer 2/3 pyramidal neurons in primate V1. *Vis. Neurosci.* **15**. [2]

Yoshimura, Y., J. L. M. Dantzker, and E. M. Callaway. 2005. Excitatory cortical neurons form fine-scale functional networks. *Nature* **433**:868–873. [6]

Young, M. P. 1992. Objective analysis of the topological organization of the primate cortical visual system. *Nature* **358**:152–155. [16]

Yuille, A., and D. Kersten. 2006. Vision as Bayesian inference: Analysis by synthesis? *Trends Cogn. Sci.* **10(7)**:301–308. [15]

Zentall, T. R. 2004. Action imitation in birds. *Learn. Behav.* **32**:15–23. [5]

Zentall, T. R., T. S. Clement, R. S. Bhatt, and J. Allen. 2001. Episodic-like memory in pigeons. *Psychon. Bull. Rev.* **8**:685–690. [4]

Zhang, L. I., H. W. Tao, C. E. Holt, W. A. Harris, and M. Poo. 1998. A critical window for cooperation and competition among developing retinotectal synapses. *Nature* **395**:37–44. [11]

Zilles, K., N. Palomero-Gallagher, C. Grefkes, et al. 2002. Architectonics of the human cerebral cortex and transmitter receptor fingerprints: Reconciling functional neuroanatomy and neurochemistry. *Eur. Neuropsychopharm.* **12**:587–599. [2]

Subject Index

Adaptive Resonance Theory 293
after-activity 87, 91
Aha! effect 153
alexithymia 259, 260
alpha 118, 130, 239–241
  oscillations 92–94, 286–288
  rhythms 120, 212
Alzheimer's disease 165, 166, 246, 252, 257, 262
amblyopia 246, 248, 254, 257, 260, 262, 284, 298
anti-Hebbian learning 97
apical dendrites 15, 31
arousal 256, 286, 291
assemblies 160–164. *See also* cell assemblies
  synaptic 15, 144–147
assembly coding 160
association field 17, 186, 188, 189
  dynamic 169, 187–190
associative learning 46, 54, 56, 202
associative memory 84, 87, 89, 91, 99, 151
attention 3, 5, 7, 8, 10–17, 21, 46, 92, 137, 164–167, 196, 208, 210, 253, 254, 268, 279, 280, 286
  focal 151, 152, 203
  models 211
  neuronal bases of 17
  role of frequency bands 288, 289
  selective 45, 200, 291, 297
  shifts 200
attentional bias 21, 246, 261, 262
attentional blink 92, 98
attractor 139, 144–146
  convergence 90, 91
  dynamics 94–97, 144
  network paradigm 86–92
  neural 139, 144–147
  states 96–98, 204, 255, 264, 270
    quasi-stable 87, 88, 92, 97
audiovisual binding 288
auditory learning 69, 76

autism 165, 166, 246, 251, 252, 256, 257, 261–265, 298
avian brain 28, 44, 72–74, 82
awareness 5, 237, 240, 255, 258, 279, 281, 286
  role of frequency bands 288, 289

basal ganglia 13, 64, 72, 73, 77, 79, 82
basket cells 89, 93–95, 208, 209
Bayesian inference 5, 23, 230, 231
beta 119–122, 239–242, 251, 283, 286–290
  oscillations 92, 94, 121, 137, 162
  rhythms 15, 17, 123, 124, 130
  synchrony 250, 253, 296
beta phenomenon 186
Betz neurons 30, 31
biased competition 18, 21, 201, 210
bias-variance dilemma 154
binding 159, 161–164, 167, 170–172, 213, 270, 271
  audiovisual 288
  by synchrony hypothesis 7, 201
  context-dependent 279
  Gestalt-like 189
  perceptual 172, 191
  temporal 7, 205
binocular rivalry 196
BOLD 204, 241
borderline personality disorder 259, 260, 263
brain
  avian 28, 44, 72–74, 82
  avian–mammal comparison 74–77
  invertebrate 172
  macaque monkey 30
  mammalian 25–28, 44, 60, 75–78, 172
  size 31, 64, 65, 75
brain–behavior relationships 59–82
brainstem 26, 27, 69, 78, 79, 135
Brunelleschi, Filippo 218, 219
burst synchronization 92, 98

canonical cortical microcircuit 9, 13, 134, 135
Cartesian Theater hypothesis 236, 237
causal understanding 51
cell assemblies 15, 60, 67, 86–88, 146–148, 205, 265
  defined 103, 121
  dynamic reconfiguration of 171
  oscillation frequency of 110
  stimulus-driven coordination 169–191
central executive 8, 22, 23, 177, 291
central pattern generator 68–70, 172
Chorus of Fragments model 205
Chorus of Prototypes algorithm 205
chronophotography 224
chunking 242
classical conditioning 45, 48
closure. See contour closure
coding 3, 4, 6, 37–40, 96, 194, 199
  activation 93
  assembly 160
  biological movement 223–226, 274
  defined 38
  labeled-line 159
  population 4, 7, 272
  precision 241
  predictive 6, 277, 278
  relational 175
  retinal 199, 200
  sequence 35–42
  temporal 38, 160, 166–168, 179, 205
cognitive behavior
  avian–mammalian comparison 43–58, 74–77
coherence 95, 200, 279–281, 288, 293
  global 20–22
  spike-field 163, 164, 208
coherent infomax 6, 270–273
color perception 213, 298
compositionality 170, 171, 202
composition, rules of 151, 155
concatenation 115, 117, 162, 168
  local circuit features 127–130
conduction velocity 30, 65, 108, 109, 184, 208
conjunction-sensitive neurons 160
connectivity 83, 88, 93, 95, 121, 134, 144, 157, 166, 183, 213, 279
  convergence 181, 182, 236
  global 96
  intracortical 186–188, 196
  local 90, 120, 189
  nonsynaptic axo-axonic 124, 125
  recurrent 87, 90
  synaptic 78, 85, 86, 98, 115, 119
consciousness 5, 12, 17, 92, 98, 165, 253, 258, 289
context 1–5, 19, 20, 25, 45, 46, 188, 219, 225, 232, 294
contextual disambiguation 5, 6, 9, 10, 15, 16, 19–21, 264, 274, 295
contextual fields 13, 253
contextual gain modulation 3, 6, 8, 17, 274
contextual modulation 5, 16, 19, 270, 271
continuous processing 210
contour closure 220, 222, 225, 284
  superiority effect 221–224
contour integration 188, 246–252, 274, 285, 291
contrast sensitivity 225–227
convergence
  attractor 90, 91
  connectivity 181, 182, 236
coordinating interactions 5–7, 19, 170, 194, 271, 291
coordination
  cross-frequency 125–128
  defined 150–152, 194, 235
  evaluation of 152, 153
  modeling 83–98
  prespecified 6, 7, 11, 12, 45
  taxonomy of 198
correspondence problem 71, 72
cortical circuits 25–34, 133–148
  local 6, 66, 116–132
cortical microcircuits. See microcircuits
cortical modules 25, 29, 34
corvids 44, 45, 54–56, 75
  physical cognition 51
coupling 8, 96, 101, 124, 137, 162, 198, 199, 202, 269, 274, 279
covert attention task 208
cross-frequency coordination 125–128
cue integration 6, 231–234

visual 231, 250
current-source density 106, 108

decoding 38–40
delta 105, 118, 125, 129, 140, 239, 286
  rhythms 130, 283
democratic cue integration 233
depth cues 230
devil's pitchfork 20, 22
diet 263
dimensionality 4, 6
  reduction in 270, 273, 274
disconnection 24, 258
discontinuous communication 210
dissociative disorders 258, 260, 263
distal dendrites 15
divergence, connectivity 181, 182
double bouquet cells 89, 90
driving interactions 13, 23, 245, 246, 253, 261
Dynamical Systems Theory 287
dynamic association field 169, 187–190
dynamic coordination
  as optimization 277–279
  defined 1, 3, 6, 7, 170, 194, 196, 245, 273, 275
  diagnosing 195
  dimensions of 275
  effects of emotion on 291, 292
  evolution of 11–13, 43–58, 59–82
  examples 173–176
  experimental approaches 212
  failures of 24, 245–266
  functional role 273
  modulating 290–294
  schematic structure 49
dynamic embedding 9, 11, 264
dynamic equilibrium 146, 148
dynamic grouping 3, 6, 9, 10, 13, 15, 271, 274
dynamic linking 3, 6, 9, 15, 17, 264
Dynamic Pattern Theory 271
dynamic routing 3, 6, 9–11, 15, 18, 200–202, 207, 264, 271, 282
dysconnection theory 24

Ebbinghaus illusion 249
ECoG data 117
emergence 237, 238, 270
emotions 78, 150, 259, 260, 291–293
encoding 38–40, 254
  compositionality 202
  flexible 159
  memory 200, 239, 279, 280, 286
  relations 159–162, 207
entorhinal–hippocampal circuit 103, 142
epilepsy 107, 137, 246, 252, 253, 257, 263, 264
episodic-like memory 45, 47–51, 56, 75
episodic memory 12, 47, 239
evolution 11–13, 44, 56, 63, 148
  convergent 44, 56, 63, 80
excitatory-inhibitory (E-I) networks 93, 94, 135–138, 140, 146–148
executive control 8, 12. See also central executive

face recognition 155, 204, 205, 211, 242
fast oscillations 92–94, 103, 120, 124, 125
fatigue 291
feature-similarity gain 210
feedback connections 25, 26, 34
feedback inhibition 93–96
feedforward inhibition 121, 122
feedforward input 14, 16, 122, 245
feedforward processing 87, 88, 97, 98, 238, 286
figure–ground segmentation 87, 287
flexible behavior 45, 46, 60, 70
flexible encoding 159
Florence Dome 215, 217–220
focal attention 151, 152, 203
food caching 47–49, 75
free energy principle 6, 277
frequency bands. See also alpha, beta, delta, gamma, theta
  cognitive functions of 239, 240
  coordination between 125
  integrating between 242, 243
  oscillations in 286–290
  roles of 287–290
Freud, Sigmund 257

frontal cortex 256, 258, 291, 293
frontal eye field (FEF) 207, 208, 288, 292, 293
functional synaptic imaging 179, 182
fundamental goals 153

GABA receptors 14, 103, 121, 136, 137, 209
Gabor signal 220, 225, 247, 270, 291
gain modulation 15, 198, 199
 contextual 3, 6, 8, 17, 274
 multiplicative 271
gamma 196, 204, 206, 208, 239–241, 251, 286–288
 oscillations 92–96, 98, 102, 103, 121, 136–140, 162, 208, 243, 282–285
 rhythms 15, 120, 122–125, 129, 151
 synchrony 201, 250, 252, 255–259, 279, 296
gap junctions 93, 120
Gestalt perception 7, 10, 12, 18, 45, 152, 186–188, 215–227, 279, 284, 287
global coherence 20–22
global connectivity 96
global emergent properties 66
global network 96, 97
global shape 186, 219, 222, 226
global workspace theories 236, 237
goal-directed behavior 9, 99, 153
grandmother cells 236
graph matching 17, 156
gray matter reduction 248, 249, 257, 262, 263
grooming behavior 45, 80
ground state 91, 93, 94, 97
grouping 160–167, 194, 249, 250, 256, 280
 dynamic 3, 6, 9, 10, 13, 15, 271, 274
 perceptual 99, 186, 271, 291, 295

hallucinations 255–259, 263
Hebbian learning 86, 91, 96
hippocampus 94, 101–114, 125, 203, 208, 265
 CA1 region 103, 108, 109
 CA3 region 103, 109

fast-conducting interneurons 209
information processing in 101–103
holism 2, 3, 169
holistic processing 87, 90, 92, 96–98
Hopfield attractor dynamics 144
Hopfield network 87, 88
horizontal cortical connections 32–34
Hox genes 64
Human Connectome Project 211
hypercolumns 89, 93, 95, 97
 winner-take-all module 90
hypnosis 258
hypothesis-driven models 86, 89

information flow 18, 38, 101, 235–237, 276, 294
information theory 38
information transfer 104–107, 238
inhibitory interneurons 8, 14, 32, 116, 188
innate rules 43, 45, 46
insight 12, 54, 55, 153
intracellular recording 177–179, 181–185
invariance 87, 155, 201

Janet, Pierre 257
Jung, Carl Gustav 260

labeled-line coding 159
lamination 26, 28, 64
language 19, 246, 253, 254, 261, 279, 281, 297
large-scale network models 84, 88, 97
lateral inhibition 66, 92, 98
layer 4 88, 97, 98, 135
layer 5 107, 117, 120
layers 2/3 87–90, 107, 117, 135
layers 5/6 135
learning 40–42, 87, 282, 285
 anti-Hebbian 97
 associative 46, 54, 56, 202
 auditory 69, 76
 Hebbian 86, 91, 96
 reinforcement 99, 232–234, 277

rules 41, 42
schema-based 152
sequence 35, 96
social 71, 72
vocal 75–78
liquid state machines 86
local circuits 6, 25, 66, 116–132
local connectivity 90, 120, 189
local field potential (LFP) 92, 104, 105, 139, 162, 177, 200, 287
  spectogram 94
  V1 170
local filters 220–222
localism 2, 3
long-range excitation 90, 96

Marey, Etienne-Jules 223, 224
m-current 118, 120
memory 35, 99, 211, 254, 256, 261. *See also* working memory
  associative 84, 87, 89, 91, 99, 151
  consolidation 105
  encoding 200, 239, 279, 280, 286
  episodic 12, 47, 239
  episodic-like 45, 47–51, 56, 75
  long-term 99, 253
  patterns 86, 87, 90, 91
  procedural 298
  recall 109, 110, 170
  what–where–when 47–49
metacognition 56, 264
Meynert neurons 30, 31
microcircuits 12–15, 89, 92–94, 112
  canonical 9, 13, 134, 135
  plasticity in 96, 97
minicolumn 90, 93, 95, 97
  structure 89
mirror neurons 71
modeling
  approaches to 85–87, 232
  computer 83–85, 158, 204, 210
  coordination 83–98
  multiscale 84, 99
monocular rivalry 216
motivation 49, 50, 56, 97
movement coding 223–226, 274
multimodal integration 88, 99

multiple sclerosis 257
multisensory integration 281, 286, 288
multisite recordings 162, 168, 196, 212

neocortex 25, 29, 105, 135, 203
  axonal spread in owl monkey 27
  emergence of 34
  lamination in 26, 64
  mammalian 25–27, 44, 75, 78
  whale 28
neoencephalon 72, 73
nesting 125, 126, 156, 162, 202
neural attractors 139, 144–147
neuromodulators 61, 66, 136–138, 145–148, 175, 197, 209, 269
  diffusion of 172
  monoamine 92, 135
neuronal adaptation 87, 91, 97
neuronal geometry 61, 63, 64
neuron density 27, 29
NMDA 14, 24, 63, 199, 253, 262, 263, 298
n:m locking 8
no-free-lunch theorems 154
nonsynaptic axo-axonic connectivity 124, 125

object recognition 17, 203, 205, 211, 217
olfactory system 37, 63, 65, 80
  insect 39, 66
optimization 277–279
orchestra leaders 172, 173, 197
orientation 188–190
oscillations 104–107, 118, 124, 141, 147, 148, 199–203, 239–242, 245, 251, 253, 260–263, 269, 270
  alpha 92–94, 286–288
  beta 92, 94, 121, 137, 162
  clocking signal 288
  evidence for 280–283
  fast 92–94, 103, 120, 124, 125
  features of 138
  gamma 92–96, 98, 102, 103, 121, 136–140, 162, 208, 243, 282–285
  information processing 101–103

oscillations (*continued*)
  synchronized 136, 137, 172, 295, 296
  theta 101–103, 109, 113, 137, 140, 142, 243, 286, 289
oscillatory patterns 162, 163, 167, 199, 203, 212

paleocortex 105
paleoencephalon 72, 73
parasynchronization 170
Parkinson's disease 137, 252, 257, 260, 261, 289
pathophysiology 14, 15, 83, 166, 246, 249, 295–299
patterns 155–157, 170, 187, 194–197, 222, 268, 269, 272, 292
  detection 151–153
  distributed 91
  memory 86, 87, 90, 91
  oscillatory 162, 163, 167, 199, 203, 212
  recognition 17, 155, 293
  recurrent 151
  rivalry 91
  spatiotemporal 8, 37, 39, 42, 183, 204, 285
perception 19, 20, 35, 40, 98, 99, 137, 169, 170, 188, 211, 229–234, 246, 277, 297. *See also* Gestalt perception
  color 213, 298
  CPG-like circuits 70
  in schizophrenia 250, 255
  motion 186, 191, 224
  visual 274
perceptual binding 172, 191
perceptual completion 87, 91
perceptual grouping 99, 186, 271, 291, 295
perceptual organization 246–259, 261, 284, 296
phase
  code 95, 141, 142, 147, 148
  locking 8, 95–98, 164, 165, 207, 296
  precession 141, 142
  relationships 162, 163, 168, 202–204
  resetting curve 137, 140
  transitions 60

phylogenetic homology 44, 61, 62
physical cognition 45, 51–55
place cells 109, 112–114, 141–143, 146
point-light display 224, 274
population coding 4, 7, 272
posttraumatic stress disorder 257
power spectral density 180
predictive coding 6, 277, 278
prefrontal cortex 8, 9, 23, 30, 33, 137, 237, 291–293
  dysfunction 256
  role of 18
prespecified coordination 6, 7, 11, 12, 45
prespecified feature hierarchies 11, 159, 250, 264
primary visual cortex 217, 218, 222, 247, 298
  local filters 220–222
  orientation column in 89
  V1 28, 169–191, 196, 197, 217, 284
  V2 196, 249, 264, 291
  V3 250, 291
  V4 207, 208, 249, 264, 291
procedural memory 298
projection neuron 39, 40, 64, 65, 78
propagation waves, reconstruction of 181–186
psychopathology 2, 8, 24, 128, 129, 290
purpose 153, 154
pyramidal cell 15, 29–31, 89, 90, 93–95, 112, 208
  Betz 30, 31
  CA3–CA1 103
  Meynert 30, 31
  windows of opportunity 8, 15, 24

quasi-stable attractor state 87, 88, 92, 97

rate code 141, 160, 167, 168, 200
reasoning 12, 22, 23, 255
receptive fields 3, 6, 13, 69, 143, 183, 187, 205, 208, 210, 217, 220, 225
  dynamic 144
  organization 171
  selectivity 4, 16
  V1 173, 174, 181

Subject Index

reciprocal information transfer 104–107
recognition 40, 156
  face 155, 204, 205, 211, 242
  object 17, 203, 205, 211, 217
  patterns 17, 155, 293
recurrent cortical connectivity 87, 97, 135–137
recurrent inhibition 95
recurrent network 87, 88, 144, 146, 147, 181, 183
recurrent processing 238
reinforcement learning 99, 232–234, 277
relatedness 160–162, 167, 195, 200, 204
REM sleep 101, 109
representational constraints 97, 222–224
retinal coding 199, 200
reverse mapping 225
rhythms 8, 13, 23, 98, 159–168, 235, 239–243, 283
  alpha 120, 212
  beta 15, 17, 123, 124, 130
  coordination of 125, 126
  delta 130, 283
  dynamic coordination of 128–130
  gamma 15, 120, 122–125, 129, 151
  generation 115–119, 123, 125, 130
  theta 120, 125
Ricci, Massimo 219, 220
rivalry 87, 91, 95, 98, 99
  binocular 196
  monocular 216
robotics 99
route planning 109

schemata 152, 153, 156
  based learning 152
schizophrenia 14, 15, 24, 165, 166, 246, 249–265, 290–297
search 278, 288
  visual 200, 242, 285
segmentation 169, 170, 172, 216–218, 222, 226
  figure–ground 87, 287
selective attention 45, 200, 291, 297
selective inhibition 95
self-organization 1, 8, 9, 18, 87, 146, 174, 197, 268

self-representation 246, 255
sensor fusion 23
sensory integration 200, 229–234, 286, 288
sensory processing 159, 160, 175, 194, 273, 287
  role of frequency bands 287, 288
sequence 35–37
  action 36
  coding 35–42
  compression slope 111
  learning 35, 96
  temporal 99
sharp waves 101, 103, 107, 108
similarity measure 150, 151, 154, 155
single-site recordings 177–181
sleep 137
  REM 101, 109
  slow-wave 101, 105, 130, 140, 172, 203
  spindles 106, 107, 137
social learning 71, 72
space–time resolution 223–228
sparse code 88, 90, 97, 98, 133, 138, 144
spatiotemporal patterns 8, 37, 39, 42, 183, 187, 204, 285
spike doublet generation 121–123, 206
spike-field coherence 163, 164, 208
spike initiation zone 64
spike timing-dependent plasticity 41, 96, 98, 161, 202, 203, 282, 283
stimulus selection 164, 165, 279, 286, 289
streaking effect 202
structural relationships 149, 155–157
subnetworks 89, 98
subsumption architecture 67
symmetry 220, 224–228, 250
  based representations 215
synaptic assemblies 15, 144–147
synaptic connectivity 78, 85, 86, 98, 115, 119
synaptic depression 87, 91, 97
synaptic echoes 181–186
synaptic inhibition 119–121
synaptic input coordination 180
synaptic plasticity 87, 97, 103, 142, 148, 204, 205, 295

synchronization 15, 22, 37, 45, 95, 96, 196, 200, 212, 213, 241, 246, 250–253, 260, 264, 279, 282, 283, 287
 across E-I networks 136, 137
 between populations 205–210
 burst 92, 98
 functions of 163–165
 manifolds 270
 mechanisms 139, 205–210
 spike 17, 121
 zero-phase lag 121, 162, 205–207
synchrony 8, 17, 23, 89, 94–96, 138, 160–162, 196, 205, 245, 261, 264, 269, 270, 278, 295, 296
 abnormal 165, 166, 296
 beta-band 250, 253, 296
 during human ontogeny 283–285
 during sensory processing 175
 establishment of 199
 gamma-band 201, 250, 252, 255–259, 279, 296
 oscillatory 66, 239–241
 reduced 248
 temporal 161, 208
 theta-band 253, 296
synergetics 272
synfire chains 86, 96, 125, 139, 170, 171
system architectures 12, 67, 75, 81, 82

task-relevant data 99
telencephalon 73, 75, 76
template matching 155, 242, 293
temporal binding 7, 205
temporal coding 38, 160, 166–168, 179, 205
temporal lags, cross-neuronal 110–113
temporal patterning 38, 39, 42, 155
temporal structure 17, 35, 38, 110, 124, 194–196, 242
 changes in interregional 213
thalamocortical spindles 106, 107, 137
thalamus 26–28, 66, 69, 79, 89, 102, 134, 137, 172
theory of mind 75, 254, 255, 264, 297, 298
therapsids 74

theta 91–94, 108, 110, 112, 119, 239, 241, 283
 burst synchronization 92, 98
 oscillations 101–103, 109, 113, 137, 140, 142, 243, 286, 289
 phase biasing 104
 rhythms 120, 125
 synchrony 253, 296
time-frequency wavelet analysis 179, 180
tool use 51, 52, 75
 New Caledonian crows 54, 55
top-down models 85, 89, 92
transfer tests 51–55
transient phase synchrony 269
transitive inference 55, 56, 75
trap tube 51–53, 74
trauma 257–259, 291
two-trap box design 52, 53

unitary event analysis 163

V1 28, 169–191, 196, 197, 217, 284
V2 196, 249, 264, 291
V3 250, 291
V4 207, 208, 249, 264, 291
visual processing 170, 181, 217, 225
visual search 200, 242, 285
visuomotor integration 281
vocal imitation 71, 75, 76
vocal learning 75–78
voltage clamp mode 178
Von Economo neuron 31, 32

white matter 31, 77, 248, 251, 252, 255, 257
 reduction 249, 257, 262
Williams syndrome 246, 248–249, 254, 257, 261, 263, 298
working memory 10, 12, 16, 18, 22, 45, 46, 99, 137, 194, 200, 246, 253, 256–259, 279, 280, 286, 289, 297
 attractor models of 96
zero-phase lag synchronization 121, 162, 205–207